Methods in Molecular Biology

Series Editor
John M. Walker
School of Life and Medical Sciences
University of Hertfordshire
Hatfield, Hertfordshire, UK

For further volumes:
http://www.springer.com/series/7651

For over 35 years, biological scientists have come to rely on the research protocols and methodologies in the critically acclaimed *Methods in Molecular Biology* series. The series was the first to introduce the step-by-step protocols approach that has become the standard in all biomedical protocol publishing. Each protocol is provided in readily-reproducible step-by-step fashion, opening with an introductory overview, a list of the materials and reagents needed to complete the experiment, and followed by a detailed procedure that is supported with a helpful notes section offering tips and tricks of the trade as well as troubleshooting advice. These hallmark features were introduced by series editor Dr. John Walker and constitute the key ingredient in each and every volume of the *Methods in Molecular Biology* series. Tested and trusted, comprehensive and reliable, all protocols from the series are indexed in PubMed.

Viroids

Methods and Protocols

Edited by

Ayala L. N. Rao, Irene Lavagi-Craddock, and Georgios Vidalakis

Department of Microbiology and Plant Pathology, University of California, Riverside, Riverside, CA, USA

Humana Press

Editors
Ayala L. N. Rao
Department of Microbiology
and Plant Pathology
University of California, Riverside
Riverside, CA, USA

Irene Lavagi-Craddock
Department of Microbiology
and Plant Pathology
University of California, Riverside
Riverside, CA, USA

Georgios Vidalakis
Department of Microbiology
and Plant Pathology
University of California, Riverside
Riverside, CA, USA

ISSN 1064-3745 ISSN 1940-6029 (electronic)
Methods in Molecular Biology
ISBN 978-1-0716-1463-1 ISBN 978-1-0716-1464-8 (eBook)
https://doi.org/10.1007/978-1-0716-1464-8

This Humana imprint is published by the registered company Springer Science+Business Media, LLC, part of Springer Nature.
The registered company address is: 1 New York Plaza, New York, NY 10004, U.S.A.

Preface

Among many microbes that are pathogenic to plants, viroids are classified as unique and are the smallest known subviral pathogens that evolved to manipulate the cellular systems of the hosts they invade. Three hallmark features distinguish viroids from conventional viruses: (1) their ability to replicate autonomously using cellular polymerases (including RNA polymerases); (2) they are not encapsidated and exist as single-stranded (ss) RNAs, covalently closed circular molecules; and (3) their ability to replicate in the nucleus. Additionally, due to extensive regions of intramolecular complementarity, they appear either as "rod-like collapsed circles" or adopt "multibranched" conformation. Prior to the advent of the recombinant DNA (rDNA) technology, detection and characterization of viroid diseases was predominantly dependent on the biology of viroids followed by electrophoresis-based technologies. The advent of rDNA technology and its high sensitivity significantly improved the detection and characterization of viroids. The invention of the polymerase chain reaction (PCR) technique revolutionized the detection and characterization of nucleic acids in general and viroids in particular. Consequently, this book has been divided into seven major parts containing a total of twenty-six chapters.

Part I provides detection methods based on the biology of viroids. Chapters 1 and 2 present biological assays to detect viroids in woody and herbaceous plants.

Part II provides detection techniques based on electrophoresis and hybridization techniques. Since viroids exist as free nucleic acid entities, methods to detect viroids using electrophoretic methods are presented in Chapters 5–8. Chapters 9–13 describe hybridization-based detection of viroids.

Part III provides a wide range of PCR-based techniques that are envisioned to provide a high degree of sensitivity. Therefore, a series of chapters from 9 to 19 describe how variations in PCR, such as RT-PCR, qPCR, multiplex RT-PCR, and others, are used in viroid research.

Part IV covers the emerging area of nucleic acid sequence-based technology, a concept that will expand in the coming years. Just as our knowledge of nucleic acid sequencing has expanded, next-generation sequencing technology was applied to viroid research as well. Chapters 20–23 describe methods in applying next-generation sequencing approaches to the identification and detection of viroids.

Parts V, VI, and VII are each composed of an individual chapter (Chapters 24, 25, and 26) dedicated to emerging techniques in viroid research such as RNA silencing, splicing techniques, and viroid structure.

Viroids: Methods and Protocols is the result of the collective hard work and effort of all three editors along with a large number of contributors who are leaders in the field of viroid research. The editors would like to extend their gratitude to all the contributors. Additionally, the editors would like to acknowledge the help of Miss Yi Tien Hsieh in supporting the completion of this project. We hope this book will be a useful and valuable addition to graduate students and researchers dealing with viroid diseases.

Riverside, CA, USA

Ayala L. N. Rao
Irene Lavagi-Craddock
Georgios Vidalakis

Contents

Part III PCR Techniques

Part IV Cloning and Sequencing Techniques

Part V RNA Silencing Techniques

PART VI SPLICING TECHNIQUES

Contributors

SOHRAB BODAGHI • *Department of Microbiology and Plant Pathology, University of California Riverside, Riverside, CA, USA*

IRAKLIS N. BOUBOURAKAS • *Rural Economy & Veterinary Services, Regional Sector of Piraeus, Attica Prefecture, Piraeus, Greece*

MENGJI CAO • *National Citrus Engineering Research Centre, Citrus Research Institute, Southwest University, Chongqing, China*

JOANA CARBAJAL-MORENO • *Department of Microbiology and Plant Pathology, University of California Riverside, Riverside, CA, USA*

KITTY CARDWELL • *Institute of Biosecurity and Microbial Forensics, Oklahoma State University, Stillwater, OK, USA*

GRANT A. CHAMBERS • *Elizabeth Macarthur Agricultural Institute, New South Wales Department of Primary Industries, Menangle, NSW, Australia*

TYLER DANG • *Department of Microbiology and Plant Pathology, University of California Riverside, Riverside, CA, USA*

JOSÉ-ANTONIO DARÒS • *Instituto de Biología Molecular y Celular de Plantas (Consejo Superior de Investigaciones Científicas-Universitat Politècnica de València), Valencia, Spain*

SHOU-WEI DING • *Department of Microbiology and Plant Pathology, University of California, Riverside, CA, USA*

J. ALLAN DODDS • *Department of Microbiology and Plant Pathology, University of California, Riverside, CA, USA*

NERIDA J. DONOVAN • *Elizabeth Macarthur Agricultural Institute, New South Wales Department of Primary Industries, Menangle, NSW, Australia*

NÚRIA DURAN-VILA • *Plant Protection and Biotechnology, Instituto Valenciano de Investigaciones Agrarias (IVIA), Moncada, Valencia, Spain*

ANDRES ESPINDOLA • *Institute of Biosecurity and Microbial Forensics, Oklahoma State University, Stillwater, OK, USA*

FRANCESCO FAGGIOLI • *Consiglio per la Ricerca in Agricoltura e l'analisi dell'Economia agraria (CREA)—Centro di Ricerca Difesa e Certificazione (CREA-DC), Rome, Italy*

JUDE GROSSER • *Horticultural Sciences Department, University of Florida-IFAS, Citrus Research and Education Center (CREC), Lake Alfred, FL, USA*

SUBHAS HAJERI • *Citrus Pest Detection Program, Central California Tristeza Eradication Agency, Tulare, CA, USA*

ROSEMARIE W. HAMMOND • *United States Department of Agriculture-Agricultural Research Service (USDA-ARS), Molecular Plant Pathology Laboratory, Beltsville, MD, USA*

JAMES A. HEICK • *Department of Microbiology and Plant Pathology, University of California, Riverside, CA, USA*

YI HSIEH • *Department of Microbiology and Plant Pathology, University of California Riverside, Riverside, CA, USA*

KRITON KALANTIDIS • *Biology Department, University of Crete, Heraklion, Greece; IMBB/FoRTH, Heraklion, Greece*

MARIA S. KAPONI • *Department of Heraklion, Hellenic Agricultural Insurance Organization (ELGA), Maiandrou & Petrakogiorgi, Crete, Greece*
KONSTANTINA KATSAROU • *Biology Department, University of Crete, Heraklion, Greece; IMBB/FoRTH, Heraklion, Greece*
OXANA KEKSTIDOU • *Scientific Directorate of Phytopathology, Laboratory of Virology, Benaki Phytopathological Institute, Athens, Greece*
ROBERT R. KRUEGER • *United States Department of Agriculture, Agricultural Research Service (USDA-ARS), National Clonal Germplasm Repository for Citrus and Dates, Riverside, CA, USA*
NIKOLETA KRYOVRYSANAKI • *Biology Department, University of Crete, Heraklion, Greece; IMBB/FoRTH, Heraklion, Greece*
PANAYOTA E. KYRIAKOPOULOU • *Department of Crop Science, Agricultural University of Athens, Athens, Greece*
IRENE LAVAGI-CRADDOCK • *Department of Microbiology and Plant Pathology, University of California Riverside, Riverside, CA, USA*
MARTA LUIGI • *Consiglio per la Ricerca in Agricoltura e l'analisi dell'Economia agraria (CREA)—Centro di Ricerca Difesa e Certificazione (CREA-DC), Rome, Italy*
JUNFEI MA • *Department of Biological Sciences, Mississippi State University, Mississippi State, MS, USA*
DEBORAH M. MATHEWS • *Department of Microbiology and Plant Pathology, University of California, Riverside, CA, USA*
SHACHINTHAKA D. DISSANAYAKA MUDIYANSELAGE • *Department of Biological Sciences, Mississippi State University, Mississippi State, MS, USA*
JAMES NG • *Department of Microbiology and Plant Pathology, University of California Riverside, Riverside, CA, USA*
BELTRÁN ORTOLÁ • *Instituto de Biología Molecular y Celular de Plantas (Consejo Superior de Investigaciones Científicas-Universitat Politècnica de València), Valencia, Spain*
FATIMA OSMAN • *Department of Plant Pathology, University of California Davis, Davis, CA, USA*
DEBORAH PAGLIACCIA • *Department of Microbiology and Plant Pathology, University of California Riverside, Riverside, CA, USA*
ALI RAZA • *Hefei National Laboratory for Physical Sciences at Microscale, School of Life Sciences, University of Science and Technology of China, Hefei, China*
TAVIA RUCKER • *Department of Microbiology and Plant Pathology, University of California Riverside, Riverside, CA, USA*
TERUO SANO • *Faculty of Agriculture and Life Science, Hirosaki University, Hirosaki, Japan*
PEDRO SERRA • *Instituto de Biología Molecular y Celular de Plantas (UPV-CSIC), Consejo Superior de Investigaciones Científicas-Universitat Politecnica de Valencia, Valencia, Spain*
HEATHER N. SMITH • *Department of Biological Sciences, Mississippi State University, Mississippi State, MS, USA*
GERHARD STEGER • *Institut für Physikalische Biologie, Heinrich-Heine-Universität Düsseldorf, Düsseldorf, Germany*
ALEXANDRA SYED • *Department of Microbiology and Plant Pathology, University of California Riverside, Riverside, CA, USA*
SHI-HUA TAN • *Department of Microbiology and Plant Pathology, University of California Riverside, Riverside, CA, USA*

GERARDO URIBE • *Department of Microbiology and Plant Pathology, University of California Riverside, Riverside, CA, USA*
CHRISTINA VARVERI • *Scientific Directorate of Phytopathology, Laboratory of Virology, Benaki Phytopathological Institute, Athens, Greece*
NIKON VASSILAKOS • *Scientific Directorate of Phytopathology, Laboratory of Virology, Benaki Phytopathological Institute, Athens, Greece*
GEORGIOS VIDALAKIS • *Department of Microbiology and Plant Pathology, University of California Riverside, Riverside, CA, USA*
JINBO WANG • *Department of Microbiology and Plant Pathology, University of California Riverside, Riverside, CA, USA; United States Department of Agriculture-APHIS-Biotechnology Regulatory Service-Biotechnology Risk Assessment Program, Riverdale, MD, USA*
YING WANG • *Department of Biological Sciences, Mississippi State University, Mississippi State, MS, USA*
QINGFA WU • *Hefei National Laboratory for Physical Sciences at Microscale, School of Life Sciences, University of Science and Technology of China, Hefei, China*
RAYMOND K. YOKOMI • *USDA-ARS San Joaquin Valley Agricultural Sciences Center, Parlier, CA, USA*

Part I

Biological Techniques

Chapter 1

Study and Detection of Citrus Viroids in Woody Hosts

Robert R. Krueger and Georgios Vidalakis

Abstract

Biological indexing is based upon the ability of certain plants, referred to as indicator plants or indicators, to produce specific symptoms when inoculated with a pathogen using mechanical means or grafting. In the case of citrus viroids, clonal indicators are grafted on to vigorous rootstocks such as rough lemon (*Citrus* × *granulata* Raf.). The 'Arizona-861-S-1' citron clonal indicator (*C. medica* L.) can detect and bioamplify all citrus viroids; however, for specific citrus variants of the hop stunt viroid (i.e., CVd-IIb and CVd-IIc), the clonal indicator 'Parson's special # 9' mandarin (*C. reticulata* Blanco) is preferred. Inoculation techniques and symptom expression are described in detail. Other supporting elements, such as greenhouse conditions and propagation techniques, are also presented.

Key words Indexing, Bioassay, Exocortis, Cachexia, Indicator, Graft inoculation

1 Introduction

Biological indexing, also referred to as "indexing" and "bioassay," is a long-established and useful method of detecting plant pathogens, particularly viruses and other graft-transmissible agents. It is based upon the ability of certain plants, referred to as indicator plants or indicators, to produce specific symptoms when inoculated with a pathogen using mechanical means or grafting [1]. Although newer rapid and sensitive laboratory-based techniques are available, biological indexing is still recommended in plant introduction and sanitation programs because it may detect pathogens or isolates not detectable by molecular-based methods using genome sequence-specific probes or primers. Biological indexing is also useful in studying pathogen/host interactions. Viroid bioindexing may be carried out in either herbaceous (*see* Chapter 2) or woody hosts. The current chapter concentrates on the latter.

The first citrus diseases later shown to be caused by viroids were xyloporosis (later synonymized with cachexia) in the early 1930s and exocortis in the late 1940s [2]. Detection or diagnosis of cachexia was initially done visually in the field or using various

Ayala L. N. Rao et al. (eds.), *Viroids: Methods and Protocols*, Methods in Molecular Biology, vol. 2316,
https://doi.org/10.1007/978-1-0716-1464-8_1,

indicator genotypes, preferably 'Orlando' tangelo (*Citrus* × *aurantium* L.). Later, under specific conditions, 'Parson's special #9' (*C. reticulata* Blanco) mandarin budded onto a vigorous rootstock such as rough lemon (*C.* × *granulata* Raf.) or Volkamer lemon (*C.* × *granulata* Raf.) became the preferred indicator [3]. Detection or diagnosis of exocortis was also originally done visually in the field. Later, 'Etrog' citron (*C. medica* L.) was identified as a superior indicator [4], with the selections USDA 60-13 [5] and later A-861 (also known as Arizona 861) [6] being identified as the preferred selections for this purpose. Citron seedlings were initially used as indicators. This was not optimal since citron is monoembryonic. Although they have a high level of self-pollination and are often morphologically similar, citron seedlings are genetically different and can show differences in sensitivity to inoculation with viroid-infected tissue. Identification and subsequent sanitation of the more sensitive A-861-S-1 (also known as S-1) seedling selection [7], now the preferred indicator, made use of the clonal line by budding 'A-861-S-1' onto a vigorous rootstock the preferred method of detection using woody indicators. The range of symptoms displayed on the citron indicator led to the erroneous concept that there were many different strains of exocortis of varying severity. It was not until the development of molecular techniques in the 1980s and later that it was possible to identify specific viroid species [8]. At that time, symptoms specific to different viroids (other than citrus variants of the hop stunt viroid (HSVd), i.e., CVd-II) were identified, and combinations of viroids were found to have synergistic effects, producing stronger responses in the indicator [9].

Classic biological indexing for a wide range of graft-transmissible citrus pathogens has been practiced since the late 1940s [2]. Although this chapter concentrates on viroid pathogens of citrus, the general principles and practices are the same for all biological indexing activity. In classical biological indexing for graft-transmissible pathogens of citrus, indicator plants are grown in controlled, protected conditions (greenhouses), are graft-inoculated with suspect tissue (blind buds) or a positive control blind bud or not inoculated at all (negative control), and develop typical symptoms if the pathogen is present or remain asymptomatic if it is not present. Biological indexing is a system in which environment, plant host, and pathogen all play a part. The current chapter presents information mainly on the actual inoculation and symptom development in indexing for citrus viroids, with only an outline of the supporting system of greenhouses, soil mix (substrate), and plant growth. Many references for general aspects of these factors are available. Authoritative works specific to citrus are available [10, 11]. The reader is referred to these works for more complete information on the infrastructure needed for successful indexing activities with the understanding that certain features may have to be modified to deal with local conditions.

This chapter will describe the use of clonal indicators 'A-861-S-1' for the detection of citrus viroids other than CVd-II and 'Parson's special # 9' for the detection of CVd-II variants.

2 Materials

There are many different types of greenhouses and plant indicator production systems available, and most can be used successfully for biological indexing. Therefore, in addition to the materials list, a few comments are also provided.

2.1 Greenhouse Indexing Facilities

1. Temperature. Temperature control is very important. Indexing for most citrus pathogens is done under "cool" conditions (24–27 °C day/18–21 °C night), whereas indexing for viroids and a few other pathogens is done under "hot" conditions (32–40 °C day/24–27 °C night). These temperatures need to be maintained for up to 6 months for the detection of most viroids in the citron indicator and for up to a year for the detection of cachexia in 'Parson's special # 9'. Multiple greenhouses or greenhouses with multiple chambers are thus necessary for a complete indexing program. In most cases, there is also a third intermediate temperature room (30–35 °C day/10–24 °C night) for growing indicator plants and rootstocks. The requirement for accurate temperature control has several implications. In many regions, it is more difficult to maintain cool greenhouse temperatures during warm weather periods than it is to maintain high greenhouses temperatures during cool weather periods. Consequently, indexing under cool temperatures is usually initiated in the fall, whereas indexing for viroids can be initiated during most times of year (*see* **Note 1**).
2. Light. Illumination conditions are also important. High light intensities are needed for good symptom expression, so the light transmissibility of the greenhouse covering is important (*see* **Note 2**). In the case of the cool temperature pathogens, it might be necessary to use a shade system to maintain suitable temperatures. This is not as important for high-temperature viroid indexing, but a shade system can in some cases increase the efficiency of heating due to the lower volume of air heated. Supplemental lighting is useful for symptom expression in some cool temperature pathogens and also increases the rate of plant growth when producing indicators or rootstocks.
3. Pest control. An important consideration in greenhouse use is arthropod pest control. Insect and mite damage can obscure symptom development as well as decrease plant growth. Greenhouses should be "insect proof" and have a vestibule to help exclude arthropod pests. A pest control program is necessary for proper indexing (*see* **Note 3**).

2.2 Production of Indicator Plants

1. Potting mixture. Classical biological indexing is based upon a modification of the "UC System" [12]. This system, originally developed in the 1950s, is based upon the use of a sterilized potting mixture ("soil mixture," "substrate") coupled with fertigation and strict sanitation practices. Citrus is best grown in a modified "UC Mix" [13]. Although this is the preferred substrate, if it is not possible to produce locally, a commercial soil mix may be used [14].
2. Plant nutrition. Fertilizer is supplied through the irrigation system via an injector. The classical system utilized hand-mixed complete fertilizer solutions, but now complete commercial mixes are available and work well (*see* **Note 4**). Sanitation practices include using sterilized plastic pots, elevating pots off floor surfaces, maintaining clean bench tops, performing all practices in enclosed areas, etc.
3. Rootstock seeds. In the case of indexing for citrus viroids, indicators generally consist of budded plants. A rootstock of rough lemon (preferred) or *C. volkameriana* is budded with 'S-1' citron or 'Parson's special # 9'. Thus it is necessary to maintain (or purchase) both a source of seeds and a supply of pathogen-tested source buds. If collected from on-site trees, seeds should be sterilized by immersing the seeds in a 5% solution of commercial sodium hypochlorite for 10 min, rinsing with warm water, placing in a hot water bath for 10 min at 52 °C, rinsing with ambient temperature water, and treating with a fungicide (*see* **Notes 5–7**). The seeds can be stored in plastic containers or doubled plastic bags at 4–6 °C. Since even with these precautions, germination decreases, fresh seeds should be procured every year.
4. Rootstock seedlings. The seeds are generally planted first into "flats" (trays approximately 12–15 cm deep) containing the sterilized soil mix. When they are approximately 8–15 cm tall (generally 11–19 weeks after planting), the seedlings are removed from the tray, off-types are discarded, and the seedlings are transplanted singly into #1 nursery pots. The rootstocks are usable when about 1 m tall (approximately 0.6–1.0 cm caliper or diameter), which is generally about 30 weeks after transplanting (*see* **Note 8**). The rootstocks should be trained to a single leader and all suckers removed to promote rapid upright growth. The rootstocks and other indicators are grown in the intermediate-temperature room described above (less commonly in the high-temperature room used for the actual indexing).
5. Plant indicator budsource. Use of clonal indicators necessitates a source of buds. Budsource trees must be free of all known graft-transmissible pathogens and must be maintained in a

protective structure (greenhouse or screenhouse). The budsource trees must be managed so that there are sufficient usable buds at the time of inoculation (*see* **Note 9**).

6. Positive control budsource. A source of positive control budwood is also required and should be maintained under protection preferably separated from the clean budsources.
7. Grafting tool sanitation. Tool sanitation is very important in biological indexing. Clippers, knives, and other tools in contact with the plant should be sterilized by immersion in a 25% dilution of commercial sodium hypochlorite (*see* **Note 7**). This is particularly important when dealing with known positive controls as well as the suspect tissue. The sodium hypochlorite causes oxidation of steel. Therefore, at the end of the day's use, the blades should be treated with an oil-based compound such as WD-40 or honing oil and wiped clean. More frequent sharpening is also necessary.

3 Methods

3.1 Experimental Design

1. It is very important that both mild and severe isolates of positive controls be used.
2. There are generally 4–6 replicates for each test.
3. The number of tests determines to some extent the rest of the experimental design. If there are only a few tests, it may be necessary to have a greater number of positive trees per test than if there are a larger number of candidate trees. If there are a larger number of tests, then the number of positive trees per test may be less.
4. Since conditions in different areas of the greenhouse may vary, it is important to have the positive and negative controls spaced so that they are represented in these different conditions (*see* **Note 10**).
5. It is important not to crowd the pots too close together and to avoid being too close to cool pads or exhaust fans (*see* **Note 11**).

3.2 Collection and Storage of Inoculum

1. Cutting tools should be sanitized before and after collecting material from an individual plant (*see* **Note 12**).
2. Although there are various methods of inoculation into indicator plants, the most common and easily done is insertion of stem tissue (blind bud) into a cut in the bark of the indicator plant.
3. The tissue for inoculation should be taken from "budsticks" of the same sort used for propagation (Fig. 1). Budsticks are

Fig. 1 (**a**) A typical budstick shown before (left) and after (right) leaf and thorn removal (Toni Siebert-Wooldridge). (**b**) A typical budstick and excised bud for propagation (below buding knife) and blind-bud for graft-inoculation (above buding knife) (Georgios Vidalakis)

shoots from the current year's growth and are still green without any secondary development (i.e., they are not woody but firm).

4. The diameter will vary somewhat depending upon the type of citrus but generally should be in about 0.4–0.8 cm.
5. Length is determined by the person collecting the inoculum but generally is 10–30 cm in length.
6. After removing the budwood from the tree and removing leaves and thorns from the budstick, budwood should be placed in a plastic bag with appropriate labelling.
7. If budwood is transported from one site to another, it should be placed in an ice chest.
8. Budwood may be stored for up to 2 weeks in plastic bags at 4–6 °C, but it is best if used within a few days of collection.
9. Collection of the clonal indicator ('A-861-S-1' or 'Parson's special #9') is done similarly to the collection of inoculum.

3.3 Inoculation

1. Prevention of cross-contamination is very important during inoculation.
2. Cutting tools should be sanitized before and after inoculation of each set of test replicates and after the positive controls (*see* **Note 12**).

Fig. 2 Making the T cut (Toni Siebert-Wooldridge)

3. If a group of several plants is to be inoculated in a day or a short time span, it is recommended that all the healthy controls be done first and all the positive controls be done last.
4. Use of the clonal indicators involves two insertion events: (a) budding with the indicator bud and (b) inoculating with the test tissue.
5. Budding the indicator before any inoculation is recommended, as is starting with the negative (healthy) control, followed by the test replicates, and finally the known positive control.
6. Insertion of the clonal indicator and the test inoculum is accomplished in the same manner. Use of a sharp knife is essential (*see* **Note 13**).
7. A "T cut" is made into the stem of the rootstock (Fig. 2) and the bark pulled away from the stem. The clonal indicator or inoculum is then sliced from the budstick using a single, controlled motion. The tissue removed should be very thin and approximately 1 cm in length.
8. The clonal indicator must contain the actual bud, whereas the inoculum may or may not contain a bud (blind bud).
9. The tissue (bud or blind bud-stem tissue between buds) is then placed in the T cut (Figs. 3 and 4) and wrapped with polyethylene budding tape (Fig. 5) (*see* **Note 14**).
10. The clonal indicator, either 'A-861-S-1' for citrus viroids other than HSVd citrus variants CVd-II or 'Parson's special # 9' for

Fig. 3 Inserting the bud (Toni Siebert-Wooldridge)

Fig. 4 Fully inserted bud (Toni Siebert-Wooldridge)

CVd-II, should be inserted first, approximately 20–25 cm above the soil surface (*see* **Note 15**).

11. Two pieces of inoculum are then inserted anywhere below the clonal indicator bud.

Fig. 5 Wrapping the bud (Toni Siebert-Wooldridge)

12. Negative (healthy) controls are budded with the clonal indicator but are not inoculated.
13. The rootstock is then either bent or cut off to force the clonal indicator bud to push (*see* **Note 16**).
14. It is also possible to wait until after survival is determined to force the bud, but it is more efficient to perform this action at the same time as inoculation.
15. After 2–3 weeks, the budding tape should be unwrapped and survival of both the clonal indicator and the inoculum assessed. Survival of one of the two insertions is adequate.
16. Dead tissue is brown and dry, whereas live tissue, indicating a successful insertion, is green and fresh.
17. Any plants with all inoculum and/or clonal indicator insertions dead should be re-inserted, into either the same rootstock or a fresh rootstock.

3.4 Post-inoculation Care and Time for Symptom Expression

1. Temperatures in the greenhouse must be maintained at 32–40 °C day/24–27 °C night for the duration of the test for good symptom expression.
2. The plants must receive adequate nutrition and appropriate irrigation, and arthropod pests must be controlled.
3. Both the 'A-861-S-1' and 'Parson's special # 9' clonal indicators are trained to a single leader after pushing.

Fig. 6 Training inoculated plants to a single leader on training stake (Robert Krueger)

4. When the shoot is approximately 15 cm, it is tied to a training stake (Fig. 6).
5. 'A-861-S-1' has more of a tendency than 'Parson's special # 9' to grow upright and so minimal suckering is needed with the former.
6. When the shoot reaches about 1 m in height, it is cut back to about 10 cm and again trained to a single 1-m-tall leader.
7. In 'A-861-S-1', exocortis and strong reacting viroid combinations generally begin to show symptoms in 4–10 weeks, whereas milder viroids and viroid combinations take 3–6 months.
8. In practice, this means that after the second 1 m shoot has been trained, the experiment can usually be terminated (*see* **Note 17**).
9. If the mild reacting viroid positives have not shown symptoms, another leader should be trained and possibly the temperature increased.
10. In 'Parson's special # 9', strong reacting CVd-II isolates usually need 6–9 months to show symptoms, and milder reacting isolates need 9–12 months or more.
11. Generally, it is safest to use the 12-month period as the standard.

3.5 Symptom Expression

1. Observation and recording of symptom expression (classically referred to as "reading the index") is the key to using biological indexing for diagnostics.
2. Detection of strong reacting isolates is easier than for mild reacting isolates. Detection of mild reactions is an indication of a successful index.
3. If strong reacting positive controls are detected but mild reacting ones are not, the index should not be accepted and should be repeated.
4. Sometimes environmental conditions, nutrient imbalances, or arthropod pests can cause abnormal appearances in indicator plants and sometimes mask symptoms. This is one reason for the negative controls: any abnormalities that are not caused by the pathogen will also appear in the negative control.

3.5.1 Symptom Expression of Viroids in 'A-861-S-1'

1. As mentioned previously, before the existence of multiple citrus viroid species was known, it was believed that there were a range of severe to mild isolates of exocortis. It is now known that the range of symptom expression in citron is due to various viroids or combinations thereof and that certain mild symptom expressions are associated with specific citrus viroids (Table 1) [9].
2. The following description of citrus viroid symptoms will refer to "mild," "moderate," and "severe" reactions and in some cases to specific viroids that have been associated with them. However, it is not possible to definitively identify these viroids based on symptomology; sequential PAGE (*see* Chapter 3) or molecular tests (*see* Chapters 13–15) are necessary for that.
3. Symptom expression in 'Arizona-861-S-1' can be thought of as a suite of symptoms: stunting, epinasty, petiole necrosis, midvein necrosis, tip browning, and bark cracking (Table 1).
4. Exocortis and strong reacting isolates of citrus viroids produce most or all of these symptoms, while milder isolates produce fewer and of lower intensity.
5. The classic severe reaction in citron includes most of these symptoms (e.g., severe stunting and leaf epinasty and general petiole and midvein necrosis) (Fig. 7, Table 1).
6. Milder reacting viroids or isolates may show epinasty but to a lesser degree. Very mild reactions may include only midvein necrosis, resulting in a mild "bent leaf" symptom, and sometimes stem cracks appear in association with citrus bent leaf viroid, citrus viroid V, and other apscaviroids (Fig. 8, Table 1).

Table 1
Citrus viroid symptom expression in 'Etrog' citron (*Citrus medica* L.) Arizona 861-S-1

Citrus viroid and genus	Symptoms				
	Stunting	Midvein necrosis	Leaf epinasty	Petiole necrosis	Leaf tip browning
Pospiviroid					
CEVd	Severe	General	Severe	General	±
Cocadviroid					
CBCVd	±	Local	Random	–	–
Hostuviroid					
HSVd (CVd-IIa)	None	–	–	–	Faint
HSVd (CVd-IIb)	None	–	–	–	–
Apscaviroid					
CBLVd	±	Local	Random	–	–
CDVd	±	Local	Random	–	–
CVd-V	Mild	Local	–	–	–
CVd-VI	±	–	Mild	Mild	–
CVd-VII	Mild	Local	Random	Mild	–

CEVd citrus exocortis viroid, *CBLVd* citrus bent leaf viroid, *HSVd* hop stunt viroid, *CVd* citrus viroid, *CDVd* citrus dwarfing viroid, *CBCVd* citrus bark cracking viroid

7. Other mild reactions may include only slight petiole necrosis or wrinkling (Fig. 9) with possible tip browning (Fig. 10) and local midrib necrosis (Fig. 11, Table 1).
8. Isolates containing citrus dwarfing viroid and citrus bark cracking viroid generally produce stronger reactions but less severe than exocortis and some mixed infections. Symptoms in this case include variable degrees of epinasty (Fig. 7) and a greater degree of petiole necrosis (Fig. 9).
9. Always remember the range of symptom expression ranges from severely stunted with severe epinasty to barely detectable tip browning and very slight and erratic petiole browning.
10. When the biological indexing system is working correctly, the negative controls will show none of these symptoms.
11. When there are issues with nutrients, the negative controls sometimes show tip browning, and in that case, the tip browning symptom cannot be considered reliable.

Fig. 7 Classic severe symptoms of CEVd in citron showing severe stunting and leaf epinasty (non-inoculated control is shown on the right) (**a**), severe epinasty of leaves (**b**), and severe general petiole and midvein necrosis (**c**) (Nuria Duran-Vila, Subhas Hajeri and Georgios Vidalakis)

3.5.2 Symptom Expression of HSVd Citrus Variants CVd-II in 'Parson's Special # 9'

1. 'A-861-S-1' is asymptomatic when infected with HSVd citrus variants CVd-IIb and CVd-IIc. Therefore, if biological indexing for these pathogens is desired, 'Parson's special # 9' is used as an indicator (*see* **Notes 18** and **19**).
2. Symptom expression is more difficult to observe than in 'A-861-S-1'. The basic symptom is gumming at the bud union.
3. Starting at about 6 months after inoculation, small cuts may be made on the bud union of the positive controls, and initial symptoms can be observed (Fig. 12), after which the bark is replaced and wrapped with budding tape.
4. When 75% of the positive controls show symptoms, all the bark around the bud unions can be peeled back on all plants in the experiment and symptoms observed and recorded.
5. The gumming symptoms appear only at the bud union and the union at the first cutback and rarely elsewhere on the plant.

Fig. 8 Mild leaf epinasty with local necrosis of midvein (**a** and **b**) and stem cracks (**c**) (Nuria Duran-Vila and Georgios Vidalakis)

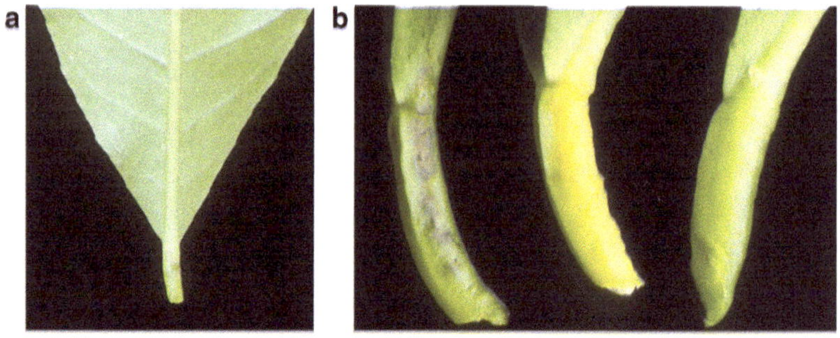

Fig. 9 Mild reaction showing petiole browning (**a**) and wrinkle (non-inoculated control is shown on the right) (**b**) (Nuria Duran-Vila)

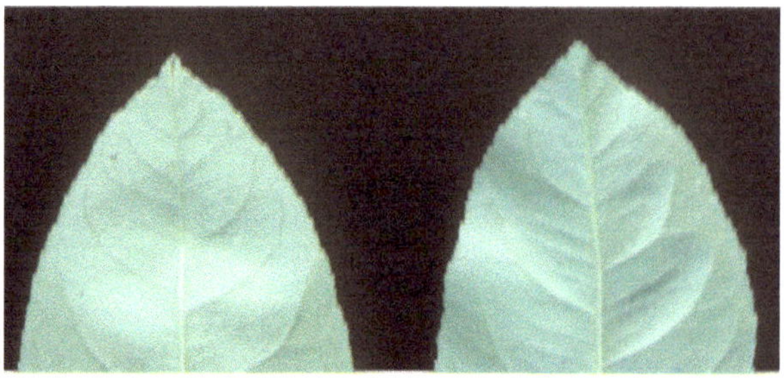

Fig. 10 Very mild reaction showing slight tip browning on leaf (left) compared to control (right) (Nuria Duran-Vila)

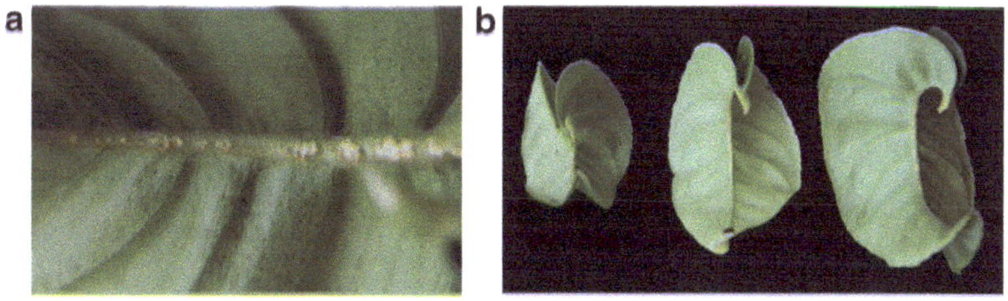

Fig. 11 Mild reaction showing midrib browning and local necrosis (**a**) resulting in leaf epinasty (**b**) (Nuria Duran-Vila)

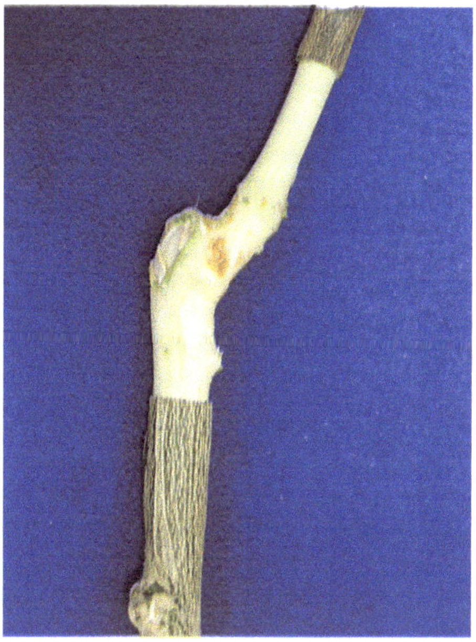

Fig. 12 The first mild gumming symptom usually seen in the bud-union area (Georgios Vidalakis)

Symptom expression is based upon the amount of gumming present (Fig. 13).

3.5.3 Record Keeping

1. As with everything science-related, record keeping is of extreme importance in biological indexing. Each researcher will have their own preferred method.
2. It is recommended that in the case of the reactions in 'A-861-S-1'and 'Parson's special # 9′, the specific symptoms be recorded rather than a general "positive," for instance, "epinasty positive, weak," "petiole wrinkling positive," "mid-rib necrosis, negative," and "gumming severe."

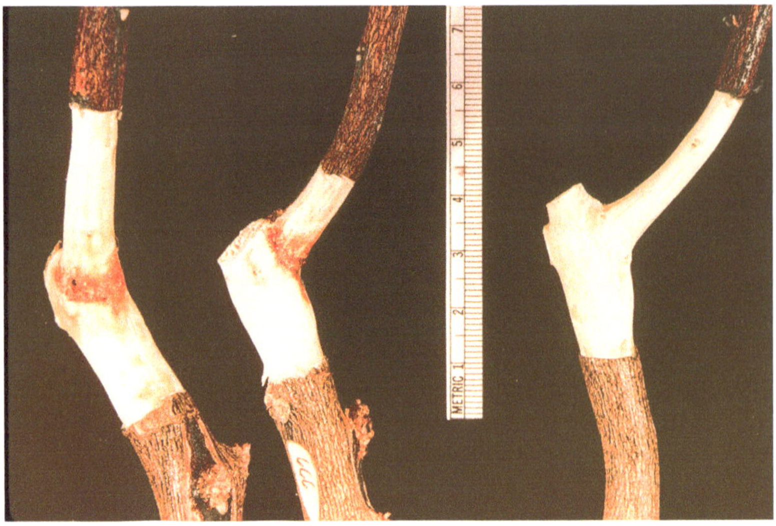

Fig. 13 Gumming and pitting reactions of 'Parson's special # 9' mandarin grafted on rough lemon (left) and of the CVd-IIb (cachexia) isolates of hop stunt viroid (non-inoculated control is shown on the right) (Nuria Duran-Vila)

3. A severity scale from 0.5 to 5 or from + to +++++, for mild to severe reactions, respectively, can be used next to the symptom name.
4. Records should be kept secure indefinitely.

3.5.4 Disposal of Plant Materials

1. Use of positive controls necessitates the safe destruction via autoclave, steam, or incineration. This is usually a condition of the permit.
2. Negative controls can generally be disposed of conventionally (i.e., landfill or compost).
3. Test plants that test positive need to be disposed of in the same manner as the positive controls.
4. Test plants that test negative can usually be disposed of conventionally but in some cases may need to be disposed in a certain way as per permit conditions or standard operating procedures and quality management standards of each program.

4 Notes

1. The condition and capacity of the greenhouse heating and cooling systems need to be taken into account. If a greenhouse lacks a heating system, then the initiation of viroid indexing will probably need to be during the spring or summer.

2. In this regard, glass is generally the best material but currently is not commonly used for greenhouse construction. Polycarbonate is superior to fiberglass as the latter can become discolored and brittle over time.
3. This really means chemical control as biological control never completely eliminates arthropod pests. There should be a "zero-tolerance" policy with regard to arthropod pests.
4. Although not well documented, growth is best with complete fertilizer mixes when actual nitrogen is adjusted to 200 ppm and magnesium to 50 ppm.
5. If there is excess pulp on the seeds when extracted from the fruit, it should be removed using a pectinase such as Clariseb RL 2×.
6. Although commercial fungicides may be used, may programs treat extracted, sterilized seeds by dipping in a 1% solution of 8-hydroxyquinoline sulfate for 3 min, followed by drying at ambient temperature.
7. "Commercial sodium hypochlorite" refers to the standard 5.25% formulation. Some brands have different formulations, and the dilution may have to be adjusted to account for this.
8. Although there is a range of usable size rootstocks, if they become too large and old, the success rate of inoculation and budding is reduced. Therefore, scheduling of rootstock planting needs to be done with the date of inoculation in mind. Many programs plant several batches of rootstocks and indicators each year.
9. Although it is best for each program to maintain its own sources of seeds and buds, this may not be possible in all cases. Limited amounts of indicator seeds and clean source buds are available for research purposes from the USDA-ARS National Clonal Germplasm Repository for Citrus and Dates https://www.ars.usda.gov/main/site_main.htm?modecode=20-36-10-00.
10. In actual fact, differences in the indicator reaction in most well-maintained greenhouses are minimal.
11. A space of 20 cm × 20 cm per pot is recommended.
12. Collection of multiple samples is facilitated by having two tools available to alternate between source plants.
13. Although many different types of knives may be used successfully, specialized budding knives work the best.
14. Specialized tapes for budding that disintegrate after wrapping are available. These are suitable for use with the clonal indicator but should not be used for the inoculum.

15. Two clonal indicator buds may be inserted if there are problems with survival.
16. Bending is the preferred method when using the 'Parson's special # 9' clonal indicator, whereas cutting off works well with the 'S-1' clonal indicator.
17. Current practice calls for tissue from this second flush to be collected for sequential PAGE analysis of citrus viroids (Chapter 3) or other experimental purposes.
18. The 'A-861-S-1' selection is the most sensitive indicator for all viroids except CVd-IIb and CVd-IIc. Consequently, it is the preferred indicator to use and is used clonally as described. In some cases, exocortis or severe isolates will kill the 'S-861-S-1' bud before it can push or show symptoms. In such a case, seedlings of 'A-861-S-1' can be used. The seedlings should be trained to a single leader and kept under the same conditions as the clonal indicator. In general, seedlings should not be used as they are more likely to miss mild isolates.
19. Although 'A-861-S-1' does not show symptoms when infected with only CVd-IIb or CVd-IIc, it does bioamplify those viroids. Therefore, they can be detected via sequential page or molecular-based methods after inoculation into the citron. Consequently, many programs that still use the citron index do not do the 'Parson's special # 9' index but rather follow up the citron index with laboratory diagnostics for the detection of CVd-IIb, CVd-IIc, and other viroids potentially present.

Acknowledgments

The authors acknowledge Dr. Chester N. Roistacher, retired, Citrus Clonal Protection Program (CCPP), University of California, Riverside, who literally wrote the book on biological indexing for graft-transmissible pathogens of citrus [10]. Chet has been an inspiration and friend for many years, and we are truly grateful to have had him as a part of our lives. We also thank Dr. Nuria Duran-Vila, Dr. Subhas Hajeri, and Ms. Toni Siebert-Wooldridge for providing pictures for this chapter. Our thanks also to Dr. Nerida Donovan and Grant Chambers for sharing symptoms information of the newly discovered CVd-VII.

References

1. Legrand P (2015) Biological assays for plant viruses and other graft-transmissible pathogens diagnoses: a review. EPPO Bull 45 (2):240–251. https://doi.org/10.1111/epp.12222
2. Roistacher CN (1995) A historical review of the major graft-transmissible diseases of citrus.

Food and Agriculture Organization, Regional Office for the Near East, Cairo, Egypt
3. Roistacher C, Blue R, Calavan E (1973) A new test for citrus cachexia. Citrograph 58 (7):261–262
4. Frolich EF, Calavan EC, Carpenter JB, Christianson DW, Roistacher CN (1965) Differences in response of citron selections to exocortis virus infection. Paper presented at the third conference International Organization Citrus Virol., Sao Paulo, Brazil, 1963
5. Calavan E, Frolich E, Carpenter J, Roistacher C, Christiansen D (1964) Rapid indexing for exocortis of citrus. Phytopathology 54(11):1359–1362
6. Garnsey S, Cohen M (1965) Response of various citron selections to exocortis infection in Florida. In: Proc. Fla. Sta. Hort. Soc. pp 41–48
7. Roistacher C, Calavan E, Blue R, Navarro L, Gonzales R (1977) A new more sensitive citron indicator for detection of mild isolates of citrus exocortis viroid (CEV) [viral diseases]. Plant Dis Rep 61:135–139
8. Duran-Vila N (2017) Citrus exocortis viroid. In: Hadidi A, Randles JW, Palukaitis P, Flores R (eds) Viroids and satellites. Academic Press, London, pp 169–179. https://doi.org/10.1016/b978-0-12-801498-1.00016-4
9. Duran-Vila N, Pina J, Ballester J, Juarez J, Roistacher CN, Rivera-Bustamante R, Semancik J (1988) The citrus exocortis disease: a complex of viroid RNAs. Paper presented at the tenth conference of the International Organization of Citrus Virologists, Valencia, Spain, 1986
10. Roistacher CN (1991) Graft-transmissible diseases of citrus: handbook for detection and diagnosis. Food and Agriculture Organization of the United Nations, Rome, Italy
11. Roistacher CN (1998) Indexing for viruses in citrus. In: Hadidi A, Khetarpal RK, Koganezawa H (eds) Plant virus disease control. American Phytopathological Society Press, St. Paul, MN, pp 301–319
12. Baker KF, Chandler PAE (1957) UC system for producing healthy container-grown plants. Manual 23. University of California, Division of Agricultural Sciences, Oakland, CA
13. Nauer EM, Roistacher CN, Labanauskas CK (1968) Growing citrus in modified UC potting mixes. Citrograph 53(12):1–3
14. Dekkers MGH, Lee RF, Roistacher CN (2000) Evaluation of commercial potting mixes for optimizing growth of citrus in containers. Proc Florida State Hortic Soc 113:130–133

Chapter 2

Detection and Characterization of Viroids via Biological Assays on Herbaceous Hosts

Rosemarie W. Hammond

Abstract

The characterization of the elusive disease agent of the potato spindle tuber disease, potato spindle tuber viroid (PSTVd), was aided by the ability to obtain large amounts of infected tomato tissue in a simple bioassay where PSTVd was easily mechanically transmissible to an alternate herbaceous host in which it thrived and produced dramatic symptoms in a relatively short period (Diener, Viroids. Handbook of plant virus infections: comparative diagnosis. Elsevier/North-Holland, Amsterdam, pp 913–934, 1981; Diener, Virology 45:411–428, 1971; Raymer and O'Brien, Am Pot J, 39:401–408, 1962). Reactions in the primary, or secondary, herbaceous indicator host can range from asymptomatic to severe depending upon the viroid strain, host species, and environmental conditions and can provide evidence of a viroid infection, but do not permit identification of the viroid in question. Further characterization by molecular hybridization, RT-PCR, and sequence analysis is used to determine the etiology of the disease agent. In this chapter, methods are described for mechanical inoculation of viroids to herbaceous hosts to determine the viroid nature of diseases and the experimental host range of the viroid or to shorten the time required for obtaining relatively large amounts of viroid for subsequent purification and characterization.

Key words Viroids, Biological indicator hosts, Herbaceous host, Infectious plasmid DNA, RNA transcripts, Agroinfiltration, Mechanical inoculation

1 Introduction

The identification of herbaceous indicator hosts for viroids has scientific and practical value as a tool for viroid discovery and purification. Herbaceous hosts are better diagnostic indicators for certain viroids than woody plants, with the short incubation times required for viroid accumulation, relatively high titers of viroid accumulation, and display of symptoms [1]. All viroids are transmissible by mechanical means, either by standard abrasion of leaf tissue with infected sap or by grafting or razor slashing, the latter two most common in woody hosts. For mechanical transmission to herbaceous hosts, rub inoculation of infected sap, viroid-containing plasmid DNAs, or in vitro synthesized RNA transcripts and agroinfiltration are the most common methods for

Ayala L. N. Rao et al. (eds.), *Viroids: Methods and Protocols*, Methods in Molecular Biology, vol. 2316,
https://doi.org/10.1007/978-1-0716-1464-8_2, © Springer Science+Business Media, LLC, part of Springer Nature 2022

experimental transmission. Matousek et al. [2] utilized biolistic bombardment of plasmid DNAs to identify new experimental weed hosts for the pospiviroids PSTVd and hop stunt viroid (HSVd). Nie and Singh summarized primary and indicator hosts for viroid bioassays that have been employed in viroid research [3] (*see* **Note 1**).

Individual viroids vary in their ability to replicate in different plant species. PSTVd can infect 160 species, primarily in the family Solanaceae [4], while several viroids share indicator hosts and elicit similar symptoms in tomato (*Solanum lycopersicon*), chrysanthemum (*Chrysanthemum morifolium*), and cucumber (*Cucumis sativus*) and *Gynura aurantiaca* herbaceous hosts [5–10].

Grafting was reported to be more reliable than sap in mums for the transmission of chrysanthemum stunt viroid (CSVd) in chrysanthemum [11].

Herbaceous hosts have also been identified for viroids whose natural hosts are woody species. Several experimental herbaceous hosts were discovered for citrus viroid V, whose natural host range is restricted to citrus, using infectious cDNA clones and agroinoculation, in addition to mechanical inoculation with infected sap, allowing further studies on its biological properties [12]. In addition, apple scar skin viroid (ASSVd: type member of the apscaviroids), whose natural host range is woody hosts, initially could only be indexed by graft inoculation and bud inoculation onto susceptible woody hosts. Nine experimental herbaceous hosts of ASSVd were identified, including cucumber and tobacco, three of which displayed visual symptoms, using an infectious ASSVd cDNA clone [13], facilitating further studies of structure/function relationships.

2 Materials

2.1 Materials for In Vitro Transcription and Inoculation

1. Plasmid DNAs or transcripts for inoculation.
2. QIAprep Spin Miniprep Kit (Qiagen).
3. MinElute Reaction Cleanup Kit (Qiagen).
4. Restriction enzymes for cloning and linearization of plasmids.
5. T7, T3, or SP6 RNA polymerase transcription kits (Ambion).
6. Inoculation buffer: 50 mM K_2HPO_4, pH 7.0, for transcript inoculation and for mechanical transmission from sap.
7. Environmentally controlled growth chamber.
8. Climate-controlled greenhouse.

2.2 Agrobacterium Strain for Agroinoculation

1. *Agrobacterium tumefaciens* containing viroid dimeric inserts cloned into a binary vector.

2.3 Materials Required for Rub Inoculation

1. 500–600 mesh carborundum.
2. Autoclaved mortar and pestle.

2.4 Materials Required for Grafting

1. Razor blades or miter-cut grafting knife.
2. Parafilm.
3. Silicon grafting clips.

2.5 Transformation of Agrobacteria and Agroinoculation

1. LBg media (2.5 g yeast extract, 5 g Bacto Tryptone, 2.5 g NaCl, 2.5 g glucose, pH 7.0 with 5 N NaOH in 500 mL water).
2. YEP media (10 g yeast extract, 10 g Bacto Peptone, 5 g NaCl in 1 L water, pH 7.0 with 5 N NaOH).
3. Chemically competent *Escherichia coli* cells (commercially available).
4. Chemically competent *Agrobacterium tumefaciens* prepared by growing a 50 mL culture of bacterial cells at 28 °C in LBg or YEP media overnight, centrifuging the cells, and resuspending in cold 20 mM $CaCl_2$.
5. 10 mM MES, pH 5.6, 10 mM $MgCl_2$, 150 μM acetosyringone (agroinfiltration buffer).
6. Needleless 1.0 mL syringes.

3 Methods

3.1 Plasmid DNA or Transcript Generation and Inoculation

1. To prepare viroid transcripts from dimeric plasmid clones [13], recombinant plasmids purified from bacteria using the QIAprep Spin Miniprep Kit are linearized using an appropriate restriction enzyme. The restricted plasmid DNA can be purified using the MinElute Reaction Cleanup Kit. One μg of plasmid DNA containing a dimeric copy of the viroid can be linearized and rub inoculated following the same procedure as for (*see* Chapter 20, this publication).
2. Positive-sense RNA transcript inocula are generated from purified, restricted plasmid DNAs using T7, T3, SP6, or other RNA polymerases and transcription kits per manufacturer's instructions; the enzyme used depends upon the RNA polymerase promoter present in the plasmid DNA and the orientation of the viroid insert within the plasmid. Approximately 100–200 ng of RNA transcripts are diluted in inoculation buffer prior to rub inoculation onto plant leaves that have been lightly dusted with carborundum (e.g., 20 μL RNA transcripts are added to 80 μL of inoculation buffer, and 20 μL of the mix are inoculated onto one leaf of each of the five host

plants). The plants are maintained in an environmentally controlled growth chamber or a climate-controlled greenhouse and are grown at 20–23 °C night and 26 °C day temperature and monitored for symptom expression (dependent upon the host species (*see* **Notes 2** and **3**)).

3.2 Agroinfiltration of Viroid Gene Constructs

Agrobacterium tumefaciens containing viroid dimeric inserts are cloned into a binary vector [14]. Briefly, dimeric viroid-derived clones are prepared by restriction digestion of a monomeric cDNA clone, followed by ligation of the monomeric inserts with compatible ends and cloning into a binary plasmid DNA vector containing compatible restriction enzyme sites [14]. The resulting ligations are transformed into chemically competent *Escherichia coli* cells and plated on LB nutrient media agar plates containing the appropriate antibiotic corresponding to the resistance gene present on the bacterial plasmid. Plasmid DNAs are isolated from individual bacterial colonies using the QIAprep Spin Miniprep Kit. The plasmid DNAs are characterized by restriction enzyme digestion and nucleotide sequence analysis to ensure that a dimeric viroid (head to tail) sequence is present within the plasmid. The resulting plasmid is transformed into chemically competent *Agrobacterium tumefaciens* (e.g., strain EHA105). Once the transformed *A. tumefaciens* is prepared, it can be used for agroinfiltration as described below.

3.3 Mechanical Transmission from Sap

Inocula can be prepared by grinding viroid-infected plant tissue (4–5 mL buffer per gm of tissue) in inoculation buffer (50 mM K_2HPO_4 buffer, pH 7.0) in a sterile mortar and pestle, followed by rubbing on leaves of an herbaceous host that has been lightly dusted with 500–600 mesh carborundum using the pestle (*see* **Note 4**). The inoculated leaves are then rinsed with tap water after inoculation. The plants are grown at 20–23 °C night and 26 °C day temperature in an environmentally controlled growth chamber or greenhouse, if possible, for tomato [6]. The infectivity index can be evaluated by dilution of inoculum, time required for symptom expression, and number of plants infected at each dilution [15]. Avoid mechanical transmission of viroids by plant-to-plant contact (*see* **Note 5**).

3.4 Grafting

Graft inoculation from one herbaceous host to another can be performed (e.g., potato to tomato) [6]. In graft transmission tests, side-grafting of scions bearing one leaf were side-grafted to stocks 2–4 nodes below the apex, and grafts were bound with parafilm [6]. Alternatively, scions and stocks can both be cut using a miter-cut cutting knife and held together using the appropriately sized silicone grafting clip. In both cases, the grafted plants should be kept under high humidity for 4–6 days to allow the graft to heal. Once the graft has healed, the parafilm or grafting clip can be removed and the plants kept at appropriate temperatures for plant growth and viroid replication.

3.5 Agroinfiltration of Viroid Gene Constructs

Agrobacterium tumefaciens constructs are grown in LBg or YEP broth overnight at 28 °C, and the bacteria are pelleted down and resuspended in agroinfiltration buffer. The resuspended bacterial cells are incubated at room temperature for 1–4 h and infiltrated on the abaxial side of host leaves using 1 mL needleless syringe. Ten plants infiltrated with each construct are maintained in a greenhouse or environmentally controlled growth chamber (20–25 °C for 16-h light/8-h dark). Fourteen to 21 days' post-infiltration, total RNA can be extracted and analyzed by RT-PCR for viroid detection and characterization.

If the host is susceptible to infection, symptoms of viroid infection can be scored. The symptoms exhibited in the inoculated plant depend upon the host plant, viroid, viroid isolate, and environmental conditions and are described in the cited literature. In many cases, the inoculated host may be symptomless. The viroid can be purified and characterized using gel electrophoresis, RT-PCR, cDNA cloning, and sequence analysis (*see* **Note 6**).

4 Notes

1. Different inoculation strategies may be better for various viroid/host combinations. Walia et al. [13] reported that agroinfiltration and mechanical inoculation of dimeric RNA transcripts achieved 100% infectivity; however, in this case, the viroid sequence was known, and these methods necessitated the generation of infectious cDNA copies of the viroid.
2. Host selection and environmental conditions for biological indexing have been summarized by Semancik [16] and Singh and Ready [17].
3. Several viroids have been identified recently in symptomless ornamental plants. Verhoeven et al. [18] describe the successful mechanical transmission of isolated sap and contaminated finger tips and razor blades between plant species, including tomato, potato, and solanaceous ornamentals.
4. Use gloves and sterilized mortars and pestles when preparing inoculum from infected plants and transferring to healthy plants. Autoclave all glassware and mortars/pestles to disinfect after use.
5. Avoid plant-to-plant contact of inoculated plants to avoid unintentional mechanical transmission of viroids.
6. Multiple virus/viroid infections may be present in the plant which is used to prepare the sap for inoculation to the

experimental host. The use of biological indexing combined with molecular assays is recommended for conclusive identification of the viroid.

References

1. Diener TO (1971) Potato spindle tuber "virus". IV. A replicating, low molecular weight RNA. Virology 45:411–428
2. Matousek J, Orctová L, Ptácek J, Patzak J, Dedic P, Steger G, Riesner G (2007) Experimental transmission of *Pospiviroid* populations to weed species characteristic of potato and hop fields. J Virol 81:11891–11899
3. Nie X, Singh RP (2017) Viroid detection and identification by bioassay. In: Hadidi A, Flores R, Randles JW, Palukaitis P (eds) Viroids and satellites. Academic Press, New York, pp 347–356
4. Singh RP (1973) Experimental host range of the potato spindle tuber 'virus'. Am Potato J 50:111–123
5. Diener TO (1981) Viroids. In: Kurstak E (ed) Handbook of plant virus infections: comparative diagnosis. Elsevier, Amsterdam, pp 913–934
6. Raymer WB, O'Brien MJ (1962) Transmission of potato spindle tuber virus to tomato. Am Potato J 39:401–408
7. Yaguchi S, Takahashi T (1984) Response of cucumber cultivars and other cucurbitaceous species to infection by hop stunt viroid. J Phytopathol 109:21–32
8. Matsushita Y, Tsuda S (2015) Host ranges of potato spindle tuber viroid, tomato chlorotic dwarf viroid, tomato apical stunt viroid, and columnea latent viroid in horticultural plants. Eur J Plant Pathol 141:193–197
9. Niblett CL, Dickson E, Horst RK, Romaine CP (1980) Additional hosts and an efficient purification procedure for four viroids. Phytopathology 70:610–615
10. Weathers LG, Greer FC (1972) Gynura as a host for exocortis virus of citrus. In: Price WC (ed) Proceedings of the fifth conference of the International Organization of Citrus Virologist. University of Florida Press, Gainesville, FL, pp 95–98
11. Hollings M, Stone OM (1973) Some properties of chrysanthemum stunt, a virus with the characteristics of an uncoated ribonucleic acid. Ann Appl Biol 73:333–348
12. Ebrahimi-Moghadam L, Zakiaghl M, Jafarpour B, Mehrvar M (2018) Identification of experimental host rage of *Citrus viroid V*. J Agric Sci Technol 20:1279–1291
13. Walia Y, Dhir S, Ram R, Zaidi AA, Hallan V (2014) Identification of the herbaceous host range of *Apple scar skin viroid* and analysis of progeny variants. Plant Pathol 63:684–690
14. Molins-Marquez J, Navarro JA, Pallas V, Gomez G (2019) Highly efficient construction of infectious viroid-derived clones. Plant Methods 15:87
15. Raymer WB, Diener TO (1969) Potato spindle tuber virus: a plant virus with properties of a free nucleic acid. I. Assay, extraction, and concentration. Virology 37:343–350
16. Semancik JS (2003) Host considerations. In: Hadidi A, Flores R, Randles JW, Semancik JS (eds) Viroids. CSIRO Publishing, Collingwood, pp 87–88
17. Singh RP, Ready KFM (2003) Biological indexing. In: Hadidi A, Flores R, Randles JW, Semancik JS (eds) Viroids. CSIRO Publishing, Collingwood, pp 89–94
18. Verhoeven JTJ, Humer L, Marn MV, Plesko IM, Roenhorst JW (2010) Mechanical transmission of *Potato spindle tuber viroid* between plants of *Brugmansia suaveoles*, *Solanum jasminoides* and potatoes and tomatoes. Eur J Plant Pathol 128:417–421

Chapter 3

Analysis on RNA Motif-Based RNA Trafficking in Plants

Heather N. Smith, Junfei Ma, and Ying Wang

Abstract

Systemic RNA trafficking widely exists in plants and is critical for integrating the healthy development and responses to environmental cues. Viroids, single-stranded circular noncoding RNAs that infect plants, have been used as a model to delineate the mechanism underlying systemic RNA trafficking. Recent work on viroids has shown that structural motifs are critical to direct RNA trafficking through distinct cellular boundaries. Here, we describe the methods for generating mutational variants using site-directed mutagenesis and infection assays to unravel the function of RNA motifs. This approach can be modified to study other RNA motif-based biological processes.

Key words RNA systemic trafficking, Viroid, Site-directed mutagenesis, DIG Northern blot

1 Introduction

Multicellular organisms evolve various approaches to integrate their development and responses to environmental changes. Intercellular RNA trafficking, including cell-to-cell and systemic RNA trafficking, is a widely existing phenomenon in plants that likely utilizes RNA molecules to communicate within an organism [1–5]. Those non-cell autonomous RNAs traffic through plasmodesmata for cell-to-cell movement and exploit phloem to move systemically within a plant [2–4]. The spreading of these genetic materials serves as an effective means for cell communications [5–8].

Recent studies on viroids have shown that RNA motifs are regulatory elements to guide RNA molecules cross distinct cellular boundaries in plants [9, 10]. Viroids, single-stranded circular noncoding RNAs that infect plants [11–13], exploit plasmodesmata and phloem to achieve cell-to-cell and systemic trafficking, respectively. Thus, viroids spreading in plants resembles those endogenous non-cell autonomous RNAs [11]. Using potato spindle tuber viroid (PSTVd) as a model, at least 11 out of 27 RNA motifs in its

Ayala L. N. Rao et al. (eds.), *Viroids: Methods and Protocols*, Methods in Molecular Biology, vol. 2316,
https://doi.org/10.1007/978-1-0716-1464-8_3,

genome have been shown to regulate PSTVd trafficking from local to systemic leaves [14]. Furthermore, critical RNA motifs have been found directing PSTVd movement across various cellular boundaries, such as from palisade to spongy mesophyll [15], from bundle sheath to phloem [16], and from bundle sheath to mesophyll [17]. These observations support a model that distinct RNA motifs exist to direct RNA movement through various cellular boundaries.

Here, we describe the methodology to analyze the function of RNA motifs in RNA systemic trafficking. Site-directed mutagenesis is used to generate saturated mutation pools of any given RNA motif. These mutational variants are used for infection assays to evaluate the function of the motif in systemic RNA trafficking. The information of the functional sequence variations can be used to support the RNA structural modeling [18, 19]. This approach can also be modified to study the function of RNA motifs in other biological processes.

2 Materials

2.1 Site-Directed Mutagenesis

1. 50× Tris-acetate-EDTA (TAE) buffer: Weigh 242 g Tris base, and mix it with 51.7 mL glacial acetic acid, 100 mL 0.5 M EDTA (pH 8.0), and 700 mL distilled water. Bring up the volume to 1 L and autoclave.
2. 1% agarose gel: 0.4 g agarose, 0.8 mL 10× TAE buffer, and 39.2 mL distilled water. Use a microwave to heat the mixture to dissolve agarose. Keep the gel solution at room temperature for 5 min, and then add 2 μL Midori Green nucleic acid staining solution (Nippon Genetics). Pour the agarose solution in a gel cassette, and place an appropriate comb.
3. 100 mg/mL ampicillin: Weigh 1 g ampicillin powder, and mix it with 9 mL ddH_2O. Bring up the volume to 10 mL using ddH_2O. Sterilize it using a 0.22 μm syringe filter.
4. LB solution: Weigh 25 g LB broth (Miller), and mix it with 900 mL ddH_2O. Bring up the volume to 1 L using ddH_2O. Autoclave the solution, and store the solution at 4 °C.
5. LB plate: Weigh 25 g LB broth (Miller) and 8 g agar, and mix them with 900 mL ddH_2O. Bring up the volume to 1 L using ddH_2O. Autoclave the mixture. Cool down to 55 °C, and add 1 mL 100 mg/mL ampicillin. Pour the solution to a petri dish, and allow the medium to cool and become solid. Store the plates at 4 °C.

2.2 Materials for DIG (Digoxigenin) Northern Blot

1. Sodium dodecyl sulfate (SDS) solution: 10% (w/v) solution in ddH_2O (*see* **Note 1**).

2. 2× Northern gel loading buffer: Mix 95 mL formamide, 25 mg xylene cyanol powder, 25 mg bromophenol blue powder, 3.6 mL 0.5 M EDTA (pH 8.0), 250 μL 10% SDS, and 1.15 mL distilled water. Store at −20 °C.
3. 10× Tris-borate-EDTA (TBE): Mix 108 g Tris base, 55 g boric acid, 40 mL 0.5 M EDTA (pH 8.0), and 900 mL distilled water. Adjust volume to 1 L by adding distilled water. Autoclave the solution before storing at room temperature.
4. Ammonium persulfate solution: 10% (w/v) solution in ddH_2O.
5. *N*,*N*,*N*,*N*′-Tetramethyl-ethylenediamine (TEMED): Store at 4 °C.
6. 5% urea denaturing PAGE gel: 1.8 mL UreaGel concentrate solution (National Diagnostics, 19:1), 6.3 mL UreaGel Diluent solution (National Diagnostics), 60 μL ammonium persulfate solution, 5 μL TEMED, 450 μL 10× TBE, and 450 μL ddH_2O. The mixture is poured into a Bio-Rad Mini-PROTEAN Tetra Cell gel cassette, and place an appropriate comb as needed. Let the gel stand at room temperature for 1 h.
7. 0.5× TBE running buffer: Mix 20 mL 10× TBE buffer with 380 mL distilled water.
8. Buffer I: Weigh 4 g NaOH, and dissolve in 900 mL distilled water. Bring up the volume to 1 L using distilled water. Do not autoclave. Store at room temperature.
9. Buffer II: Weigh 24.23 g Tris base, and dissolve in 800 mL distilled water. Adjust pH to 7.5. Bring up the volume to 1 L using distilled water. Autoclave and store the solution at room temperature.
10. 20× SSPE: Dissolve 175.3 g NaCl, 27.6 g $NaH_2PO_4 \cdot H_2O$, and 7.4 g EDTA in 800 mL distilled water. Adjust pH to 7.4 with NaOH. Bring up the volume to 1 L with distilled water. Autoclave and store the solution at room temperature.
11. Fish sperm DNA solution: 100 mg/mL solution in ddH_2O. Store at −20 °C.
12. Denhardt's blocking buffer: Mix 50 mL formamide, 25 mL 20× SSPE, 2.5 mL commercial 100× Denhardt's stock, 10 mL 10% SDS, and 1 mL 100 mg/mL fish sperm DNA. Bring up the volume to 100 mL using distilled water. Store at 4 °C.
13. Sephadex G-25 resin: Weigh 50 g Sephadex G-25, and wash three times using 500 mL 65 °C ddH_2O. The resin is then mixed with 225 mL ddH_2O and 25 mL 100% ethanol. The resin slurry is stored at 4 °C.

14. DIG-labeled PSTVd probe: pInt(−) plasmid [20] is linearized by digestion with *Spe*I at 37 °C for 1 h and 65 °C heat treatment for 20 min. To generate the DIG-labeled PSTVd probe using a T7 MAXIscript kit (Invitrogen), mix 1 μL 10× reaction buffer, 1 μL T7 polymerase, 0.5 μL 10 mM ATP, 0.5 μL 10 mM GTP, 0.5 μL 10 mM CTP, 0.2 μL 10 mM UTP, 2 μL 3.5 mM DIG-11-UTP (Millipore Sigma), 2 μL 205 ng/μL linearized plasmid, and 2.3 μL nuclease-free water. The reaction is kept at 37 °C for 2 h, followed by incubating 0.5 μL DNaseTurbo (Invitrogen) at 37 °C for 20 min. Add 10 μL nuclease-free water to the reaction, and clean up the buffer by using Sephadex G-25 resin.
15. 20× SSC: Mix 175.3 g NaCl, 88.2 g sodium citrate, and 800 mL distilled water. Adjust pH to 7.0. Bring up the volume to 1 L and autoclave.
16. 5 M NaCl: Dissolve 146 g NaCl in 400 mL distilled water. Adjust the volume to 500 mL. Autoclave and store at room temperature.
17. 0.2 M maleic acid solution: Dissolve 23.22 g maleic acid powder in 500 mL distilled water. Adjust pH to 7.5 and add distilled water to a final volume of 1 L. Autoclave and store at room temperature (*see* **Note 2**).
18. Maleic acid wash buffer: Mix 50 mL 0.2 M maleic acid solution, 3 mL 5 M NaCl, 46.7 mL distilled water, and 300 μL 100% Tween-20.
19. Maleic acid blocking buffer: Prepare 5 mL dissolving solution by mixing 2.5 mL 0.2 M maleic acid solution, 150 μL 5 M NaCl, and 2.3 mL distilled water. Mix 50 mg blocking reagent (Millipore Sigma, #11096176001) for nucleic acid hybridization and detection with the 5 mL dissolving solution. Heat at 65 °C for 30–60 min to dissolve the blocking reagent.

3 Methods

The three-dimensional (3D) structures of RNA molecules are mainly composed of helices and loops. Continuous Watson-Crick base pairings form the helices. The Watson-Crick edge, which is involved in the formation of WC base pairings, is one of the three edges in any RNA base. The other two edges are sugar and Hoogsteen edges. In RNA motifs, all three edges can be used for base interactions [21] to form non-Watson-Crick (nWC) base pairings [22]. Substituting a base in a loop motif may retain the function of the RNA motif only if the substitution keeps (1) the same base edges for interaction, (2) the same C1′–C1′ distance in 3D space, and (3) the same orientation of glycosidic bonds. The qualified base substitutions are isosteric [23]. Sequence homology search using the JAR3D program would match the input query to possible RNA motifs that have been experimentally determined and infer isosteric base interactions [24]. Readers are referred to the following references for details [15, 16, 18, 19, 24]. Below, we would focus on how to generate the mutational variants and the infection assay as a functional test. The functional data can be used to support the RNA 3D structural modeling.

3.1 Generating Mutational Variants and Infection

1. Design primers containing the desired sequence. Keep the mutation sequence in the middle of the primers, flanked by 10–15 nucleotides on each end.
2. In a 20 μL reaction system, mix 2 μL 10× Pfu polymerase buffer, 1 μL 10 mM dNTP, 0.4 μL 20 μM forward primer, 0.4 μL 20 μM reverse primer, 0.2 μL Pfu polymerase (Promega), 15 μL ddH_2O, and 1 μL 10 ng/μL plasmid containing wild-type PSTVd sequence, pRZ-Int [25], as template. Set the PCR program as following: 95 °C for 30 s, 16 cycles (95 °C for 30 s, 55 °C for 30 s, 72 °C for 4 min), and 72 °C for 1 min (*see* **Note 3**).
3. Run 6 μL PCR product in a 1% agarose gel with 1× TAE buffer. If there is a band with the expected size (~3.6 kb in this case), proceed to plasmid digestion. Add 1 μL *Dpn*I enzyme to the rest of the PCR product (~14 μL), and incubate at 37 °C for 1 h, followed by 65 °C treatment for 20 min (*see* **Note 4**).
4. Take 2 μL digested PCR product for bacteria transformation using Invitrogen TOP10 chemical competent cell. Briefly, incubate the DNA with 10 μL competent cell, and keep the mixture on ice for 5 min. Heat the mixture at 42 °C for 2 min, and supplement the mixture with 100 μL LB. Shake the culture at 37 °C for 1 h. Spread the culture on a LB plate with 100 μg/mL ampicillin.
5. Pick two colonies for miniprep and Sanger sequencing. The miniprep is performed using Qiagen plasmid miniprep kit following the user manual.
6. Choose the plasmid with the correct sequence for linearization. Digest pRZ-Int plasmid [25] with *Hind*III at 37 °C for 1 h, followed by 65 °C treatment for 20 min.
7. The linearized plasmid is used as the template for in vitro transcription using T7 MEGAscript kit (Invitrogen). The reaction mixture contains 1 μL MEGAscript reaction buffer, 1 μL MEGAscript T7 polymerase, 1 μL ATP (75 mM), 1 μL GTP (75 mM), 1 μL CTP (75 mM), 1 μL UTP (75 mM), 2 μL linearized plasmids (250 ng/μL), and 2 μL nuclease-free water. The reaction is kept at 37 °C overnight.

8. The in vitro transcription reaction is stopped by adding 1 μL DNaseTurbo and incubating at 37 °C for 20 min. The RNA is purified using MEGAclear kit (Invitrogen) following user manual. Use a NanoDrop to measure the concentration of the final product. Adjust the RNA concentration to 20 ng/μL using nuclease-free water.
9. Infect the *Nicotiana benthamiana* seedlings at the four-leaf stage. Dust the leaves with silicon carbide powder. Add 5 μL PSTVd (20 ng/μL) solution to the first and second true leaves, and use a P10 tip to scratch the leaf surface (*see* **Note 5**).

3.2 Northern Blot Detection of Viroid Trafficking

1. Take samples (~50 mg) from local and systemic leaves. Grind the leaf samples in liquid nitrogen, and then add 200 μL RNAzol RT (Molecular Research Center, Inc) and 80 μL distilled water to each sample. After incubation at room temperature for 10 min, spin the mixtures at 12,000 × *g* for 10 min at room temperature (*see* **Note 6**).
2. Take 200 μL supernatant, and add 200 μL absolute ethanol to each sample. Spin the mixtures at 12,000 × *g* for 10 min at room temperature. Remove all the liquid, and resuspend the RNA pellet with 10 μL distilled water and 10 μL 2× Northern gel loading buffer. Boil samples for 3 min, and then directly put on ice for 2 min. After a brief spin, samples are ready for gel loading.
3. Run RNA samples in a 5% urea denaturing PAGE gel with 0.5× TBE buffer. Use 200 V constant voltage to run the gel for 1 h (*see* **Note 7**).
4. After gel running, the gel is incubated in 10 mL 0.5× TBE supplement with 2 μL Midori Green nucleic acid staining solution for 5 min. The gel is then visualized with the Bio-Rad Gel Doc XR+ gel documentation system under SYBR green safe mode to obtain the loading control.
5. Incubate the gel in Buffer I for 5 min and then Buffer II for 5 min.
6. Assemble a transfer "sandwich" by laying the following items from bottom to top: pad (soaked in 0.5× TBE), filter paper, Hybond-XL nylon membrane, gel, filter paper, and pad (*see* **Note 8**).
7. Transfer the gel under constant 18 V for 20 min using a Bio-Rad Trans-Blot SD system.
8. Use a UV crosslinker to treat the membrane twice with 180 mJ/cm^2, followed by 30 min incubation in Denhardt's blocking buffer at 65 °C. Then add 5 μL DIG-labeled PSTVd probe to incubate overnight at 65 °C.

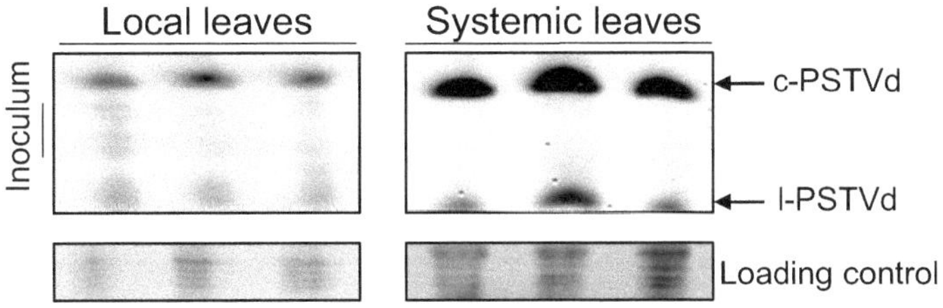

Fig. 1 A representative Northern blot result of wildtype PSTVd in local and systemic leaves. It is notable that the remaining inoculums are often visible in Northern blots using RNA from local leaves. Ethidium bromide staining of ribosomal RNAs serves as loading control. c-PSTVd and l-PSTVd depict circular and linear forms of PSTVd, respectively. (The data are adapted from Ref. [12])

9. Wash membrane twice with 2× SSC (supplemented with 0.1% SDS) by rotating for 10 min each at 65 °C, followed by two washes with 0.2× SSC (supplemented with 0.1% SDS) by rotating for 10 min each at 65 °C.
10. Rinse the membrane with 5 mL maleic acid wash buffer. Then incubate the membrane with 5 mL maleic acid-blocking buffer for 1 h. The blocking temperature should cool down from 65 °C to room temperature within this 1 h.
11. Dilute anti-DIG antibody (Millipore Sigma) 1:10,000 in maleic acid-blocking buffer, and incubate the membrane with the antibody at room temperature for 1 h.
12. Remove the antibody solution. Wash the membrane with a sufficient amount of maleic acid wash buffer at room temperature for 15 min. Repeat the wash once.
13. Remove the wash buffer, and incubate the membrane with a sufficient amount of substrate at 37 °C for 5 min.
14. Scan the membrane with a Li-Cor C-Digit scanner to visualize the result (Fig. 1).
15. Randomly pick 2–3 infected RNA samples to clone the PSTVd cDNA. For reverse transcription, 0.5 μg total RNA extracted in **step 2** of this section is dissolved in 5.5 μL nuclease-free water; then add 0.5 μL 10 mM dNTP and 0.5 μL 20 μM PSTVd-specific reverse primer (5′-cccggggatccctgaagcgctcc-3′). The mixture is heated at 65 °C for 5 min, and immediately put on ice for 2 min. The following chemicals are then added to the mixture: 2 μL 5× First-Strand buffer, 0.5 μL RNaseOUT (Invitrogen), 0.5 μL 100 mM DTT, and 0.5 μL SuperScript III reverse transcriptase (Invitrogen).
16. Take 2 μL reverse transcription product as a template to amplify PSTVd cDNA using PSTVd-specific reverse primer and PSTVd-specific forward primer (5′-ggggaaacctggagcgaa ctgg-3′) (*see* **Note 9**). The PCR product is inserted into

pCR4 vector (Invitrogen) following the user manual, followed by TOP10 chemical competent cell transformation (Invitrogen). Pick ten colonies from each plate for Sanger sequencing (*see* **Note 10**).

4 Notes

1. Do not autoclave 10% SDS solution.
2. Adjusting pH for the maleic acid solution requires a large volume of NaOH solution. Therefore, it is better not to add more than 600 mL distilled water to mix with the powder.
3. Do not allow more than 18 PCR cycles because it would introduce random mutations.
4. Other high-fidelity Taq polymerase can be used for site-directed mutagenesis PCR. If using a Phusion Taq polymerase or its derivatives, the PCR product will need to go through a Sephadex G-25 column first. After this step, *Dpn*I and its appropriate buffer are added to PCR product for digestion.
5. We have achieved the best infection rate using *N. benthamiana* seedlings at the four-leaf stage. After mechanical inoculation, it is better to keep the plants in high-moisture conditions.
6. Local leaf samples are taken 6–8 days post-inoculation. Half leaf can be sampled, and keep the other half on plants. Systemic leaf samples can be sampled as early as 14 days post-inoculation. However, sampling after 3 weeks post-inoculation would achieve a stable infection rate. It is critical to check the presence of circular form PSTVd (Fig. 1) in local leaves, which would confirm the replication capacity.
7. Before loading RNA samples, wash the wells with running buffer, using a 10 mL syringe.
8. These pads can be re-used by storing in 0.5× TBE at 4 °C to prevent molding.
9. This primer pair covers the genome region 75–116. If the desired RNA loop resides in this region, it is required to change to other PSTVd-specific primers.
10. This cloning and Sanger sequencing step is critical to ensure no mutations occurred to designed constructs.

References

1. Dlouha D, Hubacek JA (2017) Regulatory RNAs and cardiovascular disease—with a special focus on circulating microRNAs. Physiol Res 66:S21–S38
2. Ham BK, Lucas WJ (2017) Phloem-mobile RNAs as systemic signaling agents. Annu Rev Plant Biol 68:173–195

3. Nazim Uddin M, Kim JY (2013) Intercellular and systemic spread of RNA and RNAi in plants. Wiley Interdiscip Rev RNA 4:279–293
4. Reagan BC, Ganusova EE, Fernandez JC et al (2018) RNA on the move: the plasmodesmata perspective. Plant Sci 275:1–10
5. Spiegelman Z, Golan G, Wolf S (2013) Don't kill the messenger: long-distance trafficking of mRNA molecules. Plant Sci 213:1–8
6. Banerjee AK, Chatterjee M, Yu Y et al (2006) Dynamics of a mobile RNA of potato involved in a long-distance signaling pathway. Plant Cell 18:3443–3457
7. Carlsbecker A, Lee JY, Roberts CJ et al (2010) Cell signalling by microRNA165/6 directs gene dose-dependent root cell fate. Nature 465:316–321
8. Liu TY, Lin WY, Huang TK et al (2014) MicroRNA-mediated surveillance of phosphate transporters on the move. Trends Plant Sci 19:647–655
9. Ding B, Wang Y (2009) Viroids: uniquely simple and tractable models to elucidate regulation of cell-to-cell trafficking of RNA. DNA Cell Biol 28:51–56
10. Wang Y, Ding B (2010) Viroids: small probes for exploring the vast universe of RNA trafficking in plants. J Integr Plant Biol 52:28–39
11. Ding B (2009) The biology of viroid-host interactions. Annu Rev Phytopathol 47:105–131
12. Dissanayaka Mudiyanselage SD, Qu J, Tian N et al (2018) Potato spindle tuber viroid RNA-templated transcription: factors and regulation. Viruses 10:E503
13. Flores R, Minoia S, Carbonell A et al (2015) Viroids, the simplest RNA replicons: how they manipulate their hosts for being propagated and how their hosts react for containing the infection. Virus Res 209:136–145
14. Zhong X, Archual AJ, Amin AA et al (2008) A genomic map of viroid RNA motifs critical for replication and systemic trafficking. Plant Cell 20:35–47
15. Takeda R, Petrov AI, Leontis NB et al (2011) A three-dimensional RNA motif in potato spindle tuber viroid mediates trafficking from palisade mesophyll to spongy mesophyll in Nicotiana benthamiana. Plant Cell 23:258–272
16. Zhong X, Tao X, Stombaugh J et al (2007) Tertiary structure and function of an RNA motif required for plant vascular entry to initiate systemic trafficking. EMBO J 26:3836–3846
17. Qi Y, Pelissier T, Itaya A et al (2004) Direct role of a viroid RNA motif in mediating directional RNA trafficking across a specific cellular boundary. Plant Cell 16:1741–1752
18. Takeda R, Zirbel CL, Leontis NB et al (2018) Allelic RNA motifs in regulating systemic trafficking of potato spindle tuber viroid. Viruses 10:E160
19. Wang Y, Zirbel CL, Leontis NB et al (2018) RNA 3-dimensional structural motifs as a critical constraint of viroid RNA evolution. PLoS Pathog 14:e1006801
20. Qi Y, Ding B (2002) Replication of potato spindle tuber viroid in cultured cells of tobacco and Nicotiana benthamiana: the role of specific nucleotides in determining replication levels for host adaptation. Virology 302:445–456
21. Leontis NB, Westhof E (2002) The annotation of RNA motifs. Comp Funct Genomics 3:518–524
22. Leontis NB, Westhof E (2001) Geometric nomenclature and classification of RNA base pairs. RNA 7:499–512
23. Stombaugh J, Zirbel CL, Westhof E et al (2009) Frequency and isostericity of RNA base pairs. Nucleic Acids Res 37:2294–2312
24. Roll J, Zirbel CL, Sweeney B et al (2016) JAR3D Webserver: scoring and aligning RNA loop sequences to known 3D motifs. Nucleic Acids Res 44:W320–W327
25. Hu Y, Feldstein PA, Hammond J et al (1997) Destabilization of potato spindle tuber viroid by mutations in the left terminal loop. J Gen Virol 78(Pt 6):1199–1206

Chapter 4

Isolation and Transfection of Citrus Protoplasts with Citrus Exocortis Viroid

Subhas Hajeri, James Ng, Jude Grosser, and Georgios Vidalakis

Abstract

Viroids are RNA-based infectious agents that are single-stranded, covalently closed circular, non-coding, and naked. Unlike RNA viruses, which at least encode proteins for replication, encapsidation, and movement, lack of protein-coding capacity of viroids makes them completely reliant on host for replication and movement. The high genetic diversity in viroids is believed to be due to the absence of proof-reading activity of the host RNA polymerases, the large population size, and the rapid rate of replication. Protoplasts are viable plant cells that are prepared by enzymatic removal of cell walls. Plant protoplasts provide a synchronous single-cell system for studying early events of viroid infection such as replication and genetic diversity at the cellular level. A simple and efficient method to isolate and transfect citrus protoplasts with transcript RNA of citrus exocortis viroid is described in this chapter.

Key words PEG method, Suspension cell culture, Viroid replication, Genetic diversity

1 Introduction

The systemic infection process of a viroid can be divided into two major steps: (1) replication in individual cells (within nucleus or chloroplast) and (2) movement throughout the plant [1]. Viroids completely rely on the host RNA polymerase II for replication [2–4]. The plant RNA polymerase II lacks proof-reading activity, which leads to the introduction of errors into nascent RNA [5]. The high genetic diversity in viroids, like other RNA genomes, is believed to be due to the absence of proof-reading activity of the host RNA polymerases, the large population size, and the rapid rate of replication [6–8]. In fact, a viroid RNA has the highest in vivo mutation rate among all known nucleic acid replicons [1].

Cocking was the first to isolate plant protoplast enzymatically [9], transfect tomato protoplasts with tobacco mosaic virus [10], and establish the single-cell system to study plant viruses. The pioneer work showed that virus infection was synchronous and its replication was supported by the efficient uptake of viral particle/

Ayala L. N. Rao et al. (eds.), *Viroids: Methods and Protocols*, Methods in Molecular Biology, vol. 2316, https://doi.org/10.1007/978-1-0716-1464-8_4, © Springer Science+Business Media, LLC, part of Springer Nature 2022

RNAs by protoplasts [11–13]. Since then, protoplasts have been extensively used for the study of replication of plant viruses and viroids at the cellular level [14–18]. A high genetic diversity at the cellular level and a low genetic diversity at the whole plant level were observed for citrus exocortis viroid (CEVd) in citrus plant system [18]. It is conceivable that the selection pressure is much greater and the dynamics of viroid replication and movement is more complex at the plant level as compared to the cellular level in protoplasts. An obvious difference in protoplasts is the absence of both cell-to-cell and long-distance movement for viroids. Thus, protoplasts transfected with viroid molecules offer a synchronous system to study the replication and population profile of viroids at the cellular level.

2 Materials

2.1 Special-Purpose Equipment

1. Plant growth chamber (Thermo Scientific™ PR505755L).
2. Laminar flow hood Class II (Air Science VLF48).
3. Benchtop incubator shaker (New Brunswick™ I24).
4. Hemocytometer (Weber Scientific 3048-11).
5. Epifluorescent Zeiss Axioplan Microscope.

2.2 General-Purpose Equipment

1. Hotplate magnetic bar stirrer (Thermo Scientific SP88857100).
2. Refrigerated centrifuge (Eppendorf 5430 R) with Rotor F-35-6-30 and Rotor FA-45-30-11.
3. Weighing scale (Mettler Toledo™ 30029095).
4. Electronic pipet controller/dispenser (Drummond 4-000-105).
5. Manual pipet controller/dispenser (Wheaton W835148).

2.3 Consumables

1. Syringe filters 0.2 μM (Fisher Scientific 09-719C).
2. Thermo Scientific™ Nalgene™ 0.2 μM sterile disposable filter unit (Fisher Scientific 09-741-02).
3. Disposable glass Pasteur pipets (Fisher Scientific 13-678-4).
4. Disposable serological pipets 10 mL (Fisher Scientific 13-678-11E).
5. Disposable Petri dishes (Fisher Scientific 09-720-500).
6. Falcon™ cell strainers 70 μm (Fisher Scientific 08-771-2).
7. 15 mL conical centrifuge tubes (Fisher Scientific 14-959-53A).
8. 50 mL conical centrifuge tubes (Fisher Scientific 14-959-49A).
9. Falcon™ round-bottom snap cap tubes 5 mL (Fisher Scientific 14-959-11A).

2.4 Source of Viroid RNA and Citrus Protoplast Material

A cDNA clone of CEVd in a plasmid with T7 or SP6 promoters to generate in vitro transcripts [18]. A suspension cell culture, developed from callus culture of the sweet orange cultivar 'Hamlin' (*Citrus sinensis* L. Osbeck) [19–21].

2.5 Enzymes, Reagents, and Consumables

Prepare all solutions using sterile distilled deionized (dd) water and reagents suitable for plant cell culture. Prepare and store all solutions at room temperature (unless indicated otherwise). PEG-8000 solutions have to be prepared freshly.

2.5.1 Suspension Cell Culture Media (H + H)

1. Prepare stock solutions [19–23] in advance (*see* **Note 1**) (*see* Table 1).
2. MT macronutrient stock (50×): 1 M NH_4NO_3, 1 M KNO_3, 50 mM KH_2PO_4, 10 mM K_2HPO_4, and 75 mM $MgSO_4 \cdot 7H_2O$. Add 800 mL dd water to a beaker. Weigh 82.5 g NH_4NO_3, 95 g KNO_3, 18.5 g $MgSO_4 \cdot 7H_2O$, 7.5 g KH_2PO_4, and 1 g K_2HPO_4 and dissolve completely and bring the final volume to 1 L.
3. BH3 macronutrient stock (100×): 2 M KCl, 150 mM $MgSO_4 \cdot 7H_2O$, 100 mM KH_2PO_4, and 20 mM K_2HPO_4. Add 800 mL dd water to a beaker. Weigh 150 g KCl, 37 g $MgSO_4 \cdot 7H_2O$, 15 g KH_2PO_4, and 2 g K_2HPO_4 and dissolve completely and bring the final volume to 1 L.
4. MT micronutrient stock (100×): 10 mM H_3BO_3, 10 mM $MnSO_4 \cdot H_2O$, 3 mM $ZnSO_4 \cdot 7H_2O$, 1 mM KI, 0.1 mM $Na_2MoO_4 \cdot 2H_2O$, 0.01 mM $CuSO_4 \cdot 5H_2O$, and 0.01 mM $CoCl_2 \cdot 6H_2O$. Add 900 mL dd water to a beaker. Weigh 620 mg H_3BO_3, 1680 mg $MnSO_4 \cdot H_2O$, 860 mg $ZnSO_4 \cdot 7H_2O$, 83 mg KI, 25 mg $Na_2MoO_4 \cdot 2H_2O$, 2.5 mg $CuSO_4 \cdot 5H_2O$, and 2.5 mg $CoCl_2 \cdot 6H_2O$ and dissolve completely and bring the final volume to 1 L.
5. MT iron stock (200×): 20 mM Na_2.EDTA and 20 mM $FeSO_4 \cdot 7H_2O$ (EDTA). Add 900 mL dd water to a beaker. Weigh 7.45 g Na_2.EDTA and 5.57 g $FeSO_4 \cdot 7H_2O$ (EDTA) and dissolve completely and bring the final volume to 1 L.
6. MT calcium stock (66×): 200 mM $CaCl_2 \cdot 2H_2O$. Weigh 29.33 g $CaCl_2 \cdot 2H_2O$ to 900 mL dd water and dissolve completely and bring the final volume to 1 L.
7. MT vitamin stock (100×): 3 mM thiamine HCl, 5 mM pyridoxine HCl, 4 mM nicotinic acid, and 60 mM myo-inositol. Add 900 mL dd water to a beaker. Weigh 1 g thiamine HCl, 1 g pyridoxine HCl, 0.5 g nicotinic acid, and 10 g myo-inositol and dissolve completely and bring the final volume to 1 L.
8. To prepare 1 L liquid H + H media, in a sterile clean beaker, add 800 mL sterile dd water, 10 mL of MT macronutrient stock (50×), 5 mL of BH3 macronutrient stock (100×),

Table 1
Components of suspension cell culture media (H + H) [19–23]

Components	Supplier/cat. number
NH_4NO_3	Sigma-Aldrich A3795
KNO_3	Sigma-Aldrich P8291
KH_2PO_4	Sigma-Aldrich NIST200B
$MgSO_4 \cdot 7H_2O$	Sigma-Aldrich 63138
$CaCl_2 \cdot 2H_2O$	Sigma-Aldrich C5670
Na_2 EDTA	Sigma-Aldrich A6635
$FeSO_4 \cdot 7H_2O$ (EDTA)	Sigma-Aldrich F7002
$MnSO_4 \cdot H_2O$	Sigma-Aldrich 229784
$ZnSO_4 \cdot 7H_2O$	Sigma-Aldrich Z1001
H_3BO_3	Sigma-Aldrich B6768
KCl	Sigma-Aldrich P9333
KI	Sigma-Aldrich 60399
$Na_2MoO_4 \cdot 2H_2O$	Sigma-Aldrich M1651
$CuSO_4 \cdot 5H_2O$	Sigma-Aldrich C3036
$CoCl_2 \cdot 6H_2O$	Sigma-Aldrich C2911
Glutamine	Sigma-Aldrich G8540
Thiamine HCl	Sigma-Aldrich T1270
Pyridoxine HCl	Sigma-Aldrich P9755
Nicotinic acid	Sigma-Aldrich N4126
Myo-inositol	Sigma-Aldrich I7508
Malt extract	Sigma-Aldrich 70167
Sucrose	Sigma-Aldrich S0389
Agar	Sigma-Aldrich A9539

10 mL of MT micronutrient stock, 5 mL of MT iron stock (200×), 15 mL of MT calcium stock (66×), 10 mL of MT vitamin stock (100×), 35 g sucrose (100 mM), 1.55 g glutamine (10 mM), and 500 mg malt extract, and dissolve completely.

9. Bring the volume to 990 mL, adjust the pH to 5.6 with 10 N KOH, and make the final volume to 1 L.

10. Filter sterilize through sterile disposable filter unit of two 500 mL.

11. Store at room temperature in the dark (wrap with aluminum foil).
12. For maintenance of callus culture, prepare H + H medium with agar (8 g/L).

2.5.2 Stock Solution

1. Needed during preparation of enzyme solution.
2. Stock solution: 24 mM $CaCl_2$, 6.25 mM MES buffer, 1 mM NaH_2PO_4, pH 5.6.

2.5.3 Enzyme Solution

High-quality enzymes are critical for the isolation of citrus protoplasts. Prefer to obtain the enzymes from highly reliable source [18].

1. Enzyme solution: 13% D-mannitol, 2% cellulase 'Onozuka' RS, 2% macerozyme R10.
2. Dissolve the components in 60 mL of dd water (*see* **Note 2**).
3. Add 2 mL of the stock solution and adjust the pH to 5.6 with 10 N KOH.
4. Adjust the volume at 80 mL and place it in two 50 mL Oak Ridge tubes. Spin for 5 min at 12,298 × *g*.
5. Sterilize filtering through a 0.2 μM syringe filter and store at 4 °C (*see* **Note 3**).

2.5.4 Stocks of CPW Salts (100×)

1. CPW salts needed in subsequent reagent preparation.
2. CPW Stock A (100×): 20 mM KH_2PO_4, 100 mM KNO_3, 100 mM $MgSO_4 \cdot 7H_2O$, 0.01 mM KI, 0.01 mM $CuSO_4 \cdot 5H_2O$ (*see* **Notes 1** and **4**). Add 900 mL dd water to a beaker. Weigh 2.72 g KH_2PO_4 and dissolve well. Subsequently weigh each of the components 10 g KNO_3, 2.72 g $MgSO_4 \cdot 7H_2O$, 16 mg KI, and 0.3 mg $CuSO_4 \cdot 5H_2O$, and dissolve. Bring the volume to 1000 mL.
3. CPW Stock B (100×): 100 mM $CaCl_2 \cdot 2H_2O$. Dissolve 15 g of $CaCl_2 \cdot 2H_2O$ in 1000 mL water.
4. Autoclave and store both stock solutions at 4 °C.

2.5.5 1 M MES, 1 M $CaCl_2$, and 10 N KOH

1. 1 M MES: Dissolve 9.76 g of MES in 40 mL of dd H_2O (*see* **Note 5**). Adjust final volume at 50 mL. Autoclave for 20 min. Keep at room temperature.
2. 1 M $CaCl_2$: Dissolve 14.7 g of $CaCl_2$ in 80 mL of dd H_2O. Adjust final volume at 100 mL. Autoclave for 20 min. Keep at room temperature.
3. 10 N KOH: Dissolve 56.11 g of KOH in 90 mL of dd H_2O. Adjust final volume at 100 mL. Keep at room temperature.

2.5.6 25% SucroseM

1. 25% SucroseM solution needed during protoplast isolation step.
2. 25% SucroseM solution: 25% sucrose, 1× CPW Stock A, 1× CPW Stock B, 6.25 mM MES.
3. Adjust pH to 5.6 with KOH 10 N, autoclave, and store at room temperature.

2.5.7 13% MannitolM

1. 13% MannitolM solution needed during protoplast isolation step.
2. 13% MannitolM solution: 13% D-mannitol, 1× CPW Stock A, 1× CPW Stock B, 6.25 mM MES.
3. Adjust pH to 5.6 with KOH 10 N, autoclave, and store at room temperature.

2.5.8 0.6MMCPW

1. 0.6MMCPW solution needed during cell viability count and transfection step.
2. 0.6MMCPW solution: 0.6 M D-mannitol, 1× CPW Stock A, 1× CPW Stock B, 6.25 mM MES.
3. Adjust pH to 5.6 with KOH 10 N, autoclave, and store at room temperature.

2.5.9 0.6MMCPWC

1. 0.6MMCPWC solution needed during transfection step.
2. 0.6MMCPWC solution: 0.6 M D-mannitol, 1× CPW Stock A, 1× CPW Stock B, 6.25 mM MES, 10 mM $CaCl_2$.
3. Adjust pH to 5.6 with KOH 10 N, autoclave, and store at room temperature.

2.5.10 0.75MMCPWC

1. 0.75MMCPWC solution needed for preparation of PEG solution.
2. 0.75MMCPWC solution: 0.75 M D-mannitol, 1× CPW Stock A, 1× CPW Stock B, 6.25 mM MES, 10 mM $CaCl_2$.
3. Adjust pH to 5.6 with KOH 10 N, autoclave, and store at room temperature.

2.5.11 30% PEG-8000 Solution

1. 30% PEG-8000 solution needed during transfection step.
2. Place 3 g of PEG-8000 in a 15 mL plastic tube.
3. Add 0.75MMCPWC buffer to a final volume of 10 mL.
4. Mix vortexing and incubate at 55 °C for 5–10 min.
5. Vortex until PEG is completely dissolved.
6. Sterilize filtering through a 0.45 μM filter (*see* **Note 6**).

2.5.12 Citrus Protoplast Culture Media (0.6MBH3)

1. Prepare stock solutions [19–23] in advance (*see* **Note 1**) (*see* Table 2).
2. BH3 macronutrient stock (100×): 2 M KCl, 150 mM $MgSO_4 \cdot 7H_2O$, 100 mM KH_2PO_4, and 20 mM K_2HPO_4. Add 800 mL dd water to a beaker. Weigh 150 g KCl, 37 g $MgSO_4 \cdot 7H_2O$, 15 g KH_2PO_4, and 2 g K_2HPO_4 and dissolve completely and bring the final volume to 1 L.
3. BH3 KI stock (100×): 5 mM KI. Weigh 75 mg KI to 90 mL dd water and dissolve completely and bring the final volume to 100 mL.
4. BH3 multivitamin-I stock (100×): 2 mM calcium pantothenate, 10 mM ascorbic acid, 4 mM choline chloride, 0.1 mM *p*-aminobenzoic acid, 0.5 mM folic acid, 0.3 mM riboflavin, 0.04 mM biotin. Add 90 mL dd water to a beaker. Weigh 50 mg calcium pantothenate, 100 mg ascorbic acid, 50 mg choline chloride, 10 mg *p*-aminobenzoic acid, 20 mg folic acid, 10 mg riboflavin, and 1 mg biotin and dissolve completely and bring the final volume to 100 mL.
5. BH3 multivitamin-II stock (100×): 0.03 mM vitamin A (retinol), 0.02 mM vitamin D3 (cholecalciferol), and 0.01 mM vitamin B12. Add 90 mL dd water to a beaker. Weigh 1 mg vitamin A (retinol), 1 mg vitamin D3 (cholecalciferol), and 2 mg vitamin B12 (*see* **Note 7**) and dissolve completely and bring the final volume to 100 mL.
6. BH3 organic acid stock (50×): 10 mM sodium pyruvate, 10 mM citric acid, 10 mM malic acid, and 20 mM fumaric acid. Add 90 mL dd water to a beaker. Weigh 100 mg sodium pyruvate, 200 mg citric acid, 200 mg malic acid, and 200 mg fumaric acid and dissolve completely and bring the final volume to 100 mL.
7. BH3 sugar alcohol stock (100×): 140 mM fructose, 170 mM ribose, 170 mM xylose, 140 mM mannose, 140 mM rhamnose, 70 mM cellobiose, 140 mM galactose, and 140 mM glucose. Add 90 mL dd water to a beaker. Weigh 2.5 g fructose, 2.5 g ribose, 2.5 g xylose, 2.5 g mannose, 2.5 g rhamnose, 2.5 g cellobiose, 2.5 g galactose, and 2.5 g glucose and dissolve completely and bring the final volume to 100 mL.
8. To prepare 1 L of 0.6MBH3 protoplast culture media, in a sterile clean beaker, add 800 mL sterile dd water, 10 mL of BH3 macronutrient stock (100×), 10 mL of MT micronutrient stock, 5 mL of MT iron stock (200×), 15 mL of MT calcium stock (66×), 10 mL of MT vitamin stock (100×), 2 mL of BH3 multivitamin-I stock (100×), 1 mL of BH3 multivitamin-II stock (100×), 1 mL BH3 KI stock (100×), 10 mL BH3 sugar alcohol stock (100×), 20 mL BH3 organic acid stock (50×), 20 mL coconut water, 51.3 g sucrose

Table 2
Components of citrus protoplast culture media (0.6MBH3) [19–23]

Components	Supplier/cat. number
KH_2PO_4	Sigma-Aldrich NIST200B
$MgSO_4 \cdot 7H_2O$	Sigma-Aldrich 63138
$CaCl_2 \cdot 2H_2O$	Sigma-Aldrich C5670
Na_2 EDTA	Sigma-Aldrich A6635
$FeSO_4 \cdot 7H_2O$ (EDTA)	Sigma-Aldrich F7002
$MnSO_4 \cdot H_2O$	Sigma-Aldrich 229784
$ZnSO_4 \cdot 7H_2O$	Sigma-Aldrich Z1001
H_3BO_3	Sigma-Aldrich B6768
KCl	Sigma-Aldrich P9333
KI	Sigma-Aldrich 60399
$Na_2MoO_4 \cdot 2H_2O$	Sigma-Aldrich M1651
$CuSO_4 \cdot 5H_2O$	Sigma-Aldrich C3036
$CoCl_2 \cdot 6H_2O$	Sigma-Aldrich C2911
Glutamine	Sigma-Aldrich G8540
Thiamine HCl	Sigma-Aldrich T1270
Pyridoxine HCl	Sigma-Aldrich P9755
Nicotinic acid	Sigma-Aldrich N4126
Myo-inositol	Sigma-Aldrich I7508
Malt extract	Sigma-Aldrich 70167
Sucrose	Sigma-Aldrich S0389
Casein hydrolysate	Sigma-Aldrich 22090
Mannitol	Sigma-Aldrich M1902
Coconut water	Sigma-Aldrich C5915
Fructose	Sigma-Aldrich F3510
Ribose	Sigma-Aldrich R9629
Xylose	Sigma-Aldrich X3877
Mannose	Sigma-Aldrich M6020
Rhamnose	Sigma-Aldrich R3875
Cellobiose	Sigma-Aldrich 22150
Galactose	Sigma-Aldrich G5388
Glucose	Sigma-Aldrich G7021

(continued)

Table 2 (continued)

Components	Supplier/cat. number
Sodium pyruvate	Sigma-Aldrich P5280
Citric acid	Sigma-Aldrich C0706
Malic acid	Sigma-Aldrich M7397
Fumaric acid	Sigma-Aldrich F8509
Vitamin B12	Sigma-Aldrich V6629
Calcium pantothenate	Sigma-Aldrich P5155
Ascorbic acid	Sigma-Aldrich A4403
Choline chloride	Sigma-Aldrich C7527
p-Aminobenzoic acid	Sigma-Aldrich A9878
Folic acid	Sigma-Aldrich F8758
Riboflavin	Sigma-Aldrich R9504
Biotin	Sigma-Aldrich B4639
Vitamin A (retinol)	Sigma-Aldrich R7632
Vitamin D3 (cholecalciferol)	Sigma-Aldrich D1530

(150 mM), 3.1 g glutamine (20 mM), 130 g mannitol (13% or 715 mM), 250 mg casein hydrolysate, and 1 g malt extract, and dissolve completely.

9. Bring the volume to 990 mL, adjust the pH to 5.6 with 10 N KOH, and make the final volume to 1 L.
10. Filter sterilize through sterile disposable filter unit of two 500 mL.
11. Store at room temperature in the dark (wrap with aluminum foil).

3 Methods

3.1 In Vitro Transcription of CEVd RNAs

1. Isolate the plasmid containing a full-length CEVd cDNA from *Escherichia coli* strain JM109 (Promega Corp.) using QIAprep Miniprep Kit (Qiagen) following the manufacturer's instructions [22].
2. Test the presence of cloned insert DNA by a diagnostic digestion of the plasmid DNA with specific restriction enzyme (New England Biolabs Inc.) and further confirm by sequencing.
3. For the generation of plus-sense CEVd RNA transcripts, linearize the plasmid DNA with specific restriction enzyme (New England Biolabs Inc.).

Table 3
General guideline in establishing and maintaining citrus suspension cell culture

Days in 4-week cycle	Citrus suspension cell culture
0	Split into 2 flasks (2×)
7	Cell culture media change (2×)
14	Cell culture media change (2×)
21	Cell culture media change (2×)
28	– One flask: Split (2×) – Second flask: Cell culture media change (1×) (meant for protoplast isolation)
7	– Cell culture media change (2×) – No cell culture media change to the flask (meant for protoplast isolation)
14	– Cell culture media change (2×) – *Protoplast isolation*

4. Generate run-off transcripts using linearized plasmid as a template with T7 or SP6 RNA polymerase, based on insert orientation, from MEGAscript Kit (Ambion, Inc.).
5. Spectrometrically quantify the freshly prepared transcripts.
6. Aliquot transcripts into batches of 5 μg each.

3.2 Suspension Cell Culture

Follow the general guideline in establishing and maintaining citrus suspension cell culture as shown in Table 3 [22].

1. Initiate the suspension cultures by transferring approximately 2 g of callus tissue to a 125 mL Erlenmeyer flask containing 20 mL of suspension cell culture media H + H and shaking continuously on a rotary shaker at 125–150 rpm.
2. After 2 weeks, add another 20 mL of suspension cell culture media H + H to each flask to bring the total to 40 mL.
3. Maintain the rapidly growing suspension cultures by sub-culturing every 2 weeks, by splitting the contents of one flask into two and restoring the volume to 40 mL with fresh H + H media.
4. Use one of the flasks for isolating protoplast while keeping the other flask for maintenance of the suspension cell culture.

3.3 Isolation of Citrus Protoplasts

(*See* **Notes 8–10**)

3.3.1 Isolation of Citrus Protoplasts from Suspension Cell Culture

1. Cells used for protoplast isolation should be in the log phase of growth, with best results using suspension cells from 4 to 10 days into the 2 week sub-culture cycle.
2. Divide one flask of suspension culture (28th day) in four sterile disposable Petri plates.

3. Remove the H + H suspension cell culture media (*see* **Note 11**).
4. Add 7.5 mL of 0.6MBH3 protoplast culture media and 5.5 mL of enzyme solutions, and mix gently (*see* **Note 12**).
5. Seal the Petri plate with parafilm and put them on an orbital shaker and cover the Petri plate with an opaque container.
6. Incubate for overnight digestion (~16 h) in dark at 28 °C with slow shaking (20–40 rpm) (*see* **Notes 13** and **14**).
7. Filter the digested suspension cell culture through a sterile 70 μm Falcon™ cell strainer in a disposable Petri plate.
8. Pipette up strained protoplasts from the Petri plate gently and transfer to a 15 mL Falcon tube.
9. Pellet the digested cells by spinning for 5–10 min at 100 × *g* in Eppendorf 5430R centrifuge.
10. Discard the supernatant and add gently 5 mL of 25% SucroseM solution.
11. Re-suspend the protoplasts gently.
12. Overlay 2 mL of 13% MannitolM without disturbing the sucrose-protoplast layer.
13. Spin for 10 min at 100 × *g*.
14. Collect the protoplasts located in the interface between the sucrose and mannitol layers with the sterile glass Pasteur pipet.
15. Place the protoplast from all four tubes in to a new 50 mL Falcon tube.
16. Add gently 20 mL of 0.6MBH3 protoplast culture media. Re-suspend very gently.
17. Spin 5–10 min at 100 × *g*. Discard the supernatant.
18. Add gently 10 mL of 0.6MBH3 protoplast culture media. Re-suspend very gently.
19. Spin 5 min at 100 × *g*. Discard supernatant.
20. Re-suspend the pellet gently in 10 mL of 0.6MBH3 protoplast culture media, and count the number of viable protoplasts to proceed to the viroid assay.

3.3.2 Isolation of Citrus Protoplasts from Callus Culture

1. Place 8 mL 0.6MBH3 protoplast culture media and 4 mL enzyme solution in a sterile 250 mL glass flask.
2. Place one callus inside of the 250 mL glass flask with the BH3-enzyme solution. Cut the callus in small pieces with the aid of a sterile blade. Mix gently.
3. Incubate for overnight digestion (~16 h) in dark at 28 °C with slow shaking (20–40 rpm).

Table 4
Number of leaves needed to prepare protoplasts from various citrus types

Protoplast source	Number of leaves
Citron (*Citrus medica*)	4–6
Sweet orange (*C. sinensis*)	4–6
Sour orange (*C. aurantium*)	4–6
Grapefruit (*C. paradisi*)	4–6
Alemow (*C. macrophylla*)	6–8
Mexican lime (*C. aurantifolia*)	6–8
Trifoliate orange (*Poncirus trifoliata*)	24–30
Citrange Carrizo (*C. sinensis* × *P. trifoliata*)	24–30

4. Follow the **steps 7–20** from Subheading 3.3.1 to obtain the protoplasts for viroid assay.

3.3.3 Isolation of Citrus Protoplasts from Leaves

(*See* Table 4)

1. Select fully expanded young leaves. Sterilize the leaves in 200 mL of 10% bleach in water solution plus one drop of dish soap for 10 min.
2. Wash the leaves in 100 mL of water three times, 5 min each one.
3. Place 16 mL 0.6MBH3 protoplast culture media and 8 mL enzyme solution in 15 cm sterile glass Petri dish; mix swinging (*see* **Note 15**).
4. Cut the leaves in stripes, as thin as possible, and place them inside of the glass Petri dish (*see* **Note 16**).
5. Incubate for overnight digestion (~16 h) in dark at 28 °C with slow shaking (20–40 rpm).
6. Follow the **steps 7–20** from Subheading 3.3.1 to obtain the protoplasts for viroid assay.

3.4 Cell Viability Count

1. Make a dilution of the protoplasts (1:10) in 0.6MBH3 protoplast culture media (*see* **Note 17**).
2. To 20–30 μL of protoplast, add 1–2 μL fluorescein diacetate dye (*see* **Note 18**), and incubate for 1–2 min.
3. Apply the stained protoplast solution to a hemocytometer, and view under microscope equipped with UV light.
4. Live protoplast fluoresce bright green under fluorescent microscope (Fig. 1).
5. Count the number of viable protoplasts in five of the chambers.

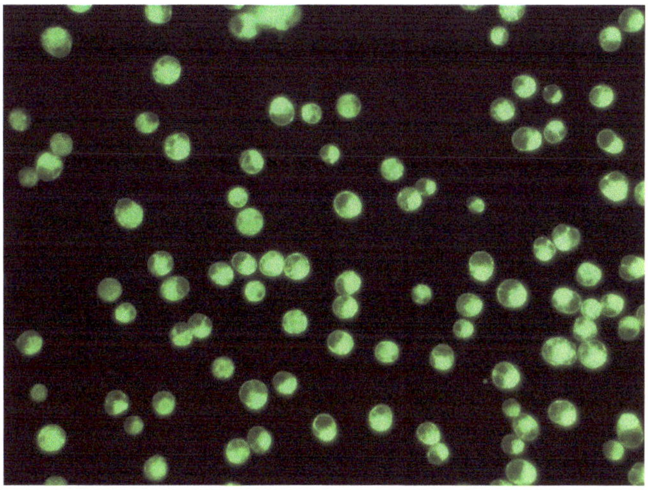

Fig. 1 Protoplasts isolated from the suspension cell culture of *Citrus sinensis* cv. Hamlin. Fluorescein diacetate-stained live protoplasts fluoresce green as observed under Epifluorescent Zeiss Axioplan Microscope (10×) equipped with 485/535 excitation/emission filter combination

6. Calculate the average protoplasts from five chambers.
7. Calculate the concentration of protoplasts using the equation:
   Average count × Dilution × Total volume × the factor of the chamber (10,000).

3.5 Transfection of Citrus Protoplasts with In Vitro Transcripts of CEVd

1. Gently re-suspend about one to two million protoplasts in 1 mL of 0.6MBH3 protoplast culture media.
2. Spin for 5 min at 100 × *g*.
3. Discard 800 μL of the supernatant.
4. Perform the viroid transfection assay [16, 22] by adding in vitro transcripts of CEVd (5 μg in 50 μL DEPC water) to one to two million citrus protoplast pellet in 5 mL snap cap tubes. Mix gently by finger tapping.
5. Add immediately 400 μL of 30% PEG-8000 solution dropwise in circular motion into the tube.
6. Mix by gentle finger tapping for 15–25 s.
7. Add immediately 4 mL of 0.6MMCPWC in circular motion.
8. Incubate on ice for at least 15 min.
9. Spin for 5 min at 100 × *g*. Discard the supernatant.
10. Wash carefully the protoplasts pellet in 4 mL of protoplast culture media (*see* **Note 19**).
11. Spin for 5 min at 100 × *g*. Discard supernatant.
12. Culture the protoplasts in 2 mL of 0.6MBH3 protoplast culture media by placing them in a parafilm sealed Petri plate for 1–5 days without shaking at 28 °C in dark.

3.6 Analysis of Replication and Population Profile of CEVd

1. Harvest protoplasts after 5 days post-transfection.
2. Isolate total RNA using the TRIzol (Invitrogen Corp.) extraction method, and treat with RQ1 RNase-Free DNase I (Promega Corp.) following the manufacturer's instructions.
3. During reverse transcription step, use forward primer (for replicative intermediate) and reverse primer (for newly generated progeny) in separate reactions.
4. Analyze the relative levels of accumulation of newly generated progeny viroid RNA and replicative intermediate by using SYBR Green I-based RT-qPCR.
5. Positive qPCR amplification from cDNA generated with forward primer indicates successful transfection and replication of viroid RNA in protoplasts.
6. Analyze the de novo generated viroid progeny population in protoplasts by generating RT-PCR products, cloning, and sequencing as described in [22].

4 Notes

1. Each chemical should be fully dissolved before adding the next chemical into solution.
2. Add 20 mL of dd water and stir with glass rod or spatula until a slurry forms. When protease powder is completely wet, add 40 mL of water. Stir the solution for 30 min at low speed until it becomes clear. Do not shake the proteolytic enzymes at high speed.
3. The enzyme solution can be active 1 month at 4 °C, but proteolytic activity could be completely lost at room temperature in a couple of hours.
4. Prepare 1 L of CPW Stock A solution because of trace amount of $CuSO_4 \cdot 5H_2O$.
5. Heat while stirring helps in faster dissolution of MES.
6. PEG-8000 solutions have to be prepared freshly.
7. Vitamin A, vitamin D, and vitamin B12 are insoluble in water. Initially dissolve in ethanol.
8. Use 10 mL wide-mouth disposable serological pipets (13-678-11E) unless otherwise mentioned.
9. Use manual pipet dispenser rather than electronic dispenser for most of the work. Use electronic dispenser only to discard the supernatants.
10. Perform re-suspension of the protoplast using glass Pasteur pipets (Fisher Scientific 13-678-4) by creating slow air bubbles in the buffer solution.

11. Do not remove the H + H suspension cell culture media completely! Otherwise the suspension cells will dry. At this point, most of the suspension cells have granular texture, while few show finer texture.
12. The 13% MannitolM solution can be used in place of 0.6MBH3 protoplast culture media.
13. Before taking out the overnight digestion, keep on shaking the plates for a couple of more minutes at 50 rpm to loosen the digested cell clumps.
14. After overnight digestion, cells look finer textured, but still granular-textured cell clumps will be seen; now suspension cells are soft and look delicate.
15. The enzyme concentration can be increased or decreased depending on the protoplast source. The 13% MannitolM solution can be used in place of 0.6BH3.
16. Add enough leave pieces to cover the surface of the Petri dish, floating on the BH3-enzyme solution.
17. The 0.6MMCPW solution can be used in place of protoplast culture media.
18. Dissolve 5 mg fluorescein diacetate in 1 mL acetone, and store at 4 °C in dark.
19. At this step, the protoplasts will be very fragile due to the PEG treatment. The protoplasts' washing or re-suspension process has to be very gentle and slow. Use manual pipet dispenser rather than electronic dispenser.

Acknowledgments

We thank the California Citrus Research Board for the partial support of this study.

References

1. Ding B (2010) Viroids: self-replicating, mobile, and fast-evolving noncoding regulatory RNAs. Wiley Interdiscip Rev RNA 1:362–375
2. Muhlbach HP, Sanger HL (1979) Viroid replication is inhibited by α-amanitin. Nature 278:185–188
3. Schindler IM, Muhlbach HP (1992) Involvement of nuclear DNA-dependent RNA polymerases in potato spindle tuber viroid replication: a reevaluation. Plant Sci 84.221–229
4. Warrilow D, Symons RH (1999) *Citrus exocortis viroid* RNA is associated with the largest subunit of RNA polymerase II in tomato *in vivo*. Arch Virol 144:2367–2375
5. Styer L (1995) RNA in eukaryotic cells is synthesized by three types of RNA polymerases. Chapter 33, RNA synthesis and splicing. In: Biochemistry, 4th edn. W.H. Freeman and Company, New York, p 853
6. Holland JJ, Spindler K, Horodyski F, Grabau E, Nichol S, Vanderpol S (1982) Rapid evolution of RNA genomes. Science 215:1577–1585

7. Domingo E, Holland JJ (1997) RNA virus mutations and fitness for survival. Annu Rev Microbiol 51:151–178
8. Garcia-Arenal F, Fraile F, Malpica JM (2001) Variability and genetic structure of plant virus populations. Annu Rev Phytopathol 39:157–186
9. Cocking EC (1960) Method for the isolation of plant protoplasts and vacuoles. Nature 187:927–929
10. Cocking EC (1966) An electron microscopic study of the initial stages of infection of isolated tomato frit protoplasts by tobacco mosaic virus. Planta 68:206–214
11. Aoki S, Takebe I (1969) Infection of tobacco mesophyll protoplasts by tobacco mosaic virus ribonucleic acid. Virology 39:439–448
12. Takebe I, Otsuki Y (1969) Infection of tobacco mesophyll protoplasts by tobacco mosaic virus. Proc Natl Acad Sci U S A 64:843–848
13. Warren GS, Hill SJL (1989) Infection of suspension-cultured cells of carrot with tobacco mosaic virus. Physiol Mol Plant Pathol 35:287–292
14. Murakishi HH, Hartmann JX, Pelcher LE, Beachy RN (1970) Improved inoculation of cultured plant cells resulting in high virus titre and crystal formation. Virology 41:365–367
15. Qi Y, Ding B (2002) Replication of Potato spindle tuber viroid in cultured cells of tobacco and *Nicotiana benthamiana*: the role of specific nucleotides in determining replication levels for host adaptation. Virology 302:445–456
16. Albiach-Martı MR, Grosser JW, Gowda S, Mawassi M, Satyanarayana T, Garnsey SM, Dawson WO (2004) *Citrus tristeza virus* replicates and forms infectious virions in protoplasts of resistant citrus relatives. Mol Breed 14:117–128
17. Sztuba-Solinska J, Bujarski JJ (2008) Insights into the single-cell reproduction cycle of members of the family *bromoviridae*: lessons from the use of protoplast systems. J Virol 82:10330–10340
18. Hajeri S, Ramadugu C, Manjunath K, Ng J, Lee R, Vidalakis G (2011) *In vivo* generated *Citrus exocortis viroid* progeny variants display a range of phenotypes with altered levels of replication, systemic accumulation and pathogenicity. Virology 417(2):400–409
19. Grosser JW, Gmitter FG (1990) Protoplast fusion and citrus improvement. Plant Breed Rev 8:339–374
20. Murashige T, Tucker DPH (1969) Growth factor requirements of citrus tissue culture. Proc First Int Citrus Symp 3:1155–1161
21. Mourão Filho FAA (1995) Protoplast fusion of citrus for rootstock and scion improvement with emphasis on wide hybridization. Doctoral dissertation, University of Florida, Gainesville
22. Hajeri (2010) Study of molecular and biological properties of *Citrus exocortis viroid* and *Dweet mottle virus*. Doctoral dissertation, University of California, Riverside
23. Grosser JW, Gmitter FG (2011) Protoplast fusion in the production of tetraploids and triploids: applications in scion and rootstock breeding. Plant Cell Tiss Org 104:343–357

Part II

Electrophoresis and Hybridization Techniques

Chapter 5

High-Throughput RNA Extraction from Citrus Tissues for the Detection of Viroids

Tyler Dang, Fatima Osman, Jinbo Wang, Tavia Rucker, Sohrab Bodaghi, Shih-hua Tan, Deborah Pagliaccia, Irene Lavagi-Craddock, and Georgios Vidalakis

Abstract

High-throughput nucleic acid extraction is critical for the implementation of modern viroid detection assays. Successful large-scale nursery, field surveys, and other regulatory, quarantine, or research diagnostic programs are increasingly dependent on high-throughput tissue pulverization and nucleic acid extraction protocols. Magnetic bead-based approaches using semi-automated robotic equipment allow high-throughput extraction and purification of high-quality uniform total nucleic acids for each individual sample. Here, we describe a streamlined and optimized protocol for citrus tissue processing and RNA extraction that can be used for downstream applications such as viroid detection by reverse transcription-quantitative polymerase chain reaction.

Key words Magnetic bead extraction, Automation, Purified total RNA

1 Introduction

Current molecular diagnostic assays heavily rely on the extraction and purification of nucleic acids. Nucleic acids' quality (purity, concentration, and integrity) [1, 2] and uniformity among samples processed are key for successful diagnostics. High-throughput nucleic acid extraction procedures are therefore critical for regulatory programs and modern diagnostic laboratories processing high volume of samples. High-throughput approaches allow to process a higher number of samples in a shorter period of time, thus lowering labor cost and producing nucleic acids extracts ready for downstream applications. The combination of high-throughput tissue pulverization, using bead beaters such as the Spex GenoGrinder® 2010, and nucleic acid purification, using magnetic beads-based approaches, such as MagMAX™ Express-96 Magnetic Particle Processor, allows high-throughput extraction of total nucleic

Ayala L. N. Rao et al. (eds.), *Viroids: Methods and Protocols*, Methods in Molecular Biology, vol. 2316,
https://doi.org/10.1007/978-1-0716-1464-8_5,

acids (TNA), which include DNA and RNA [3]. MagMAX™ Express-96 utilizes paramagnetic beads that bind to nucleic acids and then undergo a series of washing steps and a final elution. This method is much easier and safer than traditional low-throughput centrifuge and column-based approaches, of individual samples due to the lack of dangerous chemicals like phenol, chloroform, or 2-mercaptoethanol that are used in traditional plant extraction protocols to remove inhibitors such as polyphenols and polysaccharides [4].

In plants, high-throughput TNA extractions have been previously described in grapevines [3]. With this approach, up to 96 samples can be processed at once, and the protocol allows the user to scale up or down depending on the number of samples at hand.

Citrus viroid detection has largely transitioned from traditional methods such as biological indexing [5], hybridization [6–8], and sequential polyacrylamide gel electrophoresis (sPAGE) [9]-based detection to the current gold standard of high-throughput nucleic acid-based detection methods: reverse transcription (RT) followed by polymerase chain reaction (RT-PCR) [10] or quantitative PCR (RT-qPCR) [11]. The conjunction of high-throughput RNA extraction and RT-qPCR increases the laboratory capacity and delivers reliable and consistent results regarding viroids and plant health. Here, we describe a streamlined high-throughput total RNA extraction protocol that is currently used by the Citrus Clonal Protection Program (CCPP) for routine RT-qPCR citrus viroid detection (to date over 21,000 samples tested). This extraction protocol has been used for citrus viroid and virus detection from stem tissues (i.e., 1-year-old budwood) obtained from both greenhouse and field trees and has been approved by California's Department of Food and Agriculture for use at the "Citrus Nursery Stock Pest Cleanliness Program."

2 Materials

2.1 Tissue Preparation

1. Resealable plastic bags.
2. Bypass pruner.
3. Single edge razor blade.
4. Spray bottle with 10% household bleach solution (*see* **Note 1**).
5. Spray bottle with 70% ethanol.
6. Paper towels.
7. 96-microcentrifuge tube rack.
8. Microcentrifuge tube opener.
9. Kraft paper (61 cm wide × 366 m length) with paper roll cutter.

10. Chipboard pads or cardboard cut to the approximate size of 7 × 13 cm.
11. 2.0 mL Eppendorf Safe-Lock Tubes.
12. Electronic balance (0.01 g accuracy).
13. Ice bucket with ice.
14. 1 L beaker filled with 10% household bleach solution.
15. 1 L beaker filled with deionized water.

2.2 Tissue Pulverization

1. Lyophilizer (freeze-dryer).
2. Single and multichannel micropipettes (P10, P20, P200, P1000) with filtered tips.
3. 4 mm stainless steel grinding balls.
4. Tweezers.
5. Refrigerated microcentrifuge.
6. Parafilm.
7. Magnetic stir rod and magnetic stir plate.
8. Guanidine buffer: Add 47.3 g of 4M guanidine thiocyanate, 2.72 g 0.2 M sodium acetate pH 5.0, 0.93 g 25 mM ethylenediaminetetraacetic acid (EDTA), and 2.5 g 2.5% polyvinylpyrrolidone 40 (PVP-40) to 50 mL of water in a glass beaker. Add a magnetic stir rod, seal the opening of beaker with a parafilm, place the beaker on a magnetic stir plate, and put a box over the beaker to fully dissolve and protect the buffer from light for at least 4 h to overnight. Adjust to pH 5.0 with 1 M hydrochloric acid, and adjust the final volume to 100 mL with water. Transfer the guanidine buffer to bottle, and protect it from light.
9. Liquid nitrogen.
10. SPEX SamplePrep GenoGrinder® 2010.
11. SPEX SamplePrep Cryo-Station.
12. Two 48-aluminum Cryo-Blocks chilled to −80 °C.
13. Two 48-aluminum Cryo-Blocks at room temperature.

2.3 RNA Extraction

1. Molecular-grade 100% isopropanol.
2. 100% ethanol.
3. Adhesive aluminum foil seal (catalog number 2923-0100).
4. Seven 96-well deep well plates (catalog number 1896-2000).
5. MagMAX™-96 Viral RNA Isolation Kit (catalog number AMB1836-5): Lysis/binding solution, Wash 1 concentrate, Wash 2 concentrate, RNA binding beads, carrier RNA, lysis/binding enhancer, and elution buffer (*see* **Note 2**).

Prepare the following solutions: Add 40 mL of 100% isopropanol to the lysis/binding solution. Do not add the carrier RNA directly to the lysis/binding solution; it will shorten the shelf life. Add 60 mL of 100% isopropanol to Wash 1 concentrate. Add 160 mL of 100% ethanol to Wash 2 concentrate. Mix all solutions thoroughly, and the final volume for the lysis/binding solution will be 80 mL and Wash 1 and 2 will be 200 mL.

6. MagMAX™ Express-96 Magnetic Particle Processor with magnetic head (*see* **Note 3**).
7. MagMAX™ Express-96 deep well tip combs (catalog number 4388487).
8. Magnetic stand for 96-well plate (AM10027).
9. Round-bottom 96-well polypropylene microtiter plate.
10. Nano spectrophotometer.

3 Methods

3.1 Tissue Preparation

1. Collect stem tissue around the canopy of the tree of interest (if field tree minimum from 4 quadrats North, South, East, and West) from the latest fully matured growth (approximately 1 year old) using a bypass pruner, and place the samples into a resealable plastic bag (*see* **Note 4**).
2. Sanitize the working surface with 10% bleach and then with 70% ethanol. Ensure the working surface is dry using paper towels.
3. Tear a sheet of kraft paper to cover the working area.
4. Use the single edge razor blade to peel the bark from the stems.
5. Finely chop the peeled bark into small pieces on the small precut chipboards.
6. Weigh and transfer 0.25 g of chopped bark tissue into a 2 mL Safe-Lock Tubes, and place the tube on ice (*see* **Note 5**).
7. Between each sample, dispose of the chipboards and kraft paper. Then sanitize the working area with 10% bleach and then 70% ethanol with paper towels.
8. When you complete the chopping of a set of samples, drop the Safe-Lock Tubes into the 1 L beaker with 10% bleach solution, and then transfer the tubes into the 1 L beaker containing water using a pair of tongs. Allow for 5–10 s of contact time in both the bleach and water for each tube with gentle stirring.
9. Dry the tubes with paper towels.
10. Move the samples into the 96-microcentrifuge tube rack, and place them into the −80 °C freezer for at least 2 h.
11. Open the sample tubes' lids, and place them into a lyophilizer for approximately 16–24 h (*see* **Note 6**).

3.2 Tissue Pulverization

1. Remove the samples from the lyophilizer, and add one 4 mm stainless steel grinding ball into each sample tube using the tweezers.
2. Transfer samples into aluminum Cryo-Blocks, and pace inside the Cryo-Station (*see* **Note 7**).
3. Fill the Cryo-Station reservoir with liquid nitrogen, and place the samples into the station for 20 min.
4. Grind samples into fine powder, using a GenoGrinder® 2010 at 1680 rpm for 20 s. Repeat the 20 s grinding step twice.
5. Transfer sample tubes to the room temperature stainless steel Cryo-Block.

3.3 RNA Extraction and Purification

1. Add 750μL of 4 M guanidine buffer to each sample (*see* **Note 8**).
2. Place the samples back into the GenoGrinder® 2010, and grind at 1680 rpm for 20 s. Repeat the 20 s grinding once.
3. Incubate the samples at 4 °C for 15 min.
4. Remove the samples from the Cryo-Block, and centrifuge the samples at 4 °C for 45 min at 17,200 × *g*.
5. Prepare the bead mix based on sample number (*see* **Note 9**).
6. Prepare the lysis plate using the deep well plates as follows: 139μL of lysis/binding solution, 22μL of bead mix, 139μL of isopropanol, and 150μL of the sample supernatant per well (*see* **Note 10**).
7. Prepare two Wash 1 plates by adding 500μL of Wash Solution 1 to each well in the deep well plates.
8. Prepare two Wash 2 plates by adding 500μL of Wash Solution 2 to each well in the deep well plates.
9. Add 100μL of elution buffer to each well in a deep well plate.
10. Insert a tip comb into a clean deep well plate.
11. Turn on the MagMAX™ Express-96 machine, and initiate protocol script "AM1836_DW_50_V2."
12. Follow the prompt, and load the deep well plates in the following order: position 7, deep well plate with tip comb; position 6, elution plate; positions 4 and 5, Wash Solution 2 plates; positions 2 and 3, Wash Solution 1 plates; and position 1, lysis plate (*see* **Note 11**).
13. When the MagMAX™ Express-96 program is finished, place the elution plate on a magnetic rack for 5 min at room temperature to collect any residual beads.
14. Use a P200 multichannel micropipette to carefully transfer approximately 90μL elute into a new 96-well microtiter plate.

Do not disturb or interfere with the brown pellet in the deep wells by trying to recover the full 100μL elute.

15. Assess the quality of the extract by measuring the absorbance with a spectrophotometer (*see* **Note 12**).
16. Place an adhesive foil tightly over the top. Label and store the 96-well microtiter plate in the −80 °C freezer (*see* **Note 13**).

4 Notes

1. Prepare 10% household bleach solution with at least 0.5% of sodium hypochlorite in the final solution. Sodium hypochlorite degrades over time and when exposed to light; therefore, the bleach solution should be prepared fresh daily and discarded at the end of the day in order to maintain its disinfectant activity. In addition, proceed with caution when cleaning surfaces or equipment exposed to guanidine thiocyanate with sodium hypochlorite. The combination of these two chemicals can produce toxic gas. It is strongly recommended to rise the contaminated surface first and then to proceed to disinfect with 10% bleach solution and finally with 70% ethanol.
2. The lysis/binding solution may precipitate over time during storage. Before prepping the solution, ensure that it is completely dissolved by placing the lysis/binding solution in a 37 °C incubator for 1 h.
3. The KingFisher™ Flex Purification System is an equivalent system that is compatible with this protocol. The user can download protocol script "AM1836_DW_50_V2" under "MagMAX™ Express-96, Virus and Bacteria (AM1835, AM1836, AM1840) (.kf2)," and install it to the machine from the Thermo Fisher Scientific website (https://www.thermofisher.com/us/en/home/life-science/dna-rna-purification-analysis/rna-extraction/rna-types/total-rna-extraction/magmax-technology/magmax-particle-processors/magmax-instrument-protocols-scripts.html).
4. When collecting stem samples from the greenhouse or field trees, remove all leaves and thorns before placing the samples in the resealable plastic bags. Keep the samples cool in an ice chest filled with ice in order to preserve the RNA quality. Immediately transfer the samples into 4 °C refrigerator upon returning to the laboratory.
5. It is recommended to prepare a second tube with 0.25 g of chopped bark tissue as a backup in case the tube breaks during the pulverization step or if samples accidentally become contaminated during the extraction process.

6. Lyophilization time will vary for different machines. For example, the Labconco FreeZone Triad Cascade Benchtop Freeze Dryer (catalog #: 7400030) takes approximately 24 h to complete a cycle, while the FreeZone Plus 2.5 L Cascade Console (catalog #: 7420031) with Short Clear Chamber and Product Shelves (catalog # 7318702 and 7317100) takes approximately 16 h. It will be the user's discretion to determine the suitable drying time using other lyophilizers.
7. The GenoGrinder 2010 requires two aluminum Cryo-Blocks (48 tubes each) to be used at once in order to maintain balance. This configuration allows the machine to grind 96 samples at once; however, it is difficult to handle all 96 samples at once. It is recommended to grind the samples in batches of 48. For example, load 24 tubes into each liquid nitrogen-cooled aluminum Cryo-Block, transfer the tubes to another rack, and then grind the next set of 48 samples.
8. Ensure that the powder from the pulverized samples does not aggregate close to the lid of the tubes. Either tap the lids of the tube until the content aggregate at the bottom or briefly centrifuge the tubes. This will help prevent the pulverized tissue samples from cross-contaminating when opening and closing tubes to add the guanidine buffer. In addition, tubes should be opened using a microcentrifuge tube opener.
9. Ensure that the binding beads are well mixed and at room temperature for 30 min. Prepare the bead mix for the required number of reactions, plus 10% overage by adding 10μL of RNA binding beads, 10μL of lysis/binding enhancer, and 2μL of carrier RNA, and then vortex to mix. The bead mixture should be prepared fresh and used within the same day.
10. To prevent samples from cross-contaminating, use filtered tips when adding the sample supernatant to the corresponding well.
11. Ensure that all plates are correctly orientated. The protocol will take approximately 20 min to complete.
12. The quality of the extracts is defined by the concentration, purity, and integrity. The concentration of nucleic acid can be assessed by the A_{260} measurement. The purity of a sample can be determined by A_{260}/A_{280} absorbance ratio. A_{260}/A_{280} ratio of 2.0 would indicate a pure sample, while a low ratio would indicate protein contaminants [12]. The integrity of the sample can be assessed by performing RT-PCR or RT-qPCR on a plant mRNA internal control that target the mitochondrial gene NADH dehydrogenase (nad5) [13].
13. For long-term storage, transfer the elution into individual 1.5μL microcentrifuge tubes, and store in the −80 °C freezer.

Acknowledgments

Support was provided by the United States Department of Agriculture (USDA) National Institute of Food and Agriculture (NIFA), Hatch (project 1020106), and the National Clean Plant Network (NCPN) which operates under the auspices of USDA Animal and Plant Health Inspection Service (APHIS) (12-8100-1544-CA; 14-, 15-, 16-8130-0419-CA; AP17PPQS&T00C118; AP18PPQS&T00C107). The findings and conclusions in this publication are those of the author(s) and should not be construed to represent any official USDA or US government determination or policy.

URLs to sponsors' websites are listed below.

1. CRB: https://www.citrusresearch.org
2. USDA NIFA: https://nifa.usda.gov
3. NCPN: http://nationalcleanplantnetwork.org.

References

1. Manchester KL (1996) Use of UV methods for measurement of protein and nucleic acid concentrations. Biotechniques 20:968–970
2. Glasel JA (1995) Validity of nucleic acid purities monitored by 260nm/280nm absorbance ratios. Biotechniques 18:62–63
3. Osman F, Olineka T, Hodzic E, Golino D, Rowhani A (2012) Comparative procedures for sample processing and quantitative PCR detection of grapevine viruses. J Virol Methods 179:303–310
4. Schrader C, Schielke A, Ellerbroek L, Johne R (2012) PCR inhibitors–occurrence, properties and removal. J Appl Microbiol 113:1014–1026
5. Roistacher CN (1991) Graft-transmissible diseases of citrus: Handbook for detection and diagnosis. International Organization of Citrus Virologists, Food and Agriculture Organization of the United Nations, Rome
6. Romero-Durbán J, Cambra M, Duran-Vila N (1995) A simple imprint-hybridization method for detection of viroids. J Virol Methods 55:37–47
7. Palacio-Bielsa A, Foissac X, Duran-Vila N (1999) Indexing of citrus viroids by imprint hybridisation. Eur J Plant Pathol 105:897–903
8. Fonseca ME, Marcellino LH, Gander E (1996) A rapid and sensitive dot-blot hybridization assay for the detection of citrus exocortis viroid in Citrus medica with digoxigenin-labelled RNA probes. J Virol Methods 57:203–207
9. Rivera-Bustamante RF, Gin R, Semancik JS (1986) Enhanced resolution of circular and linear molecular forms of viroid and viroid-like RNA by electrophoresis in a discontinuous-pH system. Anal Biochem 156:91–95
10. Ito T, Ieki H, Ozaki K (2002) Simultaneous detection of six citrus viroids and Apple stem grooving virus from citrus plants by multiplex reverse transcription polymerase chain reaction. J Virol Methods 106:235–239
11. Osman F, Dang T, Bodaghi S, Vidalakis G (2017) One-step multiplex RT-qPCR detects three citrus viroids from different genera in a wide range of hosts. J Virol Methods 245:40–52
12. Gallagher SR (2017) Quantitation of DNA and RNA with absorption and fluorescence spectroscopy. Curr Protoc Immunol 116: A.3L.1–A.3L.14
13. Saponari M, Manjunath K, Yokomi RK (2008) Quantitative detection of Citrus tristeza virus in citrus and aphids by real-time reverse transcription-PCR (TaqMan®). J Virol Methods 147:43–53

Chapter 6

Extraction and Purification of Viroids from Herbaceous Hosts

Rosemarie W. Hammond

Abstract

Protocols for extraction and purification of viroid RNAs from the tissues of infected herbaceous plant hosts are numerous. They range from lengthy, traditional protocols that require large amounts of starting tissue and take several days to perform to those based on column chromatography which is more efficient and can be performed with smaller amounts of infected tissue. The goal of all protocols is to enrich for RNA fractions that contain viroid RNAs, and the RNA extraction procedure is chosen and adjusted for the downstream method used for detection and characterization. Removal of inhibitors/impurities is generally not an issue for herbaceous hosts unless they contain and inordinate amounts of polysaccharides, tannins, and phenols. Subsequent purification of viroid circular and linear RNAs is performed using denaturing polyacrylamide gel electrophoresis. In this chapter, a specific method routinely used for viroid purification from herbaceous hosts and problems that may be encountered is described and is intended as a reference for beginners in the field.

Key words RNA purification, Denaturing polyacrylamide gel electrophoresis, Viroid circular RNAs

1 Introduction

Viroids are low-molecular-weight RNAs present in total nucleic acid extracts from plant tissues and are composed of linear and circular molecules. Viroid RNAs are just a small fraction of the total RNA, and methods for extraction and purification are designed to separate, enrich, and concentrate the RNAs for further characterization. The traditional methods for the preparation of partially purified nucleic extracts for subsequent viroid purification from herbaceous hosts made use of detergents, such as sodium dodecyl sulfate (SDS) or cetyltrimethylammonium bromide (CTAB) [1–3], required large amounts of tissue, involved a number of steps, and required several days for completion. Small RNAs were enriched using non-ionic cellulose, such as CF-11 [4] or lithium chloride (LiCl). Additional methods used high-performance liquid or gel chromatography [5, 6]. While these

Ayala L. N. Rao et al. (eds.), *Viroids: Methods and Protocols*, Methods in Molecular Biology, vol. 2316,
https://doi.org/10.1007/978-1-0716-1464-8_6,

methods are still valuable and suitable for viroid purification, more recent methods for extraction and purification of viroid RNAs utilize commercial reagents (TRI Reagent or TRIzol or Qiagen, among others) and are faster and simpler to perform. One method is described here, and a similar procedure was published earlier [7].

Polyacrylamide gel electrophoresis (PAGE) has played a key role in viroid characterization and purification; successful subsequent purification of viroid RNAs from PAGE is highly dependent upon the quality of the RNA obtained from infected tissue. Many variations of the PAGE procedure have been described [8, 9].

2 Materials

2.1 RNA Isolation

1. Mortar and pestle (autoclaved).
2. Liquid nitrogen (N_2).
3. Microcentrifuge.
4. Autoclaved microfuge tubes.
5. Micropipettors.
6. Chloroform (molecular biology grade).
7. Ethanol (95%).
8. Isopropanol (molecular biology grade).
9. 3 M sodium acetate (NaOAc), pH 5.2.
10. 4 M lithium chloride (LiCl).
11. Autoclaved, deionized water.
12. Hot plate or water bath.
13. Commercial RNA extraction reagent: TRIzol (Invitrogen, Carlsbad, CA), TRI Reagent (Molecular Research Inc., Cincinnati, OH), or Qiagen RNeasy kit (Qiagen, Hilden, Germany).

2.2 Return Polyacrylamide Gel Electrophoresis (R-PAGE)

1. Gel plates.
2. Gel apparatus.
3. Hot plate or hot water bath.
4. 29:1 acrylamide/bis-acrylamide (*N*, *N′*-methylene-bis-acrylamide). Dissolve 29 gm of acrylamide and 1 gm of bis-acrylamide in water, and adjust volume to 100 ml. Filter to remove particulates, and store protected from light at 4 °C.
5. 10% ammonium persulfate (APS).
6. TEMED (*N*,*N*, *N′*,*N′*-tetramethylethylenediamine).
7. 10× TBE buffer (0.89 M Tris base, 0.89 M boric acid, 25 mM EDTA, pH 8.3) or (purchase from Ambion or Invitrogen).

8. 5% PAGE, 1× TBE: 18 ml water, 4.2 ml 29:1 acrylamide/bis-acrylamide, 2.5 ml 10× TBE buffer.
9. 310μl 10% ammonium persulfate, 15.6μl TEMED. Add APS last to polymerize the gel. Quickly pour into assembled gel plates. The volume of the gel can be adjusted to the size of the assemble gel plates. Insert the gel comb and allow the gel to polymerize.

3 Methods

Collect fresh tissue 2–6 weeks' post-inoculation, and grind to a powder in liquid nitrogen using a pre-cooled mortar and pestle—proceed with extraction, or store the powdered tissue at −20 or −80 °C. The tissue can be from the primary herbaceous host or indicator host (e.g., cucumber, tomato, etc.). The amount of tissue to be extracted can range from 100 mg to 1 gm when young, symptomatic tissue is collected (*see* **Notes 1** and **2**).

3.1 TRIzol or TRI Reagent Method

The protocol described is based on the manufacturer's instructions.

Take a chilled mortar, and add leaf tissue (fresh or frozen), add liquid N_2 to cover the tissue, and grind the tissue to a powder. After the N_2 has evaporated, TRIzol reagent or TRI Reagent (1 ml L100 mg) can be added directly to the mortar and mixed with the tissue to make a paste, or the powdered tissue can be transferred to a microfuge tube or similar tube and TRIzol reagent added and the tissue mixed with a small pestle. The manufacturer's protocol is then followed to extract the total RNAs at room temperature to obtain the RNA pellet. The pellet is washed with 1 ml of 70% ethanol, centrifuged, and air-dried. From 100 mg tissue, the pellet is dissolved in 50μl sterile water, heated at 50–55 °C for 10 min. The nucleic acid concentration is determined using a spectrophotometer and absorbance at 260 nm, assuming a value of 1 at 40μg/ml (*see* **Notes 3** and **4**).

3.2 Enrichment for LiCl-Soluble Nucleic Acids (DNA, 4S RNA, dsRNAs, and Viroid RNAs) if Needed

To the RNA purified above, add one volume of 4MLiCl (*see* **Note 5**). Incubate at 4 °C for 4 h or overnight. Centrifuge 10 min at maximum speed at 4 °C to pellet, and discard the pellet of LiCl-insoluble nucleic acids (ribosomal RNA). Precipitate LiCl-soluble nucleic acids overnight at −20 °C with 1/10 volume of 3 M NaOAc, pH 5.2, and 2 1/2 volumes of EtOH. Centrifuge at maximum speed at 4 °C, decant, and drain the ethanol from pellet. Re-suspend in sterile distilled water. Determine the concentration in mg/ml by adsorption in a spectrophotometer at 260 nm.

These nucleic acid samples can be stored for a long term at −20 °C to −80 °C.

3.3 Purification by Return Polyacrylamide Gel Electrophoresis (R-PAGE)

1. Assemble a 14 × 16 × 0.15 cm glass sandwich, and pour a 5% PAGE, 1 × TBE gel.
2. Run 20–30μg low-molecular-weight RNA preparation in a lane, which is adequate for silver staining and, when viroid titer is high enough, should be adequate for ethidium bromide visualization of the viroid band. Prepare the sample by adding 1/10 volume of a sample dye containing buffer and dyes (xylene cyanol and bromophenol blue).
3. Load the gel and electrophorese the vertical gel from top to bottom using 1× TBE at 46 mA (270 V) at room temperature for 2.5 h (until the xylene cyanol is about 1 cm from the bottom of the gel).
4. Exchange the buffers in the chambers, and reverse the polarity of the gel. Exchange the buffers with a low salt (1:8 dilution of 1× TBE) that have been heated to 70–80 °C for the upper chamber and 80–90 °C in the lower chamber. You must also use a heated plate at 70 °C to keep the gel warm. Alternatively, you can run the gel in a vertical gel apparatus in which the gel is submerged in the buffer. In this case, circulate and heat the buffer in the chamber to 70 °C. Electrophorese from bottom to top at 46 mA for 2 h. Remove the gel and stain with silver or with ethidium bromide (if further cloning is anticipated). (*see* Fig. 1 in Owens et al. [7] for an example of electrophoresis results) (*see* **Notes 6–13**).

4 Notes

1. Caution: Phenol, chloroform, acrylamide, and urea are toxic. Wear gloves and safety goggles when handling in the chemical fume hood, and adhere to safety guidelines for disposal of hazardous waste. Oxidized phenol (pink or red) can damage nucleic acids.
2. Use suitable lab coat, disposable gloves, and eye protection when handling phenol, chloroform, TRIzol, and TRI Reagent.
3. Liquid nitrogen (N_2) can freeze your skin. When dispensing, wear protective gloves and eye protection.
4. Avoid ribonucleases (RNases) by using gloves and disposable plastic ware or treating glassware with RNaseZap (Ambion, Austin, TX) or rinsing with 0.1M NaOH and 1 mM EDTA followed by thorough rinsing with RNase-free water.
5. Chill RNA on ice before starting the LiCl fractionation.
6. Ethidium bromide stain on acrylamide gels.

7. For small amounts of tissue (100–200 mg tissue), use a Qiagen kit (okay for purification of total RNAs for RT-PCR, but not for gel electrophoresis). The methods can be scaled up to...
8. For samples that contain high amounts of polysaccharides, the RNA can be precipitated in the TRIzol/TRI Reagent protocol using isopropanol as suggested by the manufacturer and by adding an equal volume of a high-salt precipitation solution (1.2 M NaCl, 0.8 M sodium citrate). This selectively precipitates RNA while polysaccharides remain in solution.
9. 2 M LiCl precipitates large RNA molecules (>1000 nt) but does not precipitate low-molecular-weight RNAs, such as tRNA and 5S ribosomal RNA (<100 nt). Viroids (280–400 nt) are either not precipitated or precipitated depending upon their secondary structure.
10. Other methods for small-scale isolation of plant total RNAs include the use of the RNeasy Plant Mini Kit (Qiagen, Hilden, Germany) using guanidine thiocyanate and column chromatography for processing a maximum amount of 100 mg plant material per sample, yielding from 25 to 60 mg total RNA.
11. Lysing matrix beads can be used to facilitate extraction (MP Biomedicals, Eschwege, Germany).
12. Viroids with a branched secondary structure, such as chrysanthemum chlorotic mottle viroid, are insoluble in 2 M LiCl [10].
13. If you do not have access to gel assembly plates in which coolant can be circulated, gel electrophoresis can be performed in a non-denaturing polyacrylamide gel in the first dimension, after which a gel slice corresponding to the "viroid band" defined by DNA markers of 200–400 bp can be removed. The gel slice would then be placed on the top of a second, denaturing (urea) gel and electrophoresed to separate viroid RNAs from plant RNAs. For this procedure, see additional references [7–9].

Acknowledgments

Mention of trade names or commercial products in this publication is solely for providing specific information and does not imply recommendation or endorsement by the USDA.

References

1. Diener TO, Hadidi A, Owens RA (1977) Methods for studying viroids. Methods Virol 6:185–217
2. Niblett C, Dickson E, Horst RK, Romaine CP (1980) Additional hosts and an efficient purification procedure for four viroids. Phytopathology 70:610–615
3. Uyeda I, Sano T, Shikata E (1984) Purification of cucumber pale fruit viroid. Ann Phytopathol Soc Japan 50:331–338
4. Pallas V, Navarro A, Flores R (1987) Isolation of viroid-like RNA from hop different from hop stunt viroid. J Gen Virol 68:3210–3205
5. Singh RP, Kevesque D, King RR (1986) A rapid procedure for the purification of potato spindle tuber viroid—use of high performance gel permeation chromatography. Can J Plant Pathol 8:54–58
6. Colpan M, Schumacher J, Bruggemann W, Sanger HL, Riesner D (1983) Large-scale purification of viroid RNA using Cs_2SO_4 gradient centrifugation and high-performance liquid chromatography. Anal Biochem 131:257–265
7. Owens RA, Sano T, Duran-Vila N (2012) Plant viroids: isolation, characterization/ detection, and analysis. In: Watson JM, Wang M-B (eds) Antiviral resistance in plants: methods and protocols, methods in molecular biology, vol 894, pp 252–270
8. Singh RP, Boucher A (1987) Electrophoretic separation of a severe from mild strains of potato spindle tuber viroid. Phytopathology 77:1588–1591
9. Hanold D, Vadamalai G (2017) Gel electrophoresis. In: Hadidi A, Flores R, Randles JW, Palukaitis P (eds) Viroids and satellites. Academic Press, Cambridge, pp 357–367
10. Navarro B, Flores R (1997) Chrysanthemum chlorotic mottle viroid: unusual structural properties of a subgroup of self-cleaving viroids with hammerhead ribozymes. Proc Natl Acad Sci U S A 94:11622–11267

Additional Reading

Palukaitis P, Symons RH (1980) Purification and characterization of the circular and linear forms of chrysanthemum stunt viroid. J Gen Virol 46:477–489

Semancik JS (1991) Chapter 22. Viroid purification and characterization. In: Graft-transmissible diseases of citrus-Handbook for detection and diagnosis of graft-transmissible diseases of citrus. FAO. www.fao.org.3.T0601E0m.htm

Chapter 7

Detection of Viroids by sPAGE Gel Electrophoresis

Núria Duran-Vila

Abstract

Nucleic acid polyacrylamide gel electrophoresis (PAGE) has played a critical role in the identification and characterization of viroid RNAs. In addition, double PAGE has been a very efficient tool for the detection of viroids as it is sequence independent and is based on the viroid structure of covalently closed/circular molecules. sPAGE has been widely used for the identification of new viroids as well as a routine detection tool.

Key words Silver staining, Characteristic migration, Urea denaturing gel, Sequential PAGE

1 Introduction

Polyacrylamide gel electrophoresis (PAGE) has played a critical role in viroid research and further practical applications since viroids were first discovered as a new class of plant pathogenic agents [1]. A great value of PAGE is the fact that it is sequence independent and therefore allows the identification of different sequences and sequence variants. Initial approaches under non-denaturing conditions were to be followed by PAGE under denaturing conditions, as well as a combination of both, as in the case of sequential PAGE (sPAGE) and return PAGE (R-PAGE).

sPAGE involves the migration of the extracted RNAs, first in non-denaturing conditions (5% PAGE), followed by excision of a segment of the gel (presumably containing the viroids) and placement in contact with a second gel containing 8 M urea. The advantage of this approach is that it avoids the presence and interference of DNA and most linear RNAs in the second run. The approach is based on the fact that viroids are circular RNA molecules with a slow but yet distinct-characteristic migration in the second gel [2, 3].

As reported by Rivera-Bustamante et al. [4], the procedure described below, in which the second 8 M urea-containing gel is cast with TAE (pH 6.5) buffer, and electrophoresis is run in TBE

Ayala L. N. Rao et al. (eds.), *Viroids: Methods and Protocols*, Methods in Molecular Biology, vol. 2316, https://doi.org/10.1007/978-1-0716-1464-8_7, © Springer Science+Business Media, LLC, part of Springer Nature 2022

(pH 8.3) running buffer, enhances the separation of the circular viroid forms [5]. This approach is extremely useful to analyze samples containing multiple viroids, which are very commonly found in citrus and grapevines [3, 6–9]. This procedure was critical in identifying additional viroids and/or viroid variants affecting citrus [10, 11] as well as their widespread presence in vegetable crops and wild species [12–14].

2 Materials

All solutions must be prepared with autoclaved deionized water and analytical grade reagents.

1. Electrophoresis chamber.
2. Gel casting system (glass plates, 1.5-mm-wide spacers, 1.5-mm-wide sample well comb).
3. Power supply (1000 V or greater).
4. UV transilluminator.
5. Platform shaker.
6. Micropipette equipped with a plastic tip.
7. Scalpel.
8. Single-edge razor blade.
9. Ethanol 95%.
10. Ammonium persulfate TEMED (*N*,*N*,*N*′,*N*′-tetramethylethylenediamine) TAE buffer (10×), pH 7.2: 0.4 M Tris-acetate, 0.01 M EDTA TBE buffer (5×), pH 8.3: (0.45 M Tris-borate, 25 mM EDTATAE buffer (3×), pH 6.5: 0.12 M Tris-acetate, 3 mM EDTA.
11. Stock solution A (acrylamide/bisacrylamide/40:1): Dissolve 30 g acrylamide and 0.75 g bisacrylamide in 100 ml distilled autoclaved water, and filter. Store at 4 °C.
12. Mix 35 ml of stock urea gel (acrylamide/bisacrylamide/40:1), 50 × l TEMED, and 0.5 ml 10% (w/v) ammonium persulfate.
13. Stock urea gel (acrylamide/bisacrylamide/40:1): Dissolve 72 g urea in 47.5 ml distilled autoclaved water and 15 ml TAE buffer (3×) pH 6.5. Add 25 ml of stock A (acrylamide/bisacrylamide/40:1), and store at room temperature.
14. Loading/tracking buffer (6×): (0.3% bromophenol blue +0.3% xylene cyanol FF in 40% (w/v) sucrose).
15. Ladder molecular weight marker (100 bp).
16. Ethidium bromide (0.5 μg/ml) in 0.1× TBE or water.
17. Silver staining solution (12 mM $AgNO_3$).

18. Ethanol-acetic acid: (50% ethanol, 10% acetic acid); (10% ethanol, 1% acetic acid).
19. Developer solution (0.75 M KOH + 0.28% HCOH).
20. 0.07 M Na_2CO_3.

3 Methods

Wear gloves when handling these solutions, and discard them safely.

3.1 Cast the First (Non-denaturing) Gel

1. Clean glass plates as necessary with glass cleaner followed by 95% ethanol.
2. Assemble glass plates (about 16 cm × 16 cm) and 1.5 mm spacers. The volumes of the solutions must be adjusted according to the equipment to be used.
3. Mix 28.6 ml water, 4 ml TAE (10×), 6.7 ml stock A, 60 × l TEMED, and 0.6 ml 10% (w/v) ammonium persulfate. Mix quickly and dispense.
4. Tilt the glass plate-spacer assembly slightly, fill with gel solution, and insert the comb. Avoid trapping air bubbles under the comb or against the glass plates.
5. Allow polymerization to proceed for 30–60 min at room temperature or for 15–30 min at 4 °C. When polymerization is complete, a refractive line will appear 1–2 mm below the surface of the acrylamide between the teeth of the comb.
6. Carefully remove the comb, first wetting it with a small amount of 1× TAE to avoid disturbing the wells.

3.2 Load Samples and Begin Electrophoresis

1. Place the gel assembly in the electrophoresis chamber, tighten the side clamps, and fill the upper and lower reservoirs with 1× TAE buffer. Remove any bubbles trapped against the bottom edge of the gel in the lower buffer reservoir.
2. Mix aliquots of samples to be analyzed with appropriate volumes of loading/tracking buffer (6×), and load samples (usually 20–30 ml) into the wells using a micropipette equipped with a plastic tip. Load the outer wells with a 100 bp ladder molecular weight marker.
3. Connect the electrodes to a power supply; set the amperage at a constant current of 60 mA. Dyes in the loading solution are very useful to monitor the migration of RNAs in the gel; xylene cyanol migrates like CEVd, whereas bromophenol blue migrates with the front (fast running molecules). Run electrophoresis, preferably at 4 °C, until the bromophenol blue marker dye is 7 cm down from the top of the gel. Running

time should be approximately 2 h. 15 min (constant amperage). While waiting for the run to complete, prepare the second urea-containing gel.

4. Turn off power, disconnect leads from power supply, and discard the buffer from the upper and lower reservoirs.
5. Lay the gel assembly on the bench top, remove the spacers (if necessary), and carefully separate the glass plates using a scalpel or single-edge razor blade.
6. Transfer the gel to an ethidium bromide staining solution (0.5 mg/ml ethidium bromide in $0.1\times$ TBE or water) in sufficient volume to allow the gel to float in it, and gently agitate for 15–30 min on a platform shaker.
7. Remove the gel from the ethidium bromide staining solution, and remove the excess solution by placing the gel in water for 10–15 min.
8. Place a piece of plastic wrap on the surface of a UV transilluminator and transfer the gel to the plastic wrap, and carefully remove trapped air bubbles before viewing on the UV transilluminator and/or captured image (Fig. 1).
9. Cut a horizontal strip defined by the 200–400 bp MW marker or smaller, depending upon the viroids. Most viroids (300 nt or larger) can be easily recovered from a strip defined by a 300–400 bp MW marker, whereas smaller viroids are mostly recovered from a strip defined by a 200–300 bp MW marker. Small strips give better results in the second gel.

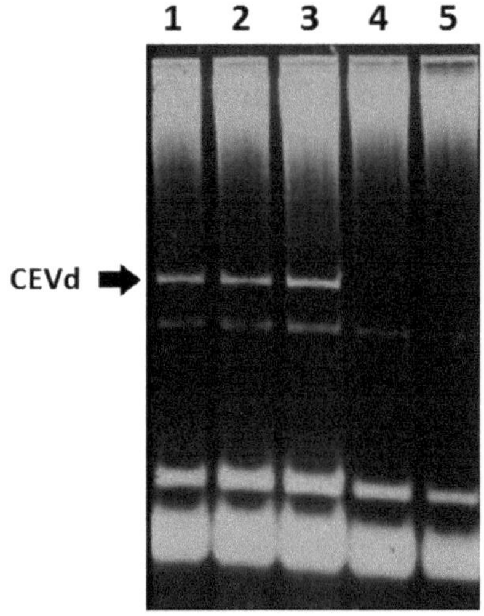

Fig. 1 Characteristic visualization of CEVd in infected tissue samples (lanes 1, 2, 3) as compared with CEVd-free samples (lanes 4, 5), after ethidium bromide staining

3.3 Cast the Second (Denaturing/Urea-Containing) Gel

1. While running the first gel, cast the second gel. Clean glass plates and assemble them as for the first gel. Place some tape on the surface of the spacers to make them slightly thicker (this will help in terms of introducing the first gel strip).
2. Mix 35 ml of stock urea gel. Mix quickly and dispense.
3. Tilt the glass plate-spacer assembly slightly, fill with gel solution up to 1–2 cm below the top end of the glass plate, and *do not* insert the comb. Avoid trapping air bubbles against the glass plates.
4. Allow polymerization to proceed for 30–60 min at room temperature (do not place at 4 °C). Make sure the top part of the gel is flat.

3.4 Apply Strip of the First Gel on the Second Gel and Begin Electrophoresis

1. Place the gel assembly in the electrophoresis apparatus, tighten the side clamps, and fill the upper and lower reservoirs with 1× TBE, pH 8.3.
2. Place the strip of the first gel on the top of the gel, making as close contact as possible. You can use a spacer to push the gel strip. Load 10 ml of the loading dye onto the outer edges of the strip. The dye will spread underneath the strip and will help confirm that there is good contact between the strip and the gel (Fig. 2).
3. Connect electrodes to a power supply; set the amperage at constant current of 16–18 mA. The two dyes in the loading solution will migrate together showing the position of the front. Run the gel until the marker dyes reach the bottom of the gel. Running time should be approximately 4 h (at constant amperage).

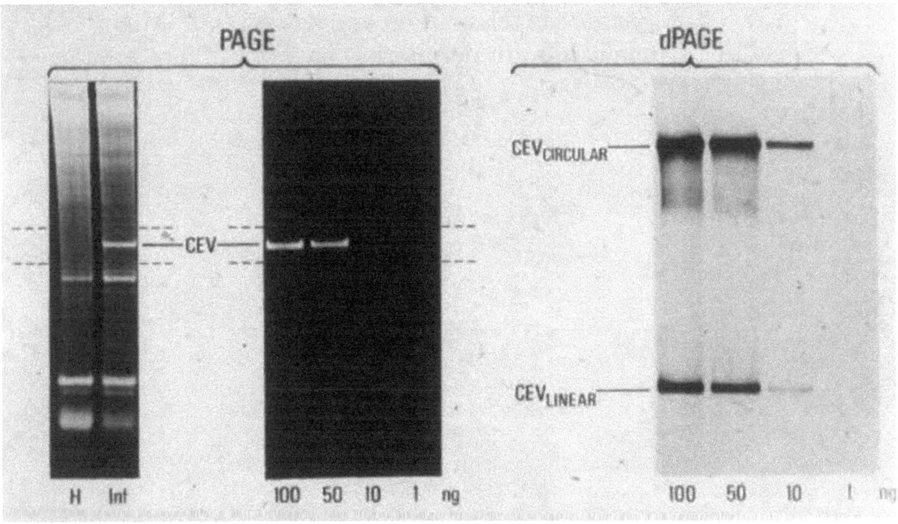

Fig. 2 Horizontal strip to be excised and placed on the top of the second urea-containing gel (dPAGE). The second PAGE shows after silver staining the characteristic mobility of the circular and linear forms of CEVd

4. Turn off power, disconnect leads from the power supply, and discard upper and lower reservoir buffer.
5. Lay the gel assembly on the bench top, remove the spacers, and carefully separate the glass plates using a scalpel or single-edge razor blade.
6. Transfer the gel to an ethidium bromide staining solution (0.5 μg/ml ethidium bromide in 0.1× TBE or water), and/or proceed to silver staining.

3.5 Visualize Small Circular RNAs

RNAs and DNAs are most commonly visualized by staining with ethidium bromide after electrophoresis. Silver staining can detect proteins as well as nucleic acids and is much more sensitive than ethidium bromide to detect the circular forms of the viroids in sPAGE.

1. Soak the gel in a solution (50% ethanol, 10% acetic acid) for at least 1 h at room temperature in gentle agitation. The gel can remain in this solution overnight or over a weekend. Long exposure in the solution will cause the gel to dehydrate and shrink, but it will get rehydrated in the following steps.
2. Discard the solution, and soak the gel in a solution (10% ethanol, 1% acetic acid) for 1 h at room temperature in gentle agitation.
3. Discard the solution, and soak the gel in a solution (12 mM $AgNO_3$) for 1 h at room temperature in gentle agitation.
4. Discard the solution and rinse the gel three times with water.
5. Rinse quickly with a small volume of developer solution (0.75 M KOH + 0.28% HCOH) and discard.
6. Soak with developer solution and keep in gentle agitation until visualizing the bands. Developing can be stopped by discarding the developer solution and adding 0.07 M Na_2CO_3, but generally it is not necessary (Fig. 3).

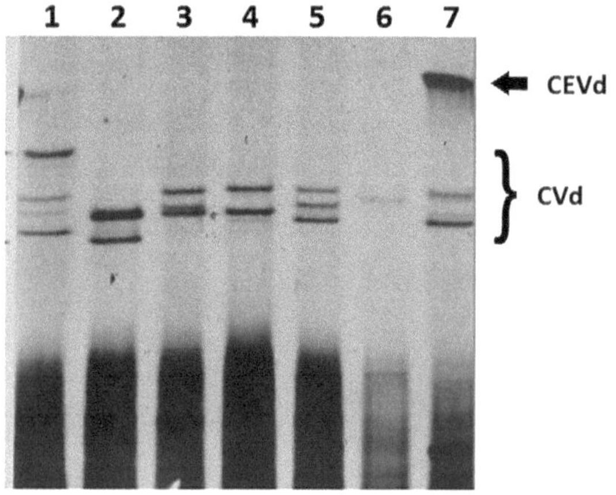

Fig. 3 Visualization of the circular forms of viroids infecting citrus. This approach allowed the identification of several viroids infecting citrus and grapevine

4 Notes

Since PAGE does not require nucleotide sequence information, it has been the most useful tool for the isolation of viroids infecting plant species. Some critical aspects are as follows: (a) the tissue grinding approach is critical because viroids are located in the nucleus or in chloroplasts; (b) LiCl separation may allow the recovery of a high concentration of viroids in the samples to be analyzed; and (c) different hosts are able to accumulate different amounts of viroids to be tested and therefore have an effect on the detection capacity.

References

1. Morris TJ, Smith EM (1977) Potato spindle tuber disease: procedures for the detection of viroid RNA and certification of disease-free potato tubers. Phytopathology 67:145–150
2. Flores R, Duran-Vila N, Pallás V, Semancik JS (1985) Detection of viroid and viroid-like RNAs from grapevine. J Gen Virol 66:2095–2102
3. Duran-Vila N, Flores R, Semancik JS (1986) Characterization of viroid-like RNAs associated with the citrus exocortis syndrome. Virology 150:75–84
4. Rivera-Bustamante R, Gin R, Semancik JS (1986) Enhanced resolution of circular and linear molecular forms of viroid and viroid-like RNA by electrophoresis in a discontinuous-pH system. Anal Biochem 156:91–95
5. Hanold D, Semancik JS, Owens RA (2003) Polyacrylamide gel electrophoresis. In: Hadidi A, Flores R, Randles JW, Semancik JS (eds) Viroids. CSIRO Publishing, Collingwood, pp 95–102
6. Duran-Vila N, Roistacher CN, Rivera-Bustamante R, Semancik JS (1988) A definition of citrus viroid groups and their relationship to the exocortis disease. J Gen Virol 69:3069–3080
7. Szychowski JA, Doazan JP, Leclair P, Garnier M, Credi R, Minafra A, Duran-Vila N, Wolpert JA, Semancik JS (1991) Relationship and patterns of distribution among grapevine viroids from California and Europe. Vitis 30:25–36
8. Serra P, Barbosa CJ, Daròs JA, Flores R, Duran-Vila N (2008) Citrus viroid V: molecular characterization and synergistic interactions with other members of the genus *Apscaviroid*. Virology 370:102–112
9. Serra P, Eiras M, Bani Hashemian SM, Murcia N, Kitajima EW, Daròs JA, Flores R, Duran-Vila N (2008) Citrus viroid V: occurrence, host range, diagnosis and identification of new variants. Phytopathology 98:1199–1204
10. Semancik JS, Roistacher CN, Rivera-Bustamante R, Duran-Vila N (1988) Citrus cachexia viroid, a new viroid of citrus: relationship to viroids of the exocortis disease complex. J Gen Virol 69:3059–3068
11. Foissac X, Duran-Vila N (2000) Characterization of two citrus apscaviroids isolated in Spain. Arch Virol 145:1075–1983
12. Fagoaga C, Pina JA, Duran-Vila N (1994) Occurrence of small RNAs in severely diseased vegetable crops. Plant Dis 78:749–753
13. Fagoaga C, Semancik JS, Duran-Vila N (1995) A citrus exocortis viroid variant from broad bean (*Vicia faba* L.): infectivity and pathogenesis. J Gen Virol 76:2271–2277
14. Fagoaga C, Duran-Vila N (1996) Natural occurrence of variants of the citrus exocortis viroid in vegetable crop species. Plant Pathol 45:45–53

Chapter 8

Two-Dimensional Polyacrylamide Gel Electrophoresis Analysis of Viroid RNAs

José-Antonio Daròs

Abstract

The circular and linear forms of viroid RNAs can be separated by two-dimensional polyacrylamide gel electrophoresis (PAGE) based on the selective delay in mobility that circular RNAs experience under denaturing conditions. First PAGE separates RNA preparations from viroid-infected plants, and the whole lane from this first gel is next perpendicularly loaded on top of a second gel. Separation continues under new conditions that differ in the degree of denaturation from the first. The result is a two-dimensional separation of the RNAs in which circular and linear molecules are distributed in two parallel diagonals.

Key words Circular RNA, Two-dimensional separation, Polyacrylamide gel electrophoresis, Non-denaturing electrophoresis, Denaturing electrophoresis, RNA staining, RNA hybridization

1 Introduction

Viroids are circular RNAs. This is one of the most distinctive properties of this particular class of infectious agents of higher plants [1]. Viroid circularity was deduced in early research after viroid discovery by biochemical evidence [2] and supported by electron microscopy analysis in denaturing conditions [2] and direct RNA sequencing using specific RNases [3]. For decades, viroids remained as the most conspicuous circular RNAs in nature. However, since the discovery of circular RNAs that serve as sponges of microRNAs [4, 5], this type of molecules is now known to be widespread in the eukaryotic tree of life, playing essential roles in health and disease [6]. A circular form is usually the most abundant viroid RNA in the plant-infected tissues and consequently considered the viroid progeny. Positive or plus polarity is arbitrarily attributed to these most abundant circular forms. Viroid circular RNAs arise from an RNA-to-RNA rolling-circle replication mechanism that involves transcription of multimeric RNAs, processing of some of these multimers to unit length, and circularization [7]. Viroids

Ayala L. N. Rao et al. (eds.), *Viroids: Methods and Protocols*, Methods in Molecular Biology, vol. 2316,
https://doi.org/10.1007/978-1-0716-1464-8_8, © Springer Science+Business Media, LLC, part of Springer Nature 2022

from the family *Pospiviroidae* replicate through an asymmetric rolling-circle mechanism, and a unit-length RNA of positive polarity is the only circular viroid RNA in infected tissues [8]. In this family, viroid circularization is mediated by the host DNA ligase 1 [9]. In contrast, viroids from the family *Avsunviroidae* replicate through a symmetric rolling-circle mechanism, and circular RNAs of negative polarity that play an intermediary role in replication are also detected in infected tissues [10, 11]. Viroids of this family, which replicate in the chloroplasts of infected cells, are circularized by the chloroplastic isoform of tRNA ligase [12].

The existence in infected tissues of viroid RNA molecules with two different topologies, circular and linear, opens the possibility of separating them in two dimensions by polyacrylamide gel electrophoresis (PAGE). This separation is based on a selective delay in mobility that circular RNAs experience with respect to their linear counterparts of the same size in denaturing conditions. While folded linear and circular viroid RNAs of the same size migrate at the same speed in non-denaturing conditions, when denatured, the linear molecules penetrate more efficiently the polyacrylamide than their circular counterparts, now converted into open rings. This differential electrophoretic behavior was brilliantly profited to develop a two-dimensional electrophoretic technique for the detection of circular viroids and viroid-like RNAs [13], which has been in the base of new viroid species discovery for decades until the coming of next-generation sequencing.

Here, I describe materials and methods for two-dimensional PAGE (2D-PAGE) separation of viroid RNAs. In principle, 2D separations of viroid RNAs can be obtained by any combination of two electrophoretic conditions with different degree of denaturation. The most typical combination is to separate the RNA preparation first in non-denaturing condition in Tris-acetate-ethylenediaminetetraacetic acid (EDTA; TAE) buffer and then in denaturing conditions in the presence of 8 M urea in a low ionic strength buffer, such as Tris-borate-EDTA (TBE) ×0.25 (Figs. 1, 2a, and 3). Note that a low ionic strength buffer contributes to a more efficient RNA denaturation and consequent slower migration of circular molecules. For this reason, 2D separation of circular and linear viroid RNAs can also be accomplished in denaturing conditions (8 M urea), provided the ionic strength of the electrophoresis buffer is changed. Good separations can be obtained electrophoresing RNAs first in ×1 TBE and then in ×0.25 TBE gels, both with 8 M urea (Fig. 2b). If viroid RNAs are abundant in the RNA preparations, the case of *Avocado sunblotch viroid* (ASBVd) in preparations from infected avocado trees, they can be directly observed after staining the gels with ethidium bromide (Fig. 2). If viroid RNAs are not abundant, detection requires northern blot hybridization using a viroid-specific probe, exemplified by *Eggplant latent viroid* (ELVd) that accumulates to low concentration in infected

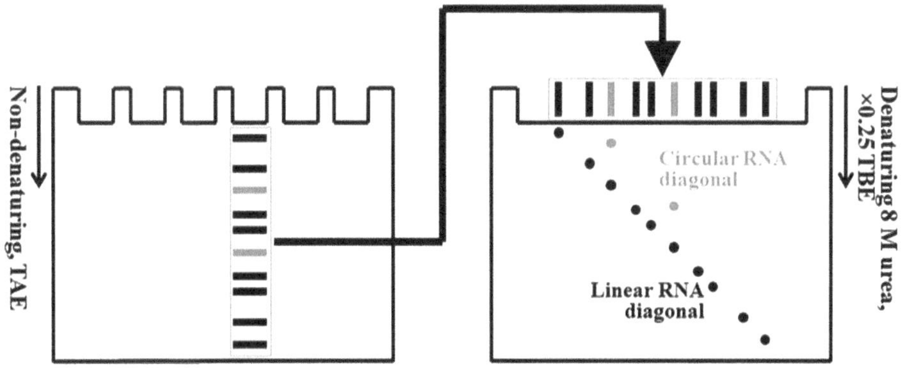

Fig. 1 Schematic representation of a 2D-PAGE separation of an RNA preparation that contains two circular RNAs (gray bands and spots) and many linear RNAs (black bands and spots). After electrophoresing the RNA in non-denaturing conditions in the TAE buffer, the whole lane is cut and applied on top of a second gel. The second electrophoresis in denaturing conditions (8 M urea, 0.25× TBE) leads to separation of circular and linear RNAs in two parallel diagonals

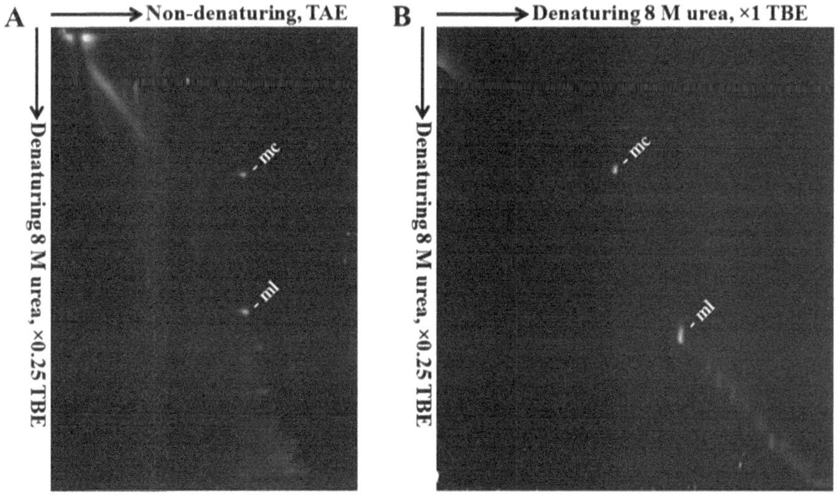

Fig. 2 2D-PAGE separation of an RNA preparation from an ASBVd-infected avocado tree. (**a**) 2D separation first under non-denaturing and second under denaturing (low ionic strength buffer) conditions, as indicated. (**b**) 2D separation first under denaturing (high ionic strength buffer) and second under denaturing (low ionic strength buffer) conditions, as indicated. RNAs were stained with ethidium bromide. Arrows indicate the direction of RNA migration. The positions of monomeric circular (mc) and linear (ml) ASBVd RNAs are indicated

eggplants (Fig. 3). Importantly, in the 2D-PAGE separation of viroid RNAs, the more denaturing run not necessarily has to follow the non-denaturing or less-denaturing. RNAs can be separated first in denaturing conditions and then in non-denaturing. In this case, circular RNAs would move faster in the second dimension. This kind of separation may be useful in some instances when viroid RNAs need to be purified from the gel, because non-denaturing gels in TAE buffer polymerized with *N*,*N*′-bis(acryloyl) cystamine are easily depolymerized, while those in TBE buffer with 8 M urea are not [14].

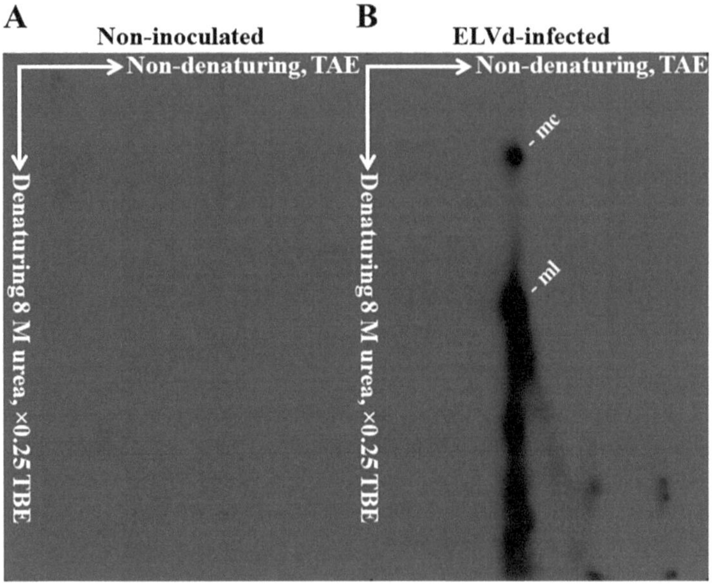

Fig. 3 2D-PAGE separation of RNA preparations from an (**a**) non-inoculated eggplant and (**b**) an ELVd-infected eggplant. RNAs were first separated under non-denaturing and then under denaturing (low ionic strength buffer) conditions, as indicated. RNAs from the second gel were transferred to a nylon membrane, and ELVd RNAs detected by hybridization with a complementary probe. Arrows indicate the direction of RNA migration. The positions of monomeric circular (mc) and linear (ml) ELVd RNAs are indicated

2 Materials

Prepare all solutions using ultrapure water and analytical grade reagents. Follow all waste disposal regulations.

2.1 Non-denaturing PAGE

1. Electrophoresis plates. We use 160-mm-wide × 150-mm-long × 3-mm-thick glass plates. One of the plates contains a 140-mm-wide × 20-mm-long notch.
2. Electrophoresis combs and spacers. We use 10-mm-wide × 150-mm-long × 2-mm-thick spacers. Combs are 2 mm thick.
3. 30% acrylamide-bisacrylamide (39:1) solution: weigh 29.25 g acrylamide and 0.75 g bisacrylamide, and dissolve in H_2O. Bring to 100 mL and store at room temperature in an amber glass bottle.
4. 10× TAE buffer: 0.4 M Tris, 0.2 M sodium acetate, 10 mM EDTA, pH 7.2. Weigh 48.5 g Tris, 27.2 g sodium acetate ($3 \cdot H_2O$), and 3.7 g disodium EDTA ($2 \cdot H_2O$), and dissolve in H_2O. Adjust pH to 7.2 with acetic acid. Approximately 20 mL are needed. Bring to 1 L and autoclave (*see* **Note 1**). Store at room temperature.

5. 10% ammonium peroxodisulfate: weigh 1 g ammonium peroxodisulfate and dissolve in H_2O. Bring to 10 mL and store at 4 °C for up to 1 month.
6. 10× non-denaturing loading buffer: 50% glycerol, 10 mM EDTA, 0.025% bromophenol blue, 0.025 xylene cyanol, pH 8.0. Weigh 2.5 mg bromophenol blue, 2.5 mg xylene cyanol, and 7.1 g 87% glycerol (density 1.228 g/mL). Add 1 mL 0.5 M EDTA, pH 8.0. Add H_2O and dissolve. Bring to a final volume of 10 mL. Divide into 1 mL aliquots and store frozen. The aliquot in use can be stored at room temperature.

2.2 Gel Staining

1. Ethidium bromide solution. We use a commercial 1% ethidium bromide solution.

2.3 Denaturing PAGE Under High Ionic Strength

1. For electrophoresis plates, combs, and spacers, *see* Subheading 2.1.
2. 10× TBE buffer: 0.89 M Tris, 0.89 M boric acid, 20 mM EDTA. Weigh 107.8 g Tris and 7.4 g disodium EDTA (2 · H_2O), and dissolve in H_2O. Add 55 g boric acid and dissolve. Bring to 1 L and autoclave (*see* **Note 1**). Store at room temperature.
3. Formamide loading buffer: 98% formamide, 10 mM Tris–HCl, pH 8.0, 1 mM EDTA, 0.025% bromophenol blue, 0.025 xylene cyanol. Weigh 2.5 mg bromophenol blue and 2.5 mg xylene cyanol. Add 100 μL 1 M Tris-HCl, pH 8.0, and 20 μL 0.5 M EDTA, pH 8.0. Add formamide and dissolve. Use deionized formamide molecular biology grade. Bring to 50 mL with formamide. Divide into 1 mL aliquots and store frozen.
4. Other reagents as described in Subheading 2.1.

2.4 Denaturing PAGE Under Low Ionic Strength

1. Electrophoresis plates, combs, and spacers as described in Subheading 2.1. Reagents as described in Subheadings 2.1 and 2.3.

2.5 RNA Blotting

1. Electrotransfer solution: 25 mM Tris, 192 mM glycine. Weigh 3 g Tris and 14.4 g glycine and dissolve in H_2O. Bring to 1 L and autoclave (*see* **Note 1**). Store at room temperature.
2. Nylon membrane. We use Whatman Nytran SuPerCharge nylon blotting membranes.
3. Blotting paper. We use 1.4-mm-thick Albet LabScience blotting paper.

2.6 Northern Hybridization

1. 20× SSC: 3 M NaCl, 0.3 M sodium citrate, pH 7.0. Weigh 175.3 g NaCl and 88.2 g tri sodium citrate (2 · H_2O). Dissolve in H_2O and adjust to pH 7.0 with HCl. Bring to 1 L and autoclave (*see* **Note 1**). Store at room temperature.

2. 10% sodium dodecyl sulfate (SDS). Weigh 10 g SDS and dissolve in H_2O. Adjust to 100 mL and store at room temperature.
3. 1% Ficoll, 1% polyvinylpyrrolidone. Weigh 1 g Ficoll 400 and 1 g polyvinylpyrrolidone (PVP; molecular weight 360,000, molecular biology grade). Sterilize by filtration and store at 4 °C.
4. 10 mg/mL DNA from fish sperm (molecular biology grade).
5. Hybridization solution. Mix 25 mL deionized formamide (molecular biology grade), 12.5 mL 20× SSC, 5 mL 1% Ficoll and 1% PVP solution, 5 mL 10% SDS, 0.5 mL 10 mg/mL fish sperm DNA, and 2 mL H_2O. Store at room temperature for 1 month.

3 Methods

For 2D-PAGE separation of viroid RNAs, two consecutive PAGE separations in different conditions must be selected.

3.1 First Dimension PAGE in Non-denaturing Conditions

1. Assemble the electrophoresis plates.
2. Mix 35.9 mL H_2O, 8.3 mL 30% acrylamide-bisacrylamide solution, and 5 mL 10× TAE buffer.
3. While gently stirring the solution, add 75 μL *N*,*N*,*N*′,*N*′-tetramethylethylenediamine (TEMED) and 0.75 mL 10% ammonium peroxodisulfate.
4. Quickly pour the solution into the assembled plates and insert a comb. Wait until acrylamide completely polymerizes.
5. Mix aliquots of the RNA preparations with the appropriate amounts of 10× non-denaturing loading buffer.
6. Remove the comb and assemble the gel into a vertical electrophoresis cuvette. Pour 1× TAE buffer in both compartments of the cuvette, and carefully wash the wells using a Pasteur pipette or a syringe (*see* **Note 2**).
7. Load the RNA samples into the gel wells (*see* **Note 3**).
8. Select electrophoresis parameters (75 mA for a 140-mm-wide × 130-mm-long × 2-mm-thick gel), and quickly start electrophoresis. Run for approximately 2 h.

3.2 RNA Staining and Cutting Gel Lanes

1. After electrophoresis, disassemble the gel.
2. Stain the RNA by incubating the gel in 200 mL of 0.5 μg/mL ethidium bromide for 15 min with agitation (125 revolutions per min, rpm).
3. Gently wash the gel three times with H_2O.

4. Visualize the RNA in the gel using a UV transilluminator equipped with 312 nm UV lamps. The gel can be photographed at this point using a gel documentation system with a camera equipped with the appropriate ethidium bromide filter.
5. Using a sharp ruler, cut the lanes of the gel corresponding to the samples that are going to be subjected to second dimension separation.

3.3 First Dimension PAGE in Denaturing (High Ionic Strength) Conditions

1. Weigh 24 g of urea, and add 17 mL H_2O, 8.3 mL 30% acrylamide-bisacrylamide solution, and 5 mL 10× TBE buffer.
2. Stir the mix until urea is completely dissolved.
3. While gently stirring, add 50 μL TEMED and 0.75 mL 10% ammonium peroxodisulfate, and quickly pour the solution into the assembled plates. Insert the comb and wait for acrylamide polymerization.
4. Mix aliquots of the RNA preparations with at least one volume of formamide loading buffer (*see* **Note 4**).
5. Heat the samples for 90 s at 95 °C and snap cool on ice.
6. Remove the comb and assemble the casted gel in a vertical electrophoresis cuvette. Fill the cuvette with 1× TBE buffer and wash the gel wells.
7. Load the denatured RNA samples, select the electrophoresis parameters (200 V for a 140 × 130 × 2 mm gel), and run the electrophoresis for approximately 2 h.
8. Stain the RNA in the gel, document the image, and cut the lanes with the RNA to be separated in a second dimension as explained in Subheading 3.2.

3.4 Second Dimension PAGE in Denaturing (Low Ionic Strength) Conditions

1. Weigh 24 g of urea, and add 20.8 mL H_2O, 8.3 mL 30% acrylamide-bisacrylamide solution, and 1.25 mL 10× TBE buffer, and stir the mix until urea is completely dissolved.
2. While gently stirring the solution, add 50 μL TEMED and 0.75 mL 10% ammonium peroxodisulfate.
3. Pour the solution quickly into the assembled plates, leaving approximately 1 cm distance between the solution and the border of the notched glass plate (*see* **Note 5**). Settle until complete acrylamide polymerization.
4. Assemble the gel in a vertical electrophoresis cuvette.
5. Using a spatula, insert the polyacrylamide band; previously cut from the first dimension gel, into the top space of the second dimension gel until it touches the second dimension gel (*see* **Note 6**).

6. Fill the cuvette with 0.25× TBE buffer, and run the electrophoresis for 2.5 h at 25 mA (140 × 130 × 2 mm gel).
7. After the electrophoresis, disassemble the gel, stain the RNA, and document the result, as explained in Subheading 3.2.

3.5 Blotting the 2D-Separated RNA to a Membrane

1. Using a sharp ruler, trim the gel, discarding those parts with no RNA.
2. Equilibrate the gel for 10 min by rocking (125 rpm) in 200 mL of electrotransfer buffer. Measure the final dimensions of the gel.
3. Cut a piece of positively charged nylon membrane and six pieces of blotting paper of the same dimensions as the gel, and equilibrate a few minutes in electrotransfer buffer.
4. Assemble the electroblotting sandwich on the plus electrode plate of an electroblotting apparatus, avoiding air bubbles.
5. First, layer three pieces of blotting paper. Then, layer the membrane. Carefully layer the gel on the membrane. Finally, layer the three remaining pieces of blotting paper, and cover with the negative electrode plate.
6. Run the electrotransfer for 1 h at 1 mA per each cm^2 of the membrane.
7. Disassemble the blotting sandwich, and cross-link the RNA to the membrane by irradiating with a dose of 0.12 J/cm^2 of UV light (254 nm).
8. The membrane can be subjected to hybridization right away or air-dried and stored at room temperature to proceed in another moment.

3.6 Viroid RNA Hybridization

1. Insert the membrane (RNA side up) into a hybridization tube, and add 10 mL of hybridization solution (*see* **Note 7**).
2. Prehybridize for at least 1 h at 70 °C (*see* **Note 8**), using a hybridization oven with constant rotation.
3. Discard the prehybridization solution and add 5 mL of fresh hybridization solution. Incubate for a few minutes to allow the solution to reach 70 °C.
4. Mix the right amount of probe (*see* **Note 9**) with 150 μL of hybridization solution. Denature for 1.5 min at 95 °C and snap cool on ice.
5. Add the probe to the hybridization tube. Avoid pipetting the probe directly on the membrane. Hybridize overnight.
6. Properly discard the hybridization solution containing the probe.

7. Wash the membrane three times for 10 min with 2× SSC and 0.1% SDS at room temperature and once for 15 min with 0.1× SSC and 0.1% SDS.
8. Use an exposure cassette with intensifying screens and an X-ray film to reveal the result of the hybridization.

4 Notes

1. There is no need to use sterile 10× TAE, 10× TBE, 20× SSC, and electrotransfer buffers, but autoclaving after preparation improves the conservation of these solutions at room temperature.
2. For good sample loading, wash out the small polyacrylamide fragments that usually remain in the wells after removing the comb.
3. Electrophoretic separation in the center of the gel is always better than in the border. Unless you plan to use all the wells, load preferentially those in the center of the gel.
4. A higher proportion of formamide loading buffer in the mix would favor the denaturation of RNA. However, the final sample volume must also be kept as low as possible to improve separation.
5. This distance can be altered depending on the width of the polyacrylamide band and cut in the first dimension gel.
6. There is no problem if part of the band keeps outside of the notched glass plate.
7. The volume of prehybridization and hybridization solutions would depend on the dimensions of the tube. The volume of the hybridization solution must be kept as low as possible to increase probe concentration, although the volume of liquid must be sufficient to wet the membrane evenly.
8. Prehybridization and hybridization temperatures depend on the nature of the probe. Since viroids are highly self-complementary RNAs, 70 °C in the presence of 50% formamide is recommended in the case of RNA probes. Use 55 °C in the case of DNA probes.
9. In the case of ^{32}P-labelled RNA probes, use around one million counts per min of the probe. Lower amounts should be, however, more than enough for good results.

Acknowledgments

I thank Verónica Aragonés for excellent technical assistance. This work was supported by grants BIO2017-83184-R and BIO2017-91865-EXP from Ministerio de Cienciae, Innovación (Spain), co-financed by European Regional Development Fund (European Commission).

References

1. Daròs JA (2016) Viroids: small noncoding infectious RNAs with the remarkable ability of autonomous replication. In: Wang A, Zhou X (eds) Current research topics in plant virology. https://doi.org/10.1007/978-3-319-32919-2_13
2. Sänger HL, Klotz G, Riesner D et al (1976) Viroids are single-stranded covalently closed circular RNA molecules existing as highly base-paired rod-like structures. Proc Natl Acad Sci U S A 73:3852–3856. https://doi.org/10.1073/pnas.73.11.3852
3. Gross HJ, Domdey H, Lossow C et al (1978) Nucleotide sequence and secondary structure of potato spindle tuber viroid. Nature 273:203–208
4. Memczak S, Jens M, Elefsinioti A et al (2013) Circular RNAs are a large class of animal RNAs with regulatory potency. Nature 495:333–338. https://doi.org/10.1038/nature11928
5. Hansen TB, Jensen TI, Clausen BH et al (2013) Natural RNA circles function as efficient microRNA sponges. Nature 495:384–388. https://doi.org/10.1038/nature11993
6. Haddad G, Lorenzen JM (2019) Biogenesis and function of circular RNAs in health and in disease. Front Pharmacol 10:428. https://doi.org/10.3389/fphar.2019.00428
7. Branch AD, Robertson HD (1984) A replication cycle for viroids and other small infectious RNAs. Science 223:450–455
8. Branch AD, Benenfeld BJ, Robertson HD (1988) Evidence for a single rolling circle in the replication of potato spindle tuber viroid. Proc Natl Acad Sci U S A 85:9128–9132
9. Nohales M-Á, Flores R, Daròs J-A (2012) Viroid RNA redirects host DNA ligase 1 to act as an RNA ligase. Proc Natl Acad Sci U S A 109:13805–13810. https://doi.org/10.1073/pnas.1206187109
10. Hutchins CJ, Keese P, Visvader JE et al (1985) Comparison of multimeric plus and minus forms of viroids and virusoids. Plant Mol Biol 4:293–304
11. Daròs JA, Marcos JF, Hernández C, Flores R (1994) Replication of avocado sunblotch viroid: evidence for a symmetric pathway with two rolling circles and hammerhead ribozyme processing. Proc Natl Acad Sci U S A 91:12813–12817. https://doi.org/10.1073/pnas.91.26.12813
12. Nohales M-Á, Molina-Serrano D, Flores R, Daròs J-A (2012) Involvement of the chloroplastic isoform of tRNA ligase in the replication of viroids belonging to the family Avsunviroidae. J Virol 86:8269–8276. https://doi.org/10.1128/JVI.00629-12
13. Schumacher J, Randles JW, Riesner D (1983) A two-dimensional electrophoretic technique for the detection of circular viroids and virusoids. Anal Biochem 135:288–295
14. Daros JA, Aragones V, Cordero T (2018) A viroid-derived system to produce large amounts of recombinant RNA in Escherichia coli. Sci Rep 8:1904. https://doi.org/10.1038/s41598-018-20314-3

Chapter 9

A Simplified Dot-Blot Hybridization Protocol for *Potato spindle tuber viroid* Detection in Solanaceae

Oxana Kekstidou, Christina Varveri, and Nikon Vassilakos

Abstract

A simplified dot-blot hybridization protocol for *Potato spindle tuber viroid* (PSTVd) detection in Solanaceae species is described here. The protocol uses an RNA DIG-labeled probe and a simplified extraction procedure that avoids the use of hazardous chemicals. PSTVd was detected in composite tomato leaf samples in a ratio of up to 1:15 of PSTVd-infected to non-infected tissue and in composite potato tuber samples in a ratio up to 1:5 of PSTVd-infected to non-infected tissue. In *Brugmansia* spp., PSTVd was detected solely in the standard sample extract preparation. The method is suitable for a reliable, large-scale sample screening especially where cost is a limiting factor.

Key words RNA hybridization, RNA DIG labeling, Viroids, DNA sequencing

1 Introduction

Potato spindle tuber viroid (PSTVd) is the type member of the *Pospiviroid* genus and an EPPO A2-listed quarantine pathogen. PSTVd infection is symptomless in solanaceous ornamentals but can cause serious losses in potato and tomato production [1]. Despite the development of highly sensitive multiplex quantitative RT-PCR protocols able to simultaneously detect several pospiviroids, PSTVd unequivocal detection requires a second detection method due to the high sequence similarity among the members of the genus. The latter is usually a conventional RT-PCR producing an amplification product of adequate length for viroid species identification through sequencing. For PSTVd first screening of a relatively high number of samples, dot-blot hybridization represents a low-cost reliable alternative method to RT-PCR/qPCR and does not require expensive laboratory equipment. The simplified dot-blot hybridization protocol for PSTVd detection in Solanaceae species presented in this chapter is based on the protocol originally presented by Vassilakos and co-workers in 2012 ([2] and references therein). The protocol uses an RNA DIG-labeled probe

Ayala L. N. Rao et al. (eds.), *Viroids: Methods and Protocols*, Methods in Molecular Biology, vol. 2316,
https://doi.org/10.1007/978-1-0716-1464-8_9,

and a simplified extraction procedure that avoids the use of hazardous chemicals. The method was shown to be suitable for PSTVd detection in tomato plants and in dormant potato tubers achieving confident PSTVd detection in composite samples that consisted of infected and non-infected tissues in a ratio up to 1:5 and 1:15 for potato tubers and tomato, respectively. In *Brugmansia* spp. however, PSTVd was detected only in the standard sample extract preparation. Overall, the method provides a reliable low-cost screening approach for PSTVd presence in different solanaceous species.

2 Materials

1. Type 1 ultrapure water; nuclease-, protease-, and bacteria-free water at the point of dispense; or water of equivalent quality (*see* **Note 1**).
2. Ice (laboratory flaker ice machine).
3. Set of micro-pipettes (preferentially autoclavable) covering a range of volumes from 0.1 μL to 1 mL and their respective tips (optionally filter).
4. Pair of clean scissors with round edges.
5. Extraction bags for ELISA tissue extraction (BIOREBA art. no. 430100 or equivalent, preferentially from heavy duty plastic with synthetic intermediate layer for optimal filtration).
6. Tissue homogenizer for ELISA extraction bags (BIOREBA HOMEX 6 with standard rack, Art. No. 400014).
7. 1.5 mL centrifuge tubes.
8. Benchtop centrifuge.
9. Water bath.
10. Heat block.
11. Incubation oven, set at 80 °C.
12. UV transilluminator (UVP) or equivalent.
13. Hybridization oven with rotation wheel and dedicated glass tubes and/or shaker with hybridization bags (*see* **Note 2**).
14. Plate shaker.
15. Plastic/glass containers, sufficient for two to three membranes, 8 × 11 cm each.
16. Vortex mixer.
17. Nylon bags (15 × 15 cm or similar dimensions).
18. Nylon bag heat sealer.

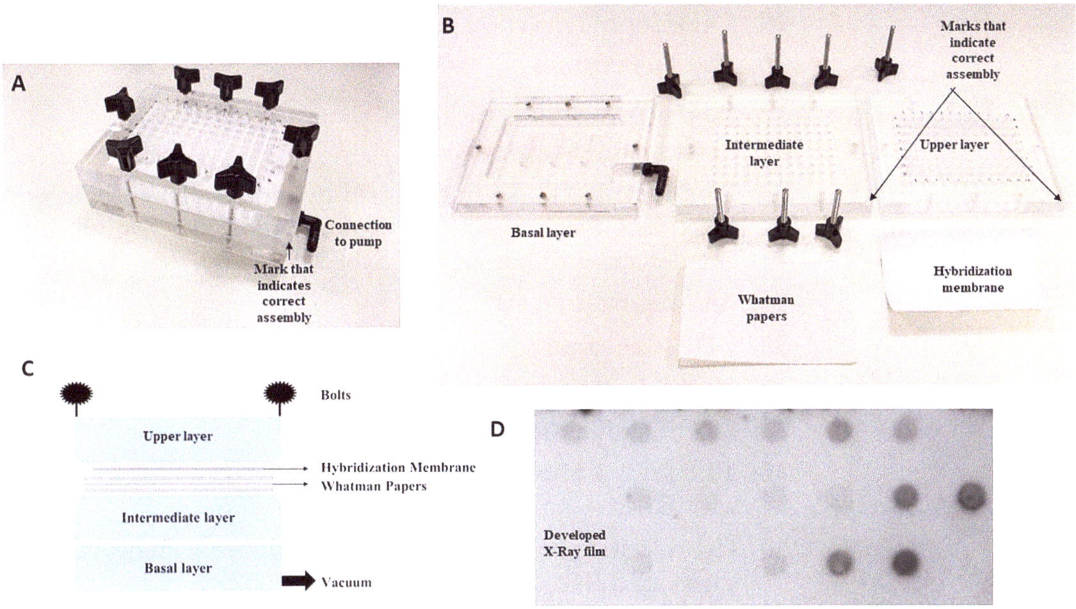

Fig. 1 Assembly and components (labeled) of the dot-blot apparatus including Whatman papers and hybridization membrane (**a**–**c**). Developed X-ray film (**d**). Positive samples appeared as dark spots in contrast to white background of the PSTVd-negative samples

19. Pathogen tested-negative plant tissues and PSTVd-infected positive controls of the same species as the tested samples. Use positive controls at the same infected to healthy ratios as the plant tissues examined.
20. Gloves and lab coat.
21. Forceps with round edges.
22. pH meter.
23. Dot-blot apparatus, provided by several brands (e.g., Hybri-Dot Manifold/Life Sciences) (Fig. 1).
24. Vacuum pump Heto SUE 300E, 75 W, rpm 2850, or of equivalent specifications.
25. A DIG-labeled full-length RNA PSTVd probe, synthesized following standard procedures as described by the manufacturer.
26. Whatman filter paper sheets.
27. Hybond-N^+ nylon membrane (Amersham Biosciences) or equivalent.
28. AMES buffer: 5.85% NaCl, 0.2% $MgCl_2$, 4.1% sodium acetate, 20% ethanol, 3% sodium dodecyl sulfate (SDS). Add the salts sequentially to 8/10 of the water, wait for them to dissolve, and then add the ethanol and finally the SDS. Adjust to pH 6.0 with hydrochloric acid or sodium hydroxide as required.

29. Chloroform (analytical grade).
30. Absolute ethanol.
31. *N*-Lauryl-sarcosine.
32. Formamide, analytical grade.
33. Blocking reagent for nucleic acid hybridization and detection (decreases the background in non-radioactive hybridization and detection of nucleic acid hybrids).
34. Anti-digoxigenin-AP.
35. CDP-Star substrate.
36. Magnetic stirrer and magnets.
37. 20% SDS, filter sterilized.
38. 20× SSC: 3 M NaCl, 0.3 M sodium citrate; adjust to pH 7.0 with HCl.
39. Hybridization buffer (HB): 5× SSC, 50% formamide, 0.1% *N*-lauryl-sarcosine, 2% blocking reagent, 0.02% SDS.
40. Wash solution 1: 2× SSC, 0.1% (w/v) SDS.
41. Wash solution 2: 0.1× SSC, 0.1% (w/v) SDS.
42. Buffer 1: Maleic acid buffer pH 7.5: 100 mM maleic acid, 150 mM NaCl, 0.3% Tween 20. pH is adjusted with NaOH pellets up to 7 and then with 5 N NaOH solution. When preparing 10× maleic acid buffer pH 7.5, add the ingredients gradually while keeping pH close to 7 with NaOH pellets.
43. Buffer 2: 1% blocking reagent in Buffer 1.
44. Equilibration buffer 1×: 100 mM Tris–HCl pH 8, 100 mM NaCl, optional 50 mM $MgCl_2$ from 1 M $MgCl_2$ stock solution. A 10× stock solution can be prepared for convenience: 1 M Tris, 1 M NaCl.
45. Safety lamp for film developing.
46. X-ray highly sensitive autoradiography film.
47. X-ray film developer.
48. X-ray film fixer.
49. Film exposure cassette (i.e., Hypercassette, Amersham Life Science, RPN 1642 18 × 24 cm).

3 Methods

3.1 Sample Preparation and Loading onto the Membrane, Nucleic Acid Fixing

1. Draw onto a paper sheet a template of 96-well plate, and mark on it the position of each sample that will be loaded on the apparatus/membrane.
2. Weigh 0.5 g of plant tissue, and place it into an ELISA extraction bag. Keep bags on ice (*see* **Note 3**).

3. Add 1.5 mL (3 w/v) AMES buffer.
4. Homogenize the samples with the tissue homogenizer, and return them immediately to ice.
5. Transfer 200 μL of homogenate to 1.5 mL centrifuge tubes.
6. Incubate tubes at 37 °C for 20 min in a water bath or a heat block.
7. Add 200 μL (1 V) chloroform (for this step work in a fume hood).
8. Vortex and centrifuge at a benchtop centrifuge 16,000 × *g* (full speed) for 5 min at room temperature.
9. Prepare three pieces of Whatman filter paper No. 3 with the dimensions of the plate of the dot-blot apparatus (Hybri-Dot Manifold) to cover all wells (8 × 11 cm). Cut the nylon hybridization membrane according to the area samples cover on the plate, depending on the number of wells loaded.
10. Soak the nylon membrane in 10× SSC for 2–3 min.
11. Assemble the dot-blot apparatus, and place the membrane as follows (Fig. 1a–c):
    - The dot-blot apparatus is composed of three layers/plates of plexiglass: (a) bottom plate (vacuum reservoir), (b) a middle plate (template to cut membranes down to size), and (c) a top plate (sample loading plate). It is important to place the layers of plexiglass in the correct order.
    - Place the three pieces of Whatman paper on the middle plate, and then add the hybridization membrane.
    - Assemble the apparatus and tighten up the bolts. Connect it to the vacuum pump using an elastic tube of appropriate dimensions (not shown). Turn on the pump.
12. Load 15 μL of sample from the supernatant of each tube on the positions indicated in the template.
13. After loading the last sample, wait for 30 min.
14. Disconnect the vacuum pump and then unscrew the plates of the apparatus. Cut one corner of the hybridization membrane to mark its orientation.
15. Cross-link nucleic acids onto the membrane (a) by incubation of the hybridization membrane at 80 °C for 30 min (sample loading surface faces up) and then (b) by placing the hybridization membrane on a UV transilluminator for 3–5 min (sample loading surface faces down) (*see* **Note 4**).
16. Place the membrane in a plastic bag, and keep it at room temperature to continue hybridizing (or store at 4 °C until further processing).

3.2 Hybridization

1. Prehybridization: heat the hybridization buffer (HB) at 50 °C in the hybridization oven until it becomes transparent.
2. Place the membrane in a dry hybridization tube with the sample loading surface facing the hybridization tube interior.
3. Add HB but not directly to the membrane. The volume depends on the size of the hybridization tube, and normally HB should form a surface across the tube when horizontal (approximately 20 mL).
4. Close the hybridization tube, and rotate to wet the membrane evenly avoiding the formation of dry areas between membrane and glass surface.
5. Place the tube in the hybridization chamber, and pre-hybridize at 62 °C for 1 h rotating at 8 rev/min.
6. Dilute 1 μg of PSTVd RNA-DIG-labeled probe in 10 μL Type 1 water, in an RNase-free 1.5 mL microcentrifuge tube, and denature at 92 °C for 5 min, in a heat block. Immediately chill on ice (1 min), and then centrifuge the tube briefly to collect the liquid at the bottom using a benchtop centrifuge.
7. Transfer 1 mL of HB from the hybridization tube into the 1.5 mL microcentrifuge tube containing the denatured probe, and mix by pipetting.
8. Transfer the HB-probe mixture into the hybridization tube avoiding direct contact with the membrane.
9. Hybridize at 62 °C in the hybridization oven, with a rotation of 8 rev/min, overnight.
10. The following morning, prepare all washing solutions in advance. *Approximately 50 mL wash solution/8 × 11 cm membrane are required for each wash step.* More than one membrane can be washed in the same container adjusting the volume of the wash solution accordingly.
11. Decant the HB from the tube, transfer carefully the membrane (s) to a plastic container/box using forceps, and rinse the membrane twice with Wash solution 1 by gentle shaking for 5 min at room temperature on a plate shaker.
12. Decant Wash solution 1, and wash the membrane twice for 15 min with Wash solution 2 at 68 °C in a heat-resistant glass container by gentle shaking in a hybridization oven.
13. Rinse membrane for 5 min in Buffer 1 at room temperature with constant agitation.
14. Incubate membrane for 30 min in 70 mL Buffer 2 in a plastic container or in 10 mL Buffer 2, if a nylon bag is used under constant agitation on a shaker plate (*see* **Note 5**).

15. Dilute the anti-digoxigenin-AP (1:10,000 or according to the manufacturer's instructions) in Buffer 2 (1 μL/10 mL, for a membrane size 8 × 11 cm).
16. Insert the membrane into a nylon bag (*see* **Note 6**).
17. Add the anti-digoxigenin-AP solution, remove gently any air trapped, and seal the bag with a heat sealer.
18. Incubate for 30 min at room temperature with constant agitation.
19. Transfer the membrane to a container, and wash twice for 15 min with 100 mL Buffer 1 at room temperature with constant agitation.
20. After decanting Buffer 1, equilibrate the membrane into 20–30 mL equilibration buffer at room temperature with constant agitation for 3 min.

3.3 Chemiluminescent Detection of Alkaline Phosphatase

1. Prepare a new nylon bag matching the dimensions of the membrane.
2. Cut the edges of the bag except one, in order to be able to unfold it in two sheets (joined on the edge left uncut).
3. Lay the membrane on one sheet of the nylon bag (sample loading surface faces up), and add 1 mL CDP-Star solution to cover the membrane surface evenly as a thin layer. Then fold the nylon side of the bag gently pressing to remove trapped air, and seal the bag using the heat sealer.

3.4 Film Exposure and Development

1. Place the membrane in a film cassette, and place an autoradiography film over it in a dark room under safety lamp illumination. Mark the orientation by cutting one corner of the film.
2. Close the film cassette and expose the film for approximately 30 min (*see* **Note 7**).
3. Prepare two trays with developer and fixer according to manufacturer's instructions in the dark room under safety lamp illumination. Have two pairs of forceps for each solution without mixing those used for the fixer with those used for the developer. Remove the film from the film cassette and immerse it in the developer. Gently agitate holding the film with forceps for 30–60 s until a chemiluminescent signal corresponding to the positive control is visualized (Fig. 1d). Then transfer the film into the tray containing fixing solution for 2–3 min.
4. Rinse the film in a third tray under tap water for several minutes, and let it air-dry. Positive samples appear as dark spots in contrast to white background of the PSTVd-negative samples (Fig. 1d).

4 Notes

1. Double distilled autoclaved water can be used as well.
2. Hybridization glass tubes require a substantially lower volume of hybridization buffer and probe.
3. Samples can be composed of leaves from up to 15 tomato plants, tissues from up to 5 potato tubers (collect conical sections from the edges of each tuber), or leaves from single solanaceous ornamental plants.
4. Alternatively, cross-linking can be performed solely by heating at 80 °C for 80 min or by UV exposure, 2–6 min (optimization is required). In our hands, a combination of the two approaches worked best, eliminating background signal.
5. 1% instant milk powder can be used instead of the blocking reagent.
6. To avoid trapping of air between the membrane and the nylon surface, cut the two parallel edges of the bag to open on two sides. Then seal the two cut edges in the sealer.
7. Exposure time varies, requiring optimization.

References

1. Randles JW (2003) Economic impact of viroid diseases. In: Hadidi A, Flores AR, Randles JW, Semancik JS (eds) Viroids. CSIRO Publishing, Collingwood, pp 2–11
2. Vassilakos N, Kektsidou O, Papaeythimiou M, Varveri C (2012) Comparison of direct-RT-PCR and dot-blot hybridization for the detection of *Potato spindle tuber viroid* in natural host plant species. Eur J Plant Pathol 134(4):859–864. https://doi.org/10.1007/s10658-012-0061-y

Chapter 10

Gel Blot Hybridization for Viroids

Pedro Serra

Abstract

Northern blot analysis reveals information about RNA identity, size, and abundance. This technique has become an essential tool for the knowledge developed about viroids and also an excellent method for viroid detection. Here we describe the methodology of a Northern blot based in polyacrylamide gel electrophoresis under denaturing conditions, hybridized with a viroid full-length riboprobe labeled with chemiluminescence. Viroid detection with this approach entails positive signals, specific migration, and the differentiation of their circular and linear forms.

Key words Northern blot, Electrophoresis, Nylon membrane, Chemiluminescence probe, Electrotransference

1 Introduction

Due to the characteristic of viroids, as small circular non-protein-coding naked RNAs [1, 2], Northern blot hybridization is an essential tool for their research and detection. RNA Northern blot hybridization involves the electrophoretic partition by size of the nucleic acids recovered in a plant extract, transference of the fractioned extract to a membrane filter (Fig. 1), hybridization with an appropriated probe, and the membrane treatment for visualizing the signal produced by the hybridized probe [3] (Figs. 2, 3, and 4). The fact that Northern blot is based on a previous electrophoresis separation is a major advantage over other hybridization methods because (a) in the fractioned plant extract, the target viroid is separated from other molecules that may react positively, (b) the monomeric viroid molecule has a specific electrophoretic mobility relative to a ladder marker or other viroids, (c) circular and linear forms of the viroid can be dissociated due to different mobility in denaturing conditions, and (d) the viroid target present in a relative high amount of plant extract can be concentrated in a small fraction of the lane and therefore increase the sensitivity of its detection.

Ayala L. N. Rao et al. (eds.), *Viroids: Methods and Protocols*, Methods in Molecular Biology, vol. 2316, https://doi.org/10.1007/978-1-0716-1464-8_10, © Springer Science+Business Media, LLC, part of Springer Nature 2022

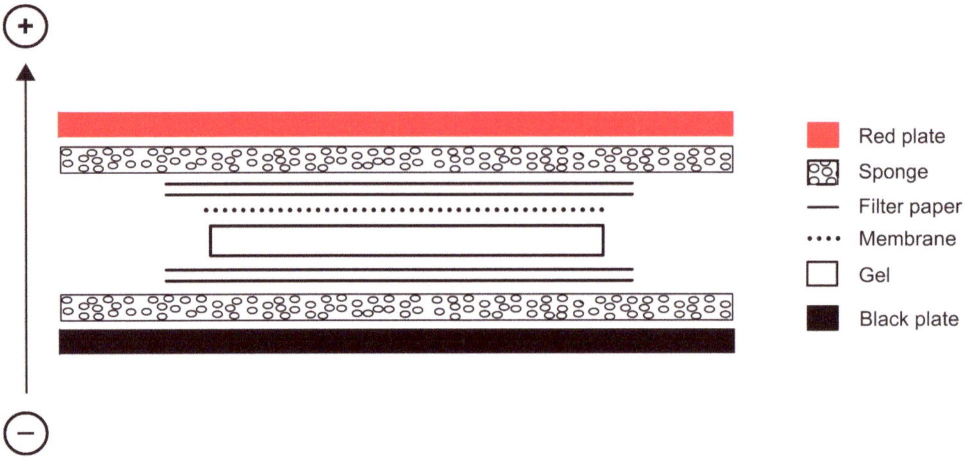

Fig. 1 Gel-membrane assembly for the electrotransfer of nucleic acids from gel to nylon membrane. The gel and the membrane are sandwiched between filter papers, sponges, and the plates of the compression cassette

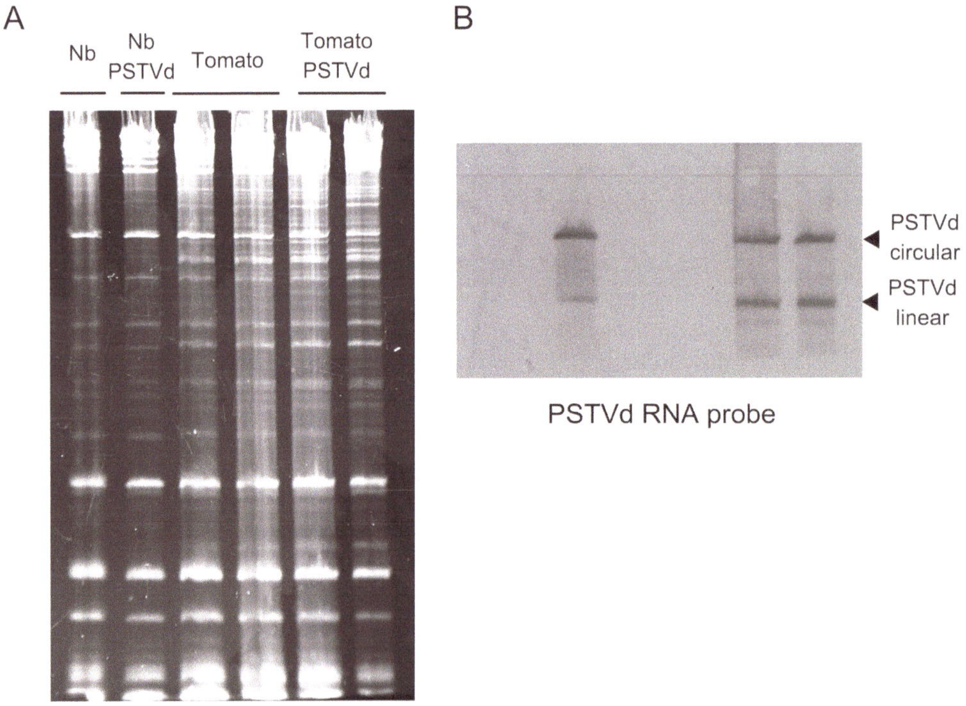

Fig. 2 Northern blot using a denaturing 8 M urea 5% polyacrylamide gel buffered in TBE and hybridized with a potato spindle tuber viroid (PSTVd) full-length riboprobe. Samples analyzed are healthy and infected plants of *Nicotiana benthamiana* and tomatoes. (**a**) Electrophoretic profile in denaturing conditions of plant extracts stained with ethidium bromide. (**b**) Northern blot result with positive PSTVd probe signals for circular and linear viroid forms in infected plants

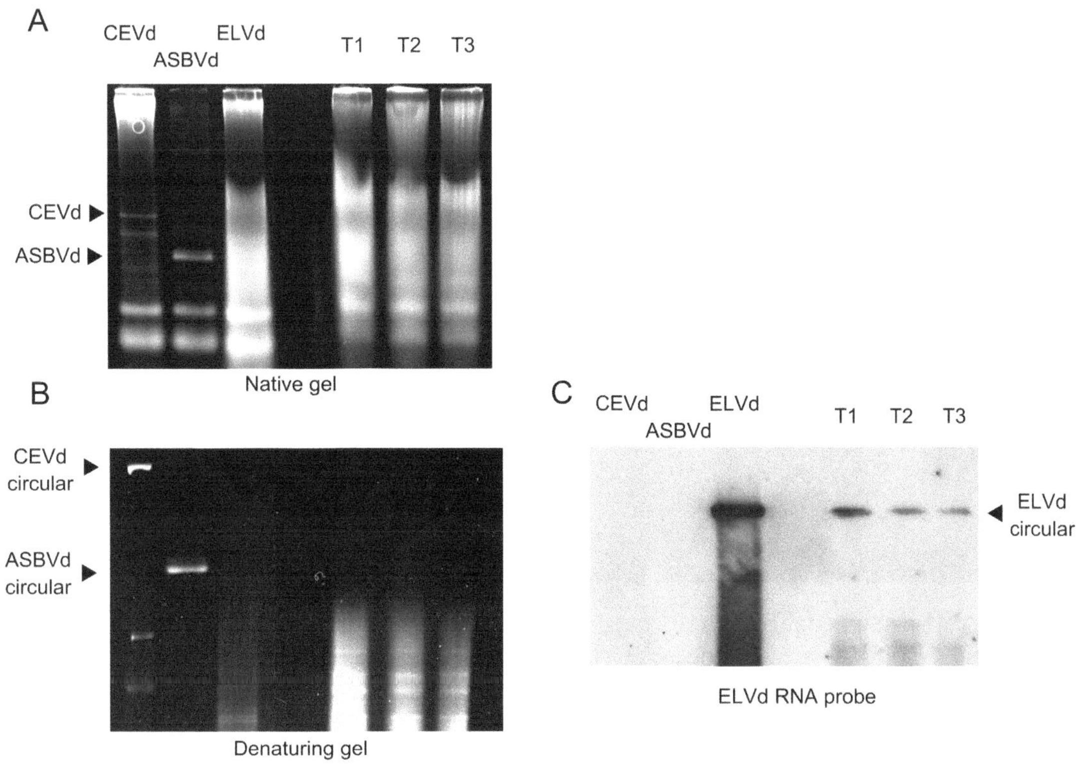

Fig. 3 Northern blot using the second dimension of a sequential polyacrylamide gel electrophoresis (sPAGE) and hybridized with an eggplant latent viroid (ELVd) full-length riboprobe. Samples analyzed are extracts from three eggplants (T1, T2, and T3) and citrus exocortis viroid (CEVd)-infected *Gynura aurantiaca*, avocado sun blotch viroid (ASBVd)-infected avocado, and an ELVd-infected eggplant as negative and positive controls. (**a**) First dimension of sPAGE (native conditions) stained with ethidium bromide. Bands corresponding to CEVd and ASBVd are indicated. (**b**) Second dimension of sPAGE (denaturing conditions) stained with ethidium bromide. Bands corresponding to CEVd and ASBVd circular forms are indicated. (**c**) Northern blot result with positive ELVd probe signals for circular viroid forms in infected plants

Northern blot provides information about the sequence, size, and type (circular or linear) of RNAs and also can be used for RNA quantification by relative accumulation. These properties have been a key element for the studies of longer- and smaller-than-unit viroid-derived molecules such as the intermediates of the replication cycle [4–6] or the degradation products of these pathogens [7]. It has also been crucial for the study of viroid-protein complexes [8–10] and for the characterization of viroids with duplicated regions [11].

Besides the multiple applications in viroid biology research, Northern blot analysis is an essential tool for diagnosis in phytopathological uses because it has a good balance in terms of sensitivity and reliability (Fig. 4). Hybridization with full-length probes can detect all the known viroids and their variants preventing the intrinsic risks associated to methods based on short fragments of

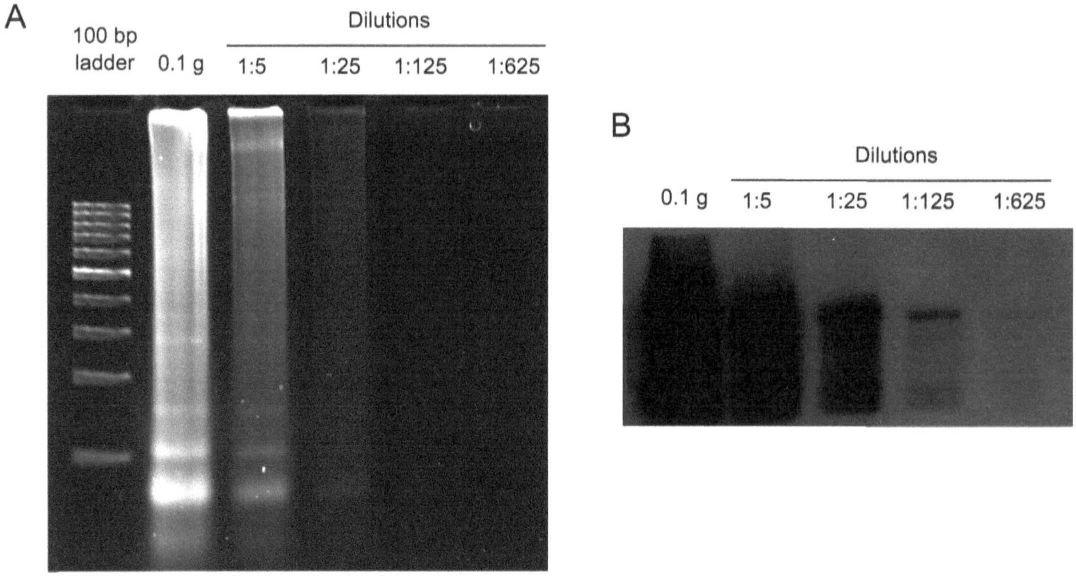

Fig. 4 Northern blot using a polyacrylamide gel electrophoresis (PAGE) and hybridized with a peach latent mottle viroid (PLMVd) full-length riboprobe. Samples analyzed correspond to 0.1 g of PLMVd-infected peach leaves and fivefold serial dilutions of the same extract. (**a**) Electrophoretic profile of the extract and dilutions stained with ethidium bromide. (**b**) Overexposed X-ray film to increase the sensibility of Northern blot analysis which reaches to detect PLMVd in the 625-fold dilution

the target RNA as it has been reported for some variants of peach latent mottle viroid (PLMVd) undetectable in routine tests using real-time quantitative RT-PCR assays [12]. In contrast, hybridization with full-length probes can be problematic in terms of discriminating viroids with high sequence similarities as in the case of potato spindle tuber viroid (PSTVd) probes that can react positively against certain viroids belonging to the *Pospiviroid* genus (personal results). It should be noted that viroids that are members of the same genus share sequence similarities in some domains as the central conserved region. In view of the above, for a reliable diagnosis, it is advisable to identify an optimal temperature of hybridization and include a positive control and an extract of a plant infected with another viroid of the same genus as a negative control.

Northern blot analysis for viroid detection may be based on agarose gels [13], but, due to the small size of these pathogens (246–433 nts), it is more suitable to use polyacrylamide gels which have a higher resolving power. Under native conditions, viroid secondary structure displays paired regions and unpaired loops along the entire RNA molecule (Figs. 3a and 4) in contrast with the completely unpaired strand under denaturing conditions (Figs. 2 and 3b). However, the native conditions in the electrophoresis do not seem to be an obstacle for the recognition and the hybridization of the probe because Northern blot analysis using native and denaturing conditions give comparable results [13].

Probes for Northern blotting can be RNA or DNA with a complementary sequence to all or part of the target viroid, labeled either with radioactive isotopes (e.g., ^{32}P) or with chemiluminescence [14, 15]. The choice of the probe type depends of the purpose of the analysis, but in general, riboprobes have a higher thermal annealing and are able to withstand more rigorous washing steps preventing some of the background noise. Many researchers prefer the chemiluminescent signals because they are faster, are more sensitive, and reduce the health hazards associated with radioactive labels. In contrast, radiolabeled probes enable extended periods of X-ray film exposures without disturbing background noise which may be helpful for the detection of weak hybridization signals.

The following is a detailed explanation of the Northern blot methodology for viroid detection using an urea denaturing 5% polyacrylamide gel buffered in TBE, electrotransferred to a nylon membrane and hybridized with a full-length riboprobe labeled with digoxigenin (UTP dig) in which alkaline phosphatase breaks down chemiluminescent substrates (CSPD) producing a detectable emission of light. Viroid detection with this method entails positive signal of the probe, specific migration site of viroid linear forms relative to an RNA ladder marker, as well as the presence of an additional band with minor electrophoretic run corresponding to circular forms of the viroid (Fig. 2).

2 Materials

Prepare all solutions using ultrapure water (prepared by purifying deionized water, to attain a sensitivity of 18 M Ω cm at 25 °C) and analytical grade reagents. Prepare all reagents at room temperature (unless otherwise indicated).

2.1 8 M Urea 5% Polyacrylamide 89 mM TBE Gel

1. Solution A 30% polyacrylamide (acrylamide 39:1 bisacrylamide): Weigh 60 g acrylamide and 1.52 g bisacrylamide (*see* **Note 1**). Transfer to a 100 mL graduated beaker, and add 180 mL water. Mix until powder is dissolved, and add water up to 200 mL. Store at 4 °C.
2. Tris-borate-EDTA (TBE) 10× buffer: Weigh 108 g Tris, 55 g boric acid, and 7.4 g EDTA·$2H_2O$. Transfer to a 1 L graduated beaker cylinder, and add 800 mL water. Mix until powder is dissolved, and add water up to 1 L. Autoclave and store at room temperature.
3. Ammonium persulfate 10% solution: Dissolve 10 g of ammonium persulfate in 100 mL water. Store at 4 °C.
4. *N,N,N,N′*-Tetramethyl-ethylenediamine (TEMED): Store at 4 °C.

5. Tris-borate-EDTA (TBE) 1× running buffer: Transfer 100 mL of TBE 10× to a 1 L cylinder, and add water up to 1 L.
6. Loading buffer (8 M urea, 50% formamide, 0.5% EDTA, 0.1% xylene cyanol, and 0.1% bromophenol blue): Weigh 24 g urea, 0.25 g EDTA · $2H_2O$, 50 mg xylene cyanol, and 50 mg bromophenol blue. Transfer to a 100 mL graduated beaker, and add 25 mL water. Mix at 60 °C until powder is dissolved, and add formamide up to 50 mL. Store at room temperature.
7. Vertical gel electrophoresis unit.
8. Electrophoresis power supply.
9. Thermoblock.
10. Ice.
11. Microcentrifuge tubes.
12. RNA plant extracts including positive and negative controls.
13. Low-range RNA ladder.
14. Ethidium bromide.
15. UV transilluminator.

2.2 Electroblotting

1. Nylon membrane.
2. Whatman 3 mm filter paper.
3. Phosphate electroblotting 20×: NaH_2 Po_4 · $2H_2O$ 0.5 M pH 6.5. Weigh 78 g of NaH_2 PO_4 · $2H_2O$, transfer to a 1 L graduated beaker, and add 800 mL water. Mix until powder is dissolved, adjust to pH 6.5 with NaOH 10 N, and add water up to 1 L. Autoclave and store at room temperature.
4. Phosphate electroblotting buffer 1×: Transfer 100 mL of TBE 10× to a 2 L graduated cylinder, and add water up to 2 L.
5. Sponges and compression cassette.
6. Plastic tray.
7. Electroblotting unit.
8. Electroblotting power supply.
9. UV crosslinker.
10. Test tube.
11. Filter paper.

2.3 Hybridization

1. Saline-sodium citrate (SSC) buffer 20×: Weigh 43 g NaCl and 88 g Na_3 citrate · $2H_2O$, transfer to a 1 L graduated beaker, and add 800 mL water. Mix until powder is dissolved, adjust to pH 7 with diluted HCl, and add water up to 1 L. Autoclave and store at room temperature.
2. Formamide.

3. Blocking solution 10%: Weigh 5 g blocking reagent, transfer to a 100 mL graduated beaker, and add 50 mL maleic acid buffer (*see* below). Mix at 60 °C until the mixture develops a milky texture without granules.
4. Sodium-dodecyl-sulfate (SDS) 20%: Dissolve 20 g of SDS into 80 mL water by stirring, add water up to a final volume of 100 mL, and store at room temperature.
5. *N*-Lauroylsarcosine 10%: Dissolve 10 g *N*-lauroylsarcosine in 80 mL water by stirring, add water up to a final volume of 100 mL, and store at room temperature.
6. Prehybridization solution: Add 25 mL formamide, 12.5 mL SSC 20×, 0.5 mL of *N*-lauroylsarcosine 10%, 0.05 mL SDS 20%, and 10 mL of blocking solution 10% into a 50 mL graduate cylinder, and add H_2O up to 50 mL. Cover the cylinder with parafilm, and mix by inverting four to six times.
7. Full-length riboprobe complementary to viroid target labeled with UTP dig.
8. Hybridization tubes.
9. Hybridization oven with rotisserie.
10. Thermoblock.
11. Microcentrifuge tube.

2.4 Membrane Treatment

1. Washing solution 1 (SSC 2×, SDS 0.1%): Add 100 mL SSC 20× and 5 mL SDS 20% in a 1 L graduated cylinder, and add water up to 1 L. Transfer to a bottle, and store at room temperature.
2. Washing solution 2 (SSC 0.1×, SDS 0.1%): Add 5 mL SSC 20× and 5 mL SDS 20% in a 1 L graduated cylinder, and add water up to 1 L. Transfer to a bottle, and store at room temperature.
3. Maleic acid buffer (maleic acid 100 mM, NaCl 150 mM pH 7.5): Weigh 11.61 g maleic acid and 8.76 g NaCl, transfer to a 1 L graduated beaker, add 800 mL water, and mix until powder is dissolved. Adjust to pH 7.5 with NaOH 10 N, and add water up to 1 L (*see* **Note 2**). Autoclave and store at room temperature.
4. Detection buffer (Tris–HCl 100 mM, NaCl 100 mM pH 9.5): Weigh 12.11 g Tris and 5.84 g NaCl, transfer to a 1 L graduated beaker, add 800 mL water, and mix until powder is dissolved. Adjust to pH 9.5 with HCl 10 N, and add water up to 1 L. Autoclave and store at room temperature.
5. Tween 20.
6. Anti-digoxigenin-AP, Fab fragments for the detection of digoxigenin-labeled compounds: Store at 4 °C.

7. CSPD (disodium 3-(4-methoxyspiro{1,2-dioxetane-3,2′-(5-′-chloro)tricyclo[3.3.1.13,7]decan}-4-yl)phenyl phosphate). Store at 4 °C.
8. Filter paper.
9. Exposure cassette.
10. X-ray films.
11. X-ray film processor.
12. Transparent plastic wrap and cellophane to secure membrane in X-ray exposure cassette.
13. X-ray light box.

3 Methods

3.1 Gel Electrophoresis

1. Clean the electrophoresis glass plates with ethanol and paper (*see* **Note 3**), place the spacers, and seal the hollows in the gel assembly cassette (*see* **Note 4**).
2. Weigh 19.2 g urea in a beaker, and transfer 6.6 mL of solution A, 4 mL TBE 10×, and 13 mL water. Mix until the urea is dissolved (*see* **Note 5**). Add 60μL TEMED and 0.6 mL ammonium persulfate 10%, mix and cast within the gel cassette, and insert a gel comb immediately without generating air bubbles. Keep at room temperature for 5 min for gel polymerization.
3. Introduce each sample and the RNA low-range ladder in different microcentrifuge tubes with the corresponding amount of 2× loading buffer, and keep for 2 min in a thermoblock at 98 °C for denaturation. Immediately place and keep the microcentrifuge tubes on ice for 5 min.
4. Remove the gel comb, place the gel cassette in the electrophoresis unit (*see* **Note 6**), and add TBE 1× running buffer. Carefully, load denatured samples in the corresponding lane, and run at 200 volts for 90 min (*see* **Note 7**).
5. Following electrophoresis, open the gel plates with the use of a spatula. The gel remains on one of the glass plates. Rinse the gel with water, and transfer carefully to a container with 200 mL water containing 10μL ethidium bromide. Keep at room temperature for 10 min.
6. Discard the ethidium bromide in an appropriated container, and wash the gel with water. Place a plastic sheet on the UV transilluminator, and transfer carefully the gel on the plastic sheet.
7. Define and cut the fraction of the gel for blotting. Remove and discard the rest of the gel, and make a mark by cutting a little piece in the right corner of the bottom to define the gel orientation (*see* **Note 8**).

3.2 Electroblotting

1. Cut a nitrocellulose membrane to the gel size, make a mark by cutting a little piece on the upper left side, and immerse it in phosphate electroblotting buffer 1× for 5 min.
2. Equilibrate the gel in phosphate electroblotting buffer 1× for 5 min.
3. Transfer 1 L phosphate electroblotting buffer 1× in a plastic tray, immerse the sponges, and remove the air bubbles by gently applying pressure.
4. Cut four Whatman filter papers 1 cM larger than the membrane, and immerse them in phosphate electroblotting buffer 1×.
5. Place the black plate of the compression cassette at the bottom of the plastic tray, and place the other elements in this order: one sponge, two Whatman filter papers, the gel in an inverted orientation (the mark must be placed at the bottom right corner), the membrane in an inverted orientation (the mark must be placed in the upper right side), and finally the two remaining filter papers. Remove any bubbles that may be present between the layers of the sandwich applying pressure with the help of a test tube. Place the remaining sponge and the red plate, and close the compression sandwich (Fig. 1).
6. Fill the electroblotting unit with phosphate buffer 1×, and insert the sandwich in the appropriated orientation. Apply a current of 1 A for 1 h (*see* **Note 9**).
7. Turn off the power supply, retire the sandwich, and open it with the red plate on the top side (*see* **Note 10**). Remove the sponge, and discard the Whatman filter papers. Place the membrane on filter paper, and let it dry.
8. Fix nucleic acids on the membrane by applying 70 mJ/cm^2 in the UV crosslinker (*see* **Note 11**).

3.3 Hybridization

1. Insert the membrane in the hybridization tube with fixed nucleic acids in the internal side, i.e., not touching the tube wall.
2. Transfer 25 mL of prehybridization solution into the hybridization tube, and keep it in the oven at constant rotation at 55 °C for 2 h (*see* **Note 12**).
3. Remove the hybridization tube from the oven, discard the prehybridization solution, and transfer 25 mL of new prehybridization solution. Keep in the oven at 60 °C.
4. Add 1μg of full-length riboprobe in a microcentrifuge tube containing 100μL of water, and denature at 98 °C for 3 min in a thermoblock. Place on ice for 5 min. Transfer the content of the microcentrifuge tube into the hybridization tube, and keep it at rotation overnight in the oven at 60 °C (*see* **Note 13**).

3.4 Membrane Treatment

1. Take the hybridization tube out of the oven, and discard the solution containing the probe (*see* **Note 14**). Add 20 mL of washing solution 1, place in the oven, and rotate at room temperature for 15 min. Discard the solution and repeat this step.
2. Discard washing solution 1, and add in the tube 20 mL of washing solution 2. Place tube in the oven, and rotate at 55 °C for 30 min.
3. Discard washing solution 2, and add 10 mL of maleic acid buffer containing 30μL of Tween 20. Place it in the oven, and rotate for 5 min at room temperature.
4. In a graduated cylinder, transfer 22 mL maleic acid buffer and 3 mL blocking solution. Cover the cylinder with parafilm, and mix by inversion. Take the hybridization tube, and discard the last step solution. Transfer 20 mL of maleic acid buffer with blocking solution, and reserve 5 mL. Place it into the oven for 30 min, and rotate at room temperature.
5. Discard the last step solution, transfer the remaining 5 mL of maleic acid solution with blocking solution, and add 1.5μL of anti-digoxigenin-AP. Place the tube into the oven, and rotate for 30 min at room temperature.
6. Discard the last step solution, and wash twice the membrane with 25 mL of maleic acid buffer in the oven with rotation for 15 min at room temperature.
7. Discard the solution, and equilibrate the membrane by adding 20 mL of detection buffer in the hybridization tube. Place it in the oven, and rotate for 30 min at room temperature.
8. Discard the last step solution, transfer 10 mL of detection buffer, and add 13μL of CSPD. Place it in the oven, and rotate for 5 min at room temperature.
9. Discard the solution, remove the membrane from the hybridization tube, and place it on filter paper and leave it to dry. Cover the membrane in a transparent plastic wrap, and secure it with cellophane into the exposure cassette (*see* **Note 15**).
10. In a dark room, put an X-ray film in the exposure cassette, and keep it at room temperature (*see* **Note 16**).
11. Transfer the X-ray film from the exposure cassette to a film processor in a dark room (*see* **Note 17**). In the developed film, one must see the results of the Northern blot (*see* **Note 18**), using X-ray light box or other source of light.

4 Notes

1. Acrylamide is highly toxic, likely to be carcinogenic. Its management requires appropriate safety procedures, and for that reason, it is mainly handled as an aqueous solution. Solution A 30% acrylamide (acrylamide 39:1 bisacrylamide) can be obtained mixing aqueous solutions.
2. Maleic acid buffer requires large amounts of NaOH to adjust pH 7.5. It is recommended supply maleic acid and NaCl with 8 g of NaOH pearl before adding water in order to increase the pH.
3. It is very important to use paper that will not peel or leave cellulose fibers because it may get into the gel and emit fluorescence in the UV exposure.
4. There are a wide variety of gel assembly cassettes. In some cases, it may be suitable use Vaseline to seal the hollows. To dispense it, fill a syringe with Vaseline, and insert a cut open pipette tip instead of a needle.
5. Mixing at 60 °C accelerates the urea dissolution, but care must be taken because heated mixture has an extremely fast polymerization.
6. In some cases, after removing the comb, many of the wells in the gel contain fibbers that may reduce the loading capacity. To remove these fibbers, it is suitable to use a Hamilton syringe to wash the wells. With the well submerged in buffer, pull back and forth on the plunger, going close to but not touching the sides of the wells. A 50 or 100μL size will work. It creates enough of a current that washes the wells but not too much so you don't damage them.
7. The values presented are guidelines; electrophoresis cells require different power settings. Conditions should be optimized for every application.
8. There are DIG-labeled RNA markers to give signals of molecular weight in Northern blot analysis. However, RNA marker reference can be transferred to membrane drilling the gel in each marker band with the help of a pipette tips. In this case, perforated marker lane must be included in the portion of gel to electroblotting (continue in *see* **Note 10**).
9. Values presented are guidelines, and electroblotting units require different power settings. Conditions should be optimized for each application.
10. To transfer the RNA marker reference to the membrane, retire the sandwich, and open it with the black plate on the top side. Retire the sponge and discard the Whatman filter papers. With

a pencil, mark the membrane through the drilled holes of the gel corresponding to marker bands. Carefully take away the gel, place the membrane on paper, and air-dry the membrane for a few minutes (continue in *see* **Note 18**).

11. Nucleic acids transferred to membrane can be observed under UV light. Likewise, electroblotted gels can also be observed in UV transilluminator to verify a decrease in luminescence regarding the gel profile before the blotting step.
12. Prehybridization time can be reduced at least up to 15 min. The time taken for prehybridization will result in cleaner results and, ultimately, time savings, especially if a probe will be reused many times.
13. Optimal hybridization temperature depends on the length and G + C base pair proportion of the duplex probe/target. In all instances, temperature for viroid detection with full-length riboprobes ranges from 60 to 70 °C.
14. Hybridization solutions with a large amount of probe (1μg) can be reused. In this case, the solution should be transferred to a 50 mL Falcon tube and stored at −20 °C. In future hybridizations, the Falcon tube containing the solution should be heated in a water bath at 60 °C for 10 min and transferred to the hybridization tube containing the prehybridized membrane.
15. Membranes must be oriented with the mark at the upper left side and nucleic acids facing the X-ray film.
16. Exposure time is variable. Optimal exposure times must not give saturated signals without disturbing background noise and must be empirically determined. First attempt with 30 min exposure time should be suitable. In the case of weak signals, insert two films in the exposure cassette, and expose overnight. The signals in the first film will be more intense than those in the second one.
17. If a film processor is not available, prepare the developing solution in a dark room with a safe light to prevent light oxidization. Add 750 mL of water to a tray. Pour in 10 mL of developer concentrate, and stir for 2 min. Prepare the fixer solution. Add 856 mL of water to a tray, and add in 95 mL fixer concentrate. Stir the mixture a few times. Fill a third tray with distilled water to be used as a stop solution and final rinse. Take the film out of the exposure cassette in the darkroom, and place it in the developer solution. Move the tray in a circular motion for 30 s to keep the developer moving across the film. Take the film out of the developer solution with long forceps. Continue using the forceps to handle the film until it is completely developed. Place the film in the distilled water for a couple of seconds to remove the remaining developer (mixing developer

and fixer will ruin the film). Place the film in the fixer solution for 30 s, and gently move the tray in a circular motion to keep the fixer flowing across the film. Remove the film from the fixer, and rinse it with water for 30–60 s. Allow the film to dry.

18. To transfer the RNA marker reference, place the developed X-ray film in the opened exposure cassette (background noise corresponds to the size of the membrane). Trace with a felt pen the pencil marks in the membrane corresponding to RNA marker bands.

References

1. Flores R, Hernández C, MartínezdeAlba E, Daròs JA, DiSerio F (2005) Viroids and viroid-host interactions. Annu Rev Phytopathol 43:117–139
2. Ding B (2010) Viroids: self-replicating, mobile, and fast-evolving noncoding regulatory RNAs. Wiley Interdiscip Rev RNA 1:362–375
3. Alawine JC, Kemp DJ, Stark GR (1977) Method for detection of specific RNAs in agarose gels by transfer to diazobenzyloxymethyl paper and hybridization with DNA probes. Proc Natl Acad Sci U S A 74:5350–5354
4. Hadidi A, Hashimoto J, Tiener D (1982) Potato spindle tuber specific double stranded RNA in extracts from infected leaves. Ann de Virol 133E:15–31
5. Hutchins CJ, Keese P, Visvader JE, Rathjen PD, McInnes JL, Symons RH (1985) Comparison of multimeric plus and minus forms of viroids and virusoids. Plant Mol Biol 4:293–230
6. Daròs JA, Marcos JF, Hernández C, Flores R (1994) Replication of avocado sunblotch viroid: evidence for asymmetric pathway with two rolling circles and hammerhead ribozyme processing. Proc Natl Acad Sci U S A 91:12813–12817
7. Flores R, Minoia S, Carbonell A, Gisel A, Delgado S, López-Carrasco A, Navarro B, Di Serio F (2015) Viroids, the simplest RNA replicons: how they manipulate their hosts for being propagated and how their hosts react for containing the infection. Virus Res 209:136–145
8. Daròs JA, Flores R (2002) A chloroplast protein binds a viroid RNA in vivo and facilitates its hammer head mediated self-cleavage. EMBO J 21:749–759
9. Gómez G, Torres H, Pallás V (2005) Identification of translocatable RNA-binding phloem proteins from melon, potential components of the long-distance RNA transport system. Plant J 41:107–116
10. Gómez G, Pallás V (2001) Identification of a ribonucleoprotein complex between a viroid RNA and a phloem protein from cucumber. Mol Plant-Microbe Interact 14:910–913
11. Fadda Z, Daròs JA, Flores R, Duran-Vila N (2003) Identification in eggplant of a variant of citrus exocortis viroid (CEVd) with a 96 nucleotide duplication in the right terminal region of the rod-like secondary structure. Virus Res 97:145–149
12. Serra P, Bertolini E, Martínez MC, Cambra M, Flores R (2017) Interference between variants of peach latent mosaic viroid reveals novel features of its fitness landscape: implications for detection. Sci Rep 7:42825
13. Murcia N, Serra P, Olmos A, Duran-Vila N (2009) A novel hybridization approach for detection of citrus viroids. Mol Cell Probes 23:95–102
14. Podleckis EV, Hammond RW, Hurtt SS, Hadidi A (1994) A comparison of digoxigenin-labeled and 32P-labeled probe: dot blot detection of plant viroids. B M Biochem 10:10–12
15. Astruc N, Marcos JF, Macquaire G, Candresse T, Pallás V (1996) Studies on the diagnosis of hop stunt viroid in fruit trees: identification of new hosts and application of a nucleic acid extraction procedure based on non-organic solvents. Eur J Plant Pathol 102:837–846

Chapter 11

Tissue Print Hybridization for Detection and Characterization of Viroids

Maria S. Kaponi, Teruo Sano, and Panayota E. Kyriakopoulou

Abstract

Simplifying sample preparation by transferring plant sap (from plant sections) directly onto a membrane is advantageous for any routine viroid detection technique, such as tissue printing. After fixing the samples on the membrane, hybridization steps similar to Northern or dot blot can be successfully implemented as long as factors like stringency and low viroid titer are properly adjusted to enable enhancement of the detection limit. The protocol described allows for indexing hundreds of field samples as a phytosanitary control measure.

Key words Tissue print hybridization, DIG-labeled probes, Nitrocellulose membrane, Nylon membrane, Viroid detection, Viroid characterization

1 Introduction

Routine viroid detection requires minimal sample manipulation, a condition met in tissue print hybridization (TPH) technique. TPH does not require nucleotide extraction from samples, as it involves the direct transfer of plant sap (from stem, cutting, leaf, or fruit sections) to a nylon or nitrocellulose membrane, with the obvious advantage of reducing time and financial costs. TPH as a viroid detection technique has been reported in many cases, including detection of chrysanthemum stunt viroid and potato, citrus, stone, and pome fruit viroids. Large-scale stone fruit viroid indexing in field conditions and disease etiology and expression as well as in viroid distribution studies within the infected plant have also been reported, whereas TPH has become the routine method for citrus indexing in Spain [1–21].

Limitations, like false negatives caused by low viroid titer in plants or seasonal fluctuations in viroid titers, can be overcome using amplification hosts, in which viroids replicate more efficiently [16]. On the other hand, false positives have been reported using TPH for PLMVd detection [22], as interactions between the viroid

Ayala L. N. Rao et al. (eds.), *Viroids: Methods and Protocols*, Methods in Molecular Biology, vol. 2316,
https://doi.org/10.1007/978-1-0716-1464-8_11,

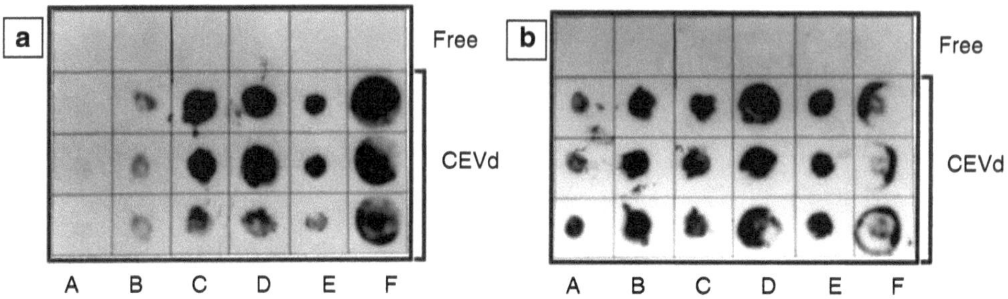

Fig. 1 (Teruo Sano). Detection of CEVd infection in tomato cross-sections by tissue print hybridization using a CEVd DIG-labelled riboprobe. (**a**), UV crosslinking. (**b**), baking at 80 °C for 2 h. Membrane treatment: A, no treatment; B, treated with distilled water; C, treated with 8 M urea; D, treated with 10× SSC; E, treated with glyoxal; F, treated with formamide

probe and plant proteins have been generating strong signals, and in cases where DNA probes were used [23–25]. In such cases, including an additional blocking step to remove host proteins, or using riboprobes and RNases at more stringent hybridization conditions, is recommended. It is also important to treat the membrane with denaturing agents or salts prior to fixing nucleic acids on it, which is better done by UV cross-linking (Fig. 1).

The presented protocols are modifications of Palacio-Bielsa et al. [16] for DNA probes and of Li et al. [26] for RNA probes. These protocols were initially used for detecting Pospiviroidae viroids (namely, apple scar skin viroid (ASSVd), apple fruit crinkle viroid (AFCVd), hop stunt viroid (HSVd), and pear blister canker viroid (PBCVd)) in tissue cross-sections of rosaceous samples imprinted on positively charged nylon membranes [27]. Both protocols were used for RNA probes and the modification of Palacio-Bielsa et al. [16] for DNA probes. TPH was initially used on positive field controls, to investigate the possibility of testing a large number of samples in the field.

The tissues used were mainly annual or tender shoots, fruit bark, stems, seeds, and leaf stalks. The initial difficulty of optimizing tissue print hybridization was about finding suitable conditions for avoiding non-specific attachment of the probes on the nylon membrane (false positives). This difficulty was overcome by an increase of the prehybridization and hybridization temperature from 50 to 55 °C for DIG-DNA probes and from 55 to 60 or 65 °C for riboprobes (depending on the protocol), by using RNase A [11, 23], and by lengthening the "blocking" time of the film from 30 min to 1–2 h. Results are shown at Fig. 2.

Implementing these protocols helped avoiding nucleic acid extraction from samples or long processing of the membrane before blotting the samples (*see* **Note 1**). A mere fresh cross-section of the sample tissue and its pressing on the membrane are enough for printing. Printing was also possible in field conditions, where some

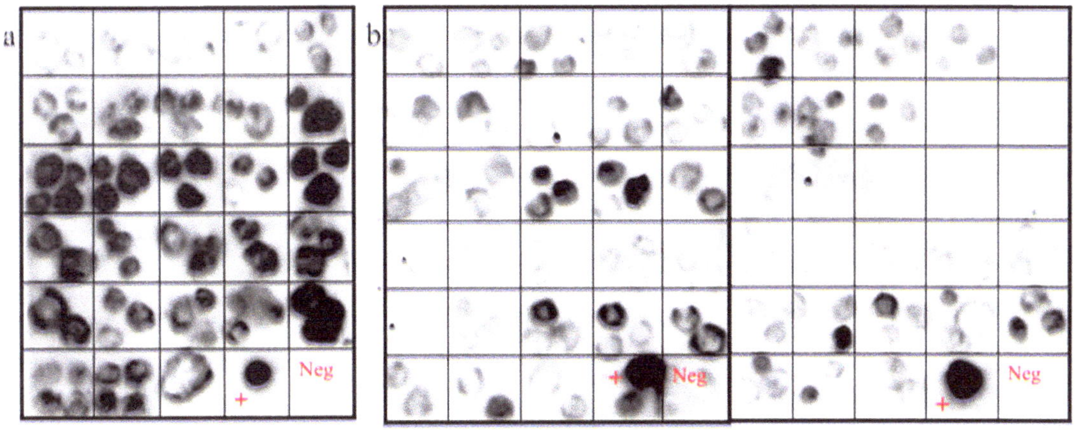

Fig. 2 (Maria Kaponi [27]). Detection of (**a**) HSVd and (**b**) ASSVd by tissue print hybridization on cross sections of stems and petioles of Rosaceous species

appropriate practices that prevent membrane contamination, such as membrane cover, gloves, and clean razor blades (immersed in bleach buffer after cutting each sample), were implemented. Finally, fixing of nucleotides on the membrane was done by irradiating it with ultraviolet light in a cross-linker up to 1 min, which allowed for long-term storage of the membrane even at ambient temperature or shipment to remote laboratories.

2 Materials

Common laboratory materials, like sterile nuclease-free Eppendorf tubes, pipettes, and tips, are not listed, since they are used in many protocol stages. All buffers are prepared using sterile water from a laboratory purification system (such as Milli-Q) or double distilled water (ddH_2O) and molecular biology or analytical grade reagents. All reagents are stored at room temperature (unless specified otherwise). All waste disposal regulations must be followed when disposing waste materials.

2.1 Protocol 1: Modification from [16]

2.1.1 DIG-DNA Probe Synthesis and Purification

1. Overlapping primer pairs, usually designed on the upper central conserved region (CCR) viroid region (*see* **Note 2**).
2. $MgCl_2$, usually contained in PCR buffer (1.5 mM final concentration) or supplemented in a separate tube with Taq polymerase (*see* **Note 2**).
3. Digoxigenin (DIG) labeling mix (DIG labeling mix, Roche Diagnostics, Mannheim, DE).
4. Unlabeled dNTPs for unlabeled control reactions.
5. Taq polymerase.

6. 2–10 ng of template (cloned cDNAs or RNA extracts of literature-cited positive controls) (*see* **Note 2**).
7. Sterile double distilled H_2O (ddH_2O), such as sterile Milli-Q water.
8. PCR thermocycler.
9. Agarose gel electrophoresis apparatus.
10. Agarose.
11. Ethidium bromide 0.5 μg/mL or other agarose gel staining dye (Midori Green, etc.).
12. TAE buffer: 1 M Tris–HCl, pH 8.0, 0.04 M Tris-acetate, 0.001 M EDTA.
13. Microwave or heat block for melting agarose.
14. Refrigerator (−20 °C or lower temperature).
15. Spectrophotometer (e.g., NanoDrop).
16. PCR products of known concentration to be used as positive, negative, and blank (non-nucleic acid) controls.
17. Chilled 100% ethanol.
18. Chilled 70% ethanol.
19. Vortex or mixer.
20. Refrigerated centrifuge apparatus.
21. Speed vacuum apparatus.
22. Sterile TE buffer: 1 M Tris–HCl, pH 7.5–8.0 1 mL, 0.5 M EDTA, pH 8.0 200 μL; adjust volume to 100 mL with double distilled water, and sterilize (*see* **Note 3**).
23. Sterile Eppendorf tubes.

2.1.2 Printing Samples on Membrane

1. Positively charged nylon membranes.
2. Plastic containers or sterile Falcon tubes.
3. Kitchen paper.
4. UV cross-linker.
5. Water bath, heat block, or boiler for denaturing controls.
6. Scissors.
7. Pencil.
8. Ruler.
9. Plastic bags and bag sealer.
10. Forceps.
11. Sterile razor blades.
12. 70% EtOH or 5% bleach solution for cleaning surfaces and blades.

13. Formaldehyde for control denaturation.
14. 2× SSC buffer: 0.3 M NaCl, 0.03 M Na citrate, pH 7.0, adjusted with HCl, or 3 M NaCl, 0.3 M Na citrate, pH 7.0, 1% Na_2SO_3.
15. Alternative to (14), 2 M ß-mercaptoethanol.

2.1.3 Prehybridization and Hybridization

1. Hybridization oven or shaking incubator.
2. Forceps.
3. Plastic containers or Falcon tubes.
4. Paper towels.
5. Sterile 45 μm syringe filter.
6. Boiler or water bath (100 °C).
7. 2 M ß-mercaptoethanol.
8. Prehybridization solution: 50% v/v deionized formamide, 20× SSPE, pH 7.4, 5% SDS.
9. 20× SSPE: 3 M NaCl, 0.2 M NaH_2PO_4 H_2O, 0.02 M EDTA, DEPC-treated H_2O; adjust pH to 7.4 with 10 N NaOH. Store at room temperature.
10. Denhardt's reagent: 1% (w/v) Ficoll 400, 1% (w/v) polyvinylpyrrolidone (PVP), 1% (w/v) bovine serum albumin (BSA, Fraction V), sterile H_2O to final volume. Dissolve the components in H_2O. Use a 0.45 μm syringe filter to sterilize and remove particulate matter. Divide into aliquots and store at −20 °C.
11. Hybridization solution: 6× SSPE, pH 7.4, 50% (v/v) deionized formamide, 1% SDS, 10% Denhardt's reagent 50×.
12. DIG-DNA probe: 20 ng/cm^2 of membrane.

2.1.4 Washes and Detection

1. RNase A (2 μg/μL), used in 2× SSC.
2. 2× SSC: 0.3 M NaCl, 0.03 M Na citrate, pH 7.0. Adjust pH with HCl, or 3 M NaCl, 0.3 M Na citrate, pH 7.0, 1% Na_2SO_3.
3. Wash 1: 2× SSC, 0.1% SDS.
4. Wash 2: 0.1× SSC, 0.1% SDS.
5. Buffer 1 (maleic acid buffer): 0.1 M maleic acid, 0.15 M NaCl, pH 7.5. Adjust pH with NaOH.
6. Tween solution: Tween 0.3% in Buffer 1.
7. Buffer 2 (blocking solution): 10% blocking reagent (Roche Diagnostics) or milk powder.
8. Anti-DIG-AP solution: 0.2 mL/cm^2 of membrane, Fab fragments (Roche Diagnostics), 1:5000 in Buffer 2.
9. Buffer 3 (equilibration solution): 0.1 M Tris-HCl, pH 9.5, 0.1 M NaCl, 0.05 M $MgCl_2$. Adjust pH with HCl.

10. CSPD or CDP-Star (disodium 2-chloro-5-[4-methoxyspiro {1,2-dioxetane-3,2′-[5′-chloro]tricycle [3.3.1.1^{3,7}] decan}-4-yl] phenyl phosphate, Roche Diagnostics), 1% in Buffer 3.
11. Autoradiography cassette.
12. X-ray film.
13. Film developer.
14. Film fixer.
15. Plastic containers for film development.
16. Dark chamber equipped with red lamp for film development.
17. Automated imaging systems, e.g., ChemiDoc XRS (Bio-Rad), instead of the aforementioned number 11–16.

2.2 Protocol 2: Modification from [26]

2.2.1 DIG-RNA Probe Synthesis and Purification

1. Template (cloned cDNAs of literature-cited positive controls).
2. Restriction enzymes.
3. Phenol/chloroform 1:1.
4. RNA transcriptase, such as T7, usually supplied with DTT, buffer, and spermidine.
5. Sterile, ribonuclease-free water of high purity, such as DEPC-treated double distilled or Milli-Q water.
6. Diethyl ether for phenol/chloroform removal.
7. Agarose.
8. Tris-acetate-EDTA (TAE) buffer: 1 M Tris–HCl, pH 8.0, 0.04 M Tris-acetate, 0.001 M EDTA.
9. Ethidium bromide 0.5 μg/mL for gel staining.
10. 10 mM NTPs.
11. 10× DIG labeling mixture (Roche).
12. RNase inhibitor, 40 U/μL.
13. DNase for template digestion after transcription.
14. 4 M LiCl for low molecular weight RNA precipitation.
15. 0.2 M EDTA or 20 mM EGTA, pH 8.0, to stop enzyme reactions.
16. PCR products of known concentration to be used as positive, negative, and blank (non-nucleic acid) controls.
17. Acrylamide-bisacrylamide stock 39:1 (for PAGE, stored at 4 °C).
18. Elution buffer: 0.5 M ammonium acetate, 1 mM EDTA, pH 8.0, or 0.5 M ammonium acetate, 10 mM magnesium acetate, 1 mM EDTA, 0.1% SDS, pH 7.5.
19. Chilled (glacial) 100% or 99.5% ethanol.

20. Chilled 70% ethanol.
21. 70% ethanol for rinsing surfaces and blades.
22. Glycogen 20 mg/mL.
23. Sterile TE buffer: 1 M Tris–HCl, pH 8.0, 1 mL, 0.5 M EDTA, pH 8.0, 200 μL; adjust volume to 100 mL with double distilled water, and sterilize (*see* **Note 3**).
24. Sodium acetate3 M, pH 5.2.
25. Incubator.
26. Vortex.
27. Hood/vertical flow chamber (diethyl ether evaporation).
28. Agarose gel electrophoresis apparatus.
29. Vertical electrophoresis apparatus for PAGE.
30. Vortex or mixer.
31. Refrigerator (−20 °C or lower temperature).
32. Microwave or heat block for melting agarose.
33. Refrigerated centrifuge apparatus.
34. Speed vacuum apparatus.
35. Sterile Eppendorf tubes.
36. Marker (for writing on plastic surfaces).
37. UV protection, such as mask, glasses, etc.
38. Aluminum foil.
39. Saran film.
40. Sterile razor blades.
41. Glass wool or filter columns.
42. Forceps.
43. Spatula.
44. Paper tissues or kitchen paper.
45. Pasteur pipetors.
46. Spectrophotometer (NanoDrop).

2.2.2 Printing Samples on Membrane

1. Same materials as in Subheading 2.1.2, with a preference to using 2× SSC buffer (**step 14** in Subheading 2.1.2) instead of 2 M ß-mercaptoethanol.
2. 50% formamide (non-deionized).
3. 17.5% formaldehyde for RNA denaturation.

2.2.3 Prehybridization and Hybridization

1. Hybridization oven or shaking incubator.
2. Forceps.
3. Plastic containers, such as bags or Falcon tubes.

4. Plastic bag sealer.
5. 2× SSC buffer: 0.3 M NaCl, 0.03 M Na citrate, pH 7.0, adjusted with HCl, or 3 M NaCl, 0.3 M Na citrate, pH 7.0, 1% Na_2SO_3.
6. Prehybridization solution (added after mixing the following components in the following order for a final volume of 6.532 mL per tube: 1.749 mL of sterile Milli-Q water, 0.625 mL of 0.5 M sodium phosphate buffer, pH 7.0, 0.313 mL of 20% SDS, 2.5 mL of non-deionized formamide, 0.075 mL of Salmon sperm DNA (sonicated: 10 mg/mL), 0.020 mL of yeast tRNA (100 mg/mL), 1.25 mL of 50% dextran sulfate.
7. Hybridization solution: Prehybridization solution plus 1 μL of 200 ng/μL or more riboprobe in prehybridization solution per membrane.
8. Magnetic stirrer.

2.2.4 Washes and Detection

1. Shaker allowing horizontal agitation.
2. Water bath.
3. Wash 2: 0.1× SSC, 0.1% SDS.
4. RNase A (10 mg/μL).
5. Blocking solution: 0.1 M maleic acid, 0.15 M NaCl, pH 7.5, 1% w/v blocking reagent (Roche Diagnostics).
6. Tween washing buffer: 0.1 M maleic acid, 0.15 M NaCl, pH 7.5, 0.3% Tween 20.
7. Anti-DIG solution: Blocking solution, 1:10,000 anti-DIG Fab fragment (Roche Diagnostics).
8. Buffer 3: 0.1 M Tris–HCl, pH 9.5, 0.1 M NaCl, 0.05 M $MgCl_2$.
9. CSPD or CDP-Star (Roche Diagnostics): 20% in Buffer 3.
10. Autoradiography cassette.
11. X-ray film.
12. Film developer.
13. Film fixer.
14. Plastic containers for film development.
15. Dark chamber equipped with red lamp for film development.
16. Automated imaging systems, e.g., ChemiDoc XRS (Bio-Rad), instead of the aforementioned number 11–16.

3 Methods

Every procedure should be carried out at room temperature unless otherwise specified.

3.1 Protocol 1: Modification from [16]

3.1.1 DIG-Labeled DNA Probe Synthesis and Purification

DIG-Labeled DNA Probe Synthesis

1. Full-length digoxigenin (DIG)-labeled DNA and RNA probes (riboprobes) can be used for the purposes of molecular hybridization. DNA probes (e.g., ASSVd, PBCVd, and HSVd) were originally synthesized by polymerase chain reaction (PCR) from cloned full-length monomeric positive viroid controls.
2. DNA probe synthesis is possible by using facing away overlapping primer pairs corresponding to the upper conserved central region (CCR) of the HSVd, PBCVd, and ASSVd genomes.
3. DIG-labeled probes are obtained by incorporation of DIG-dUTP into the PCR product, using as a target a cloned plasmid containing full-length viroid monomeric DNA.
4. PCR reactions of 50 μL contain 5 ng/μL of each primer, 1.5 mM $MgCl_2$, 0.2 mM DIG labeling mix, 5 U Taq polymerase, and 2–10 ng of template (cloned cDNAs of literature-cited positive controls) and double distilled H_2O to the final reaction volume.
5. In DIG-unlabeled control reactions, 0.25 mM unlabeled dNTPs are used instead of DIG labeling mix and the same amount of template per 50 μL of reaction (2–4 reactions).
6. PCR reactions used for DIG-DNA probe synthesis are performed by adding components in the following order (final concentration in parentheses): Sterile double distilled or Milli-Q H_2O in variable volume, 10× PCR buffer (without $MgCl_2$) 5 μL (1×), 25 mM 1–2.5 mM $MgCl_2$ in variable volume (1.5 mM), 2 mM PCR DIG labeling mix 5 μL (0.2 mM), 100 ng/μL c- or h-DNA primer 2.5 μL (5 ng/μ L), 5 U/μL Taq DNA polymerase 1 μL (5 U/μL), template DNA up to 200 ng/μL 5 μL (variable).
7. PCR includes an initial template denaturation step (95 °C/ 5 min), 35–40 amplification cycles (94 °C/45 s, 60 °C/45 s, final extension 72 °C/1 min, tested for ASSVd, PBCVd, and HSVd apricots. For other viroids, use optimum denaturation and primer annealing temperatures), and final product extension (72 °C/7 min).

DIG-Labeled DNA Probe Purification

8. DIG-DNA probe purification from primer residues and dNTPs and further comparison to PCR products of known concentration in 1.5–2% agarose gels are required before use of probes in hybridization.

9. Five (5) μL of the PCR products of each DIG-DNA probe labeling reaction are analyzed on 1.5% agarose gels, and those of expected size are purified.
10. For each 50 μL of probe, add 6.5 μL of 4 M LiCl +190 μL chilled 100% ethanol, mix well, and precipitate at −20 °C overnight (alternatively at −80 °C/30 min or −20 °C/1–2 h).
11. Centrifuge 20 min at 19,000 × *g* in −4 °C, and discard the supernatant.
12. Add 0.5 mL of 70% chilled ethanol (−20 °C), and centrifuge for 10 min at 19,000 × *g* in −4 °C.
13. Discard the supernatant, and let it dry in the speed vacuum for 10 min.
14. Resuspend the pellet in 50 μL of sterile TE buffer (*see* **Note 3**).
15. Check again the probe by 2% agarose gel electrophoresis in presence of unlabeled control, to confirm that probe was not lost during purification process (*see* **Note 4**).
16. Dilute an aliquot of the probe 1:5 or 1:10 (depending on the band intensity in the gel) in TE buffer, and measure the DNA concentration in a spectrophotometer/NanoDrop considering absorbencies at 280 and 260 nm wavelengths.
17. Aliquot the probe in 4–5 sterile Eppendorf tubes, and store at −20 °C.
18. To confirm probe sequences prior to use for viroid detection, a dot-blot hybridization control experiment can be successfully performed with denatured RNA extracts and RT-PCR positive, negative, and blank (non-nucleic acid) controls with the protocol described herein.

3.1.2 Printing Samples on Membrane

1. Appropriate positive (tissues from literature-cited and viroid-infected plants, cloned viroid monomers in plasmids, purified viroid RNA preparations, or RT-PCR amplicons) and negative controls (sterile water, viroid-free plant tissues and extracts) should be used in each membrane for tissue prints and blots.
2. After calculating the number of samples and controls needed, each membrane is separately marked, forming a grid of cells 1 cm^2 wide using EtOH-sterilized pencil and ruler.
3. Each membrane needs to be labeled by numbering one of its corners or just by making cuts on the membrane corners (1 cut for membrane #1, 2 cuts for membrane #2, etc.) to avoid membrane mix-up.
4. Nitrocellulose or positively charged nylon membrane can be blotted with 2–5 freshly cut transverse sections per 1 cm^2 cell; relatively fresh and succulent tissues can be pressed on the membrane in the laboratory or in the field. Sterilized razor

blades sanitized in 70% EtOH between samples and stems are used.

5. As controls, 1–2 μL of denatured nucleic acid preparations (*see* **Note 5**), RT-PCR products, or plant sap can be used without any pretreatment of the membranes.
6. Fix nucleic acids on membranes by cross-linking (1200×100 pJ/cm^2).
7. After cross-linking, membranes can be stored in the dark at room temperature or at 4 °C until further processing.

3.1.3 Prehybridization and Hybridization

Membrane Preparation

1. Place membranes in a sterile plastic container in 2 M ß-mercaptoethanol (15%) at room temperature for 10–15 min to activate them, destroy plant proteins, and reduce "background noise" by non-specific probe reactions.
2. Rinse membranes in sterile and distilled or Milli-Q water, dry the excess water on paper towels, and place the membranes in sterile prescribed 50 mL Falcon tubes or other plastic containers with imprints facing the inside of the tube.

Prehybridization (10 mL Prehybridization Buffer/Membrane)

3. Mix and filter prehybridization buffer using a 45 μm syringe filter, and then place in hybridization oven ($T = 55$ °C).
4. Denature 100 μL herring sperm for 10 min at $T = 100$ °C (e.g., in a boiler or water bath), and add immediately in the prehybridization buffer.
5. Incubate for 2 h in hybridization oven ($T = 55$ °C or 60 °C for RNA probes).

Hybridization (10 mL Hybridization Buffer/Membrane)

6. Discard the prehybridization buffer, and replace it with hybridization buffer.
7. Add 100 μL denatured herring sperm (denaturation for 10 min at $T = 100$ °C in boiler or water bath) to the hybridization buffer to block the membrane.
8. Calculate probe concentration based on the membrane size (20 ng/cm^2).
9. Add 100 μL boiled sterile H_2O to the probe to avoid evaporation of the probe during denaturation and not to renature the probe by lowering its temperature.
10. Denature probe for 5 min at $T = 100$ °C in a boiler or water bath, and pipette probe immediately inside hybridization buffer. Avoid pipetting the probe on top of the membrane to

avoid staining of the membrane from excess probe and probe renaturation (*see* **Note 6**).

11. Place in hybridization oven overnight ($T = 55$ °C or 60 °C for RNA probes).

3.1.4 Washes and Detection

Washes and DIG-ELISA (Buffer Volumes Depend on Membrane Size, e.g., 5 or 10 mL)

1. Discard hybridization buffer, and replace it with 5 or 10 mL of 2× SSC including RNase A (2 μg/μL) for digestion of probe excess at 37 °C for 15 min.
2. Discard carefully RNase A buffer, and replace it with 10 mL Wash 1 at room temperature for 15 min (*see* **Note 7**).
3. Repeat Wash 1.
4. Discard Wash 1, and replace with 10 mL of preheated at $T = 60$ °C Wash 2 for 1 h (*see* **Note 8**).
5. Replace Wash 2 with 10 mL Tween solution (0.3% in Buffer 1) at room temperature for 2 min.
6. Replace Tween buffer with 10 mL or more Buffer 2 at room temperature for 30 min.
7. Replace Buffer 2 with 5 mL anti-DIG-AP solution 1:5000 in Buffer 2 at room temperature for 30 min (*see* **Note 9**).
8. Discard anti-DIG solution, and wash excess anti-DIG with 10 mL Buffer 1 at room temperature for 15 min.
9. Repeat Buffer 2 wash.
10. Discard Buffer 2, and replace with 10 mL Buffer 3 at room temperature for 30 min.

Detection

11. Replace Buffer 3 with 50 μL CSPD in 5 mL of Buffer 3 for 5 min at room temperature (1% CSPD solution).
12. Seal the membrane with the diluted CSPD in a plastic bag leaving one side not sealed.
13. Carefully remove air bubbles by pressing them with fingers or a pencil out of the bag, and then seal completely.
14. Contact the imprints on the membrane with the autoradiography film by attaching firmly the bag in the cassette with imprints facing up without any air bubbles formed (*see* **step 13** just above) and placing carefully the film on the bag in the cassette for 1–3 h at dark at room temperature.
15. After that, take film from one of its corners with forceps, and submerge it in developer (e.g., Fuji Rendol) for 60–90 s at room temperature with gentle agitation in the dark chamber (*see* **Note 10**).

16. Transfer the film in fixer (e.g., Fuji Fujifix) for 3–5 min at room temperature with gentle agitation.
17. Wash film in running water for 30 s.
18. Air-dry the film and observe under fluorescent light.
19. Otherwise, attach firmly the bag containing the membrane on a black smooth surface, and place it on the ChemiDoc XRS (Bio-Rad) transilluminator. Use appropriate filters, exposure time, and software (e.g., Quantity One) to take an image.

3.2 Protocol 2: Modification from [26]

3.2.1 DIG-Labeled RNA Probe Synthesis and Purification

Use of Monomeric Full-Length Viroid Plasmids as Template for Riboprobe Synthesis

1. Riboprobe synthesis is performed by in vitro transcription from complete viroid genome monomers inserted into plasmids. These plasmids need to be cut (restricted) by endonucleases in order for the RNA polymerase to gain access to the transcription site. Restriction can be confirmed by 1–1.5% agarose-TAE electrophoresis (*see* **Note 11**).
2. Measure purified DNA plasmid concentration in spectrophotometer (NanoDrop). 2 μg of plasmid DNA (e.g., 10 μL each of 200 ng/μL) is enough for restriction.
3. The correct choice of restriction enzyme is essential; it depends on the restriction site in the plasmid, which is relevant to the position of the RNA polymerase promoter, and on the lack of restriction sites in the insert.
4. Endonuclease digestion is performed usually up to a total volume of 30 μL in the following order: Add Milli-Q H_2O up to 30 μL, e.g., 15 μL, 10× restriction buffer + BSA 3 μL, restriction enzyme (5–10 U/μL) 2 μL, e.g., 200 ng/μL plasmid DNA, 10 μL.
5. Precipitate at 37 °C overnight (2–16 h).

Restricted Plasmid Purification

6. Add 70 μL Milli-Q water to complement the restriction reaction up to 100 μL.
7. Add 100 μL phenol/chloroform 1:1 to the restriction reaction (*see* **Note 12**).
8. Vortex for 2 min.
9. Centrifuge for 3 min at 16,000 × *g*.
10. Transfer to a new tube the upper aqueous phase.
11. Add 100 μL diethyl ether (at hood/cloning chamber because of evaporation, *see* **Note 13**).
12. Vortex vigorously for 30 s.

13. Centrifuge for 1 min at 16,000 × *g* at room temperature to remove phenol dissolved in the aqueous phase and discard carefully upper phase (diethyl ether + phenol) by pipetting (*see* **Note 14**).
14. Add 2.5 volumes (250 μL) chilled 99.5% EtOH.
15. Store at −30 °C or less for at least an hour to allow nucleic acid precipitation.
16. Centrifuge for 10 min or more at 16,000 × *g*.
17. Discard ethanol supernatant by pipetting, and allow the precipitate to dry in air for 10 min or more.
18. Add 22 μL of Milli-Q water to reconstitute the precipitate, and analyze 1–2 of the reconstituted precipitates in an 1.5% agarose-TAE gel along with equal amount of non-restricted plasmid; linearized plasmid is usually moving faster than the circular plasmid DNA in agarose-TAE gels and slower than the supercoiled plasmid (*see* **Note 7**).

In Vitro Transcription for the Production of DIG-Labeled Riboprobes

T7 (Invitrogen, 50 U/μL), SP6 (Roche, 20 U/μL), or T3 (Roche, 20 U/μL) RNA polymerases can be routinely used according to their manufacturer's instructions, depending on the respective promoter on the plasmid sequence.

In vitro transcription converts the +/sense DNA strand of the plasmid into an RNA strand; the polarity of the transcribed RNA and the DNA insert is the same, so we always use minus sense (−)/antisense inserts to produce riboprobes able to hybridize with the positive sense viroid target. In vitro transcription is performed usually in a total volume of 20 μL.

19. For a T7 reaction, we add in each reaction tube in the given order Milli-Q H_2O up to 1.5 μL (up to 20 μL), 5× T7 transcription buffer (+ spermidine) 4 μL, 10 mM NTPs or 10× DIG labeling mixture (Roche) 2 μL, 0.1 M DTT 1 μL, 40 U/μL RNase inhibitor 0.5 μL, T7 RNA polymerase 1 μL, and restricted plasmid DNA 10 μL or more.
20. For a T3 or an SP6 reaction, Milli-Q H_2O volume is up to 4.5 μL (up to 20 μL), as 2 μL of T3 or SP6 transcription buffer (+ spermidine) is added, having an initial concentration of 10×.
21. Incubate at 37 °C for 2 h.

In Vitro Transcript Purification

22. Add 1 μL DNase (Promega) to the transcription reaction, and continue incubation at 37 °C for 15–20 min.

23. Add 2 μL 0.2 M EDTA or 20 mM EGTA pH 8.0 to stop the enzyme reaction.
24. Add 2.5 μL 4 M LiCl.
25. Add 75 μL 99.5% chilled EtOH.
26. Store at −30 °C overnight or at −80 °C for at least 30 min to allow nucleic acid precipitation.
27. Centrifuge for 10 min or more at 19,000 × *g* at 4 °C.
28. Dry the precipitate in air for 10 min or more.
29. Reconstitute precipitate by adding 20–100 μL ribonuclease-free water, depending on the amount of the precipitate and its viscosity.
30. Add 1 μL RNase inhibitor.
31. Run 1 μL of transcript in 1.5% agarose-TAE gel to check its quality using the restricted plasmid as control along with marker.
32. Further purify RNA using PAGE, and crush and soak method modified from [28] as follows:
33. After taking a photo, cover the PAGE gel with a plastic transparency, e.g., a saran film, and mark the desired fragments on the film during UV exposure using a marker; protect hands and eyes from the UV.
34. After marking and taking a photo, remove carefully the gel so as not to disturb the film; using a razor or a scalpel, separate all the marked area, and transfer it on an aluminum foil to mince.
35. Use forceps or a spatula to transfer the minced fragment in a new sterile 1.5 mL prenotated Eppendorf tube.
36. Clean scalpel and forceps before proceeding to the next fragment, by using a handkerchief and 70% EtOH.
37. Add 0.5 mL elution buffer in every tube, close tubes firmly, and place in a box or in a rack.
38. Put the box/rack to vigorous shaking at 37 °C overnight in an incubator.
39. The next day, transfer the liquid phase using a yellow tip to new sterile 1.5 mL prenotated Eppendorf tubes, and add 1 mL of 99.5% glacial EtOH and 1 μL of 20 mg/mL premelted glycogen (nucleic acid condensation carrier).
40. Reverse with hand a few times, and place in −20 °C or less overnight to precipitate.
41. The next day centrifuge (19,000 × *g*, 20 °C, 10 min or more).
42. Add another 0.5 volume of elution buffer to the pellet, vortex briefly, and recentrifuge.
43. Combine the two supernatants.

44. Remove any remaining fragments of acrylamide by passing the supernatant through a small column of glass wool contained in the tip of a Pasteur pipette.
45. Precipitate the riboprobe RNA with 99.5% glacial ethanol.
46. Redissolve the riboprobe RNA in 200 μL of TE.
47. Add 25 μL of 3 M sodium acetate, pH 5.2, and reprecipitate the RNA.
48. Rinse the pellet once with chilled 70% ethanol.
49. Centrifuge (19,000 × *g*, 4 °C, 5 min or more).
50. Remove ethanol and let dry in air for at least 10 min.
51. Reconstitute riboprobe precipitant in 10 μL nuclease-free Milli-Q water or TE (pH 8.0), and measure nucleic acid concentration in NanoDrop.

3.2.2 Printing Samples on Membrane

1. Same as in Subheading 3.1.2.
2. RNA controls can be spotted onto nylon membrane after mixing with 50% formamide, 17.5% formaldehyde, 1× SSC, and Milli-Q H_2O and after heating at 65 °C for 5 min for denaturation (*see* **Note 5**).

3.2.3 Prehybridization and Hybridization

Membrane Preparation

1. Place membrane in 75 mL 2× SSC for 1 min.
2. Drain membrane by holding it from a corner with forceps prior to use.

Prehybridization

3. Place membranes in a hybridization bottle or bag and seal, leaving space for pouring prehybridization solution (4 mL for a membrane of 5 cm × 10 cm, *see* **Note 15**).
4. Squeeze off air bubbles between the membrane and bottle using a long pair of forceps.
5. Incubate at 55 °C or more (*see* **Note 16**) for at least 2 h in hybridization oven, if using a bottle by rotation and if using a bag by shaking.

Hybridization

6. Add to the prehybridization buffer 1 mL of hybridization solution (1 μL heat-denatured DIG-RNA probe at 55 °C or more (*see* **Note 16**)/1 mL hybridization buffer in a sterile Eppendorf).
7. Incubate at 55 °C or more (*see* **Note 16**) overnight in hybridization oven, if using a bottle by rotation and if using a bag by shaking.

3.2.4 Washes and Detection

Washes

1. Put the membrane in a plastic ware filled with 75 mL 2× SSC. Shake gently at room temperature for 10 min. Discard 2× SSC.
2. Repeat, by changing with a new aliquot of 2× SSC (75 mL), and shake for 10 min.
3. Transfer the membrane in a plastic bag, and pour 5 mL 2× SSC with 1 μL RNase A (10 mg/mL) to digest any probe not hybridized with RNA. Seal, and incubate at room temperature under agitation for 15 min. Discard the solution carefully (*see* **Note 7**).
4. Transfer the membrane in a plastic ware filled with 75 mL 0.1× SSC –0.1% SDS (preheated at 65 °C in a shaking water bath), and shake gently at 65 °C for 10 min, to remove RNase.
5. Discard the buffer and replace with a new aliquot of 0.1× SSC –0.1% SDS (75 mL), and shake gently for 10 min at 65 °C in the water bath (*see* **Note 8**). Discard the solution.

Detection

6. Place the membrane in a plastic bag, pour 5 mL blocking solution, and incubate at room temperature under agitation for 60 min. Discard the solution.
7. Pour 5 mL anti-DIG solution (1 μL anti-DIG Fab fragment in 5 mL blocking solution) in the same bag, incubate at room temperature under horizontal agitation for 30 min, and then discard the solution.
8. Put the membrane in a plastic ware filled with 75 mL Tween washing buffer. Shake gently for 10 min at room temperature.
9. Repeat by changing washing buffer (75 mL) and shaking gently for 10 min at room temperature. Discard the solution.
10. Submerge the membrane in a plastic ware filled with 30 mL Buffer 3, and agitate for ~5 min.
11. Place the membrane carefully, not allowing formation of air bubbles, in a plastic bag pre-spotted with drops (pipette aliquots) of 500 μL diluted CSPD or CDP-Star (1/5 in Buffer 3), and seal the bag leaving one side not sealed.
12. Carefully remove air bubbles by pressing them with fingers or a pencil out of the bag, and then seal completely. "Massage" gently the membrane using your fingertips to allow uniform distribution of CSPD/CDP-Star.
13. Contact the imprints on the membrane with the autoradiography film by attaching firmly the bag in the cassette with imprints facing up without any air bubbles formed and placing carefully the film on the bag in the cassette for 1–3 h at dark at room temperature.

14. After that, in the dark chamber, take film out of the cassette from one of its corners with forceps, and submerge it in developer (e.g., Fuji Rendol) for 60–90 s at room temperature with gentle agitation (*see* **Note 10**).
15. Transfer the film in fixer (e.g., Fuji Fujifix) for 3–5 min at room temperature with gentle agitation.
16. Wash film in running water for 30 s.
17. Air-dry the film and observe under fluorescent light.
18. Otherwise, attach firmly the bag containing the membrane on a black smooth surface, and place it on the ChemiDoc XRS (Bio-Rad) transilluminator. Use appropriate filters, exposure time, and software (e.g., Quantity One) to take an image.

4 Notes

1. Generally, hybridization protocols require 2 days or more for their implementation (not including sampling, probe synthesis, and buffer preparation). In the tissue print hybridization (TPH) protocols described, DIG-DNA probes require 1 day less than DIG-RNA probes for their synthesis, and they work at lower hybridization temperatures. On the other hand, riboprobes can detect viroids that are more abundant in the plant sap at higher temperatures, diminishing false positives.
2. Optimal reaction conditions are dependent on template DNA and primers. In particular incubation times and temperatures, concentration of Mg^{2+}, but also concentration of DNA template should be optimized for optimal results for each new primer/template pair (see additional information).
3. If the probe shows a faint band prior to purification (first agarose 2% gel), it could be resuspended in a smaller volume than 50 μL.
4. When synthesizing a probe, it is important to add to the PCR reaction a control of target DNA with regular dNTPs (not DIG-labeled), in order to check the probe size (e.g., by 2% agarose gel electrophoresis and ethidium bromide staining). The probe should have a slightly higher molecular weight than regular PCR product due to the DIG incorporation.
5. Denaturation of controls prior to spotting on the membrane can be done by heating (65 °C/15 min) and by adding 37% formaldehyde (3.5 μL/20 μL of preparation) or by adding 50% formamide, 17.5% formaldehyde, 1× SSC, and Milli-Q H_2O and heating at 65 °C for 5 min, e.g., in a heat block or a water bath.

6. In order to keep the preheated probe at 100 °C in denatured conformation, place tube on ice until pipetting the probe in the hybridization buffer.
7. Be careful not to spread the solution containing RNase A in the laboratory!
8. It is wise to turn on a shaking water bath (e.g., Precision SWB 15, Fischer Scientific) at 60 or 65 °C before discarding the hybridization buffer and putting 75 + 75 mL of Wash 2 to be preheated.
9. If a membrane is small, use 0.8 μL anti-DIG-AP/4 mL Buffer 2.
10. As soon as the positive control is observed on the film, take care not to develop the film for much longer time, as false positives may appear.
11. During electrophoresis, the migration pattern of DNA is also dependent on its conformation, and according to Maniatis et al. [29], the relative mobility of the circular, nicked circular and linear forms of DNA is also influenced by the applied current strength, the running buffer ionic strength (TAE or TBE), and the density of superhelical twists.
12. Phenol chloroform is used for protein and enzyme denaturation; after centrifugation, the upper aqueous phase containing the restricted plasmid can be transferred to another tube, while the organic lower phase can be discarded.
13. Diethyl ether evaporates easily and has a characteristic intense odor.
14. In order to remove phenol dissolved in the aqueous phase, diethyl ether is mixed and then centrifuged so as to form another upper transparent phase on the aqueous one. They are separated by a somewhat curved surface. To discard this ~100 μL upper phase (diethyl ether + phenol) by pipetting, a steady hand, Eppendorf racks, and a 100/200 μL pipetor can be used in the hood. It is recommended to lift the tube on the eye level against a back light to see the phase-separating curve. If phases are disturbed or the lower phase is taken by accident, vortexing and centrifugation can be repeated.
15. Buffer volume with respect to the membrane size can be roughly calculated, e.g., 4 or 5 mL of buffer for membranes up to 25–30 cm^2, 10 mL for membranes around 50 cm^2, and 25 mL for membranes over 50–60 cm^2. Wash buffer volumes are usually larger and may reach 100 mL for a 100 cm^2 membrane.
16. Prehybridization and hybridization temperature in this protocol is normally 55 °C. Yet, higher temperatures up to 60 °C [10] or 65 °C [27] have also been used in this protocol.

Acknowledgments

ASSVd and AFCVd riboprobes were a courtesy of the laboratories Plant Virology of CRA-PAV (Centro di Ricerca per la Patologia Vegetale, Rome, Italy) and Plant Pathology (Hirosaki University, Japan), respectively.

References

1. Amari K, Cañizares MC, Myrta A, Sabanadzovic S, Di Terlizzi B, Pallás V (2001) Tracking *Hop stunt viroid* (HSVd) infection in apricot trees during a whole year by non-isotopic tissue printing hybridization. Acta Hortic 550:315–320
2. Amari K, Ruiz G, Gómez G, Sánchez-Piña MA, Pallás V, Egea J (2007) An important new apricot disease in Spain is associated with hop stunt viroid infection. Eur J Plant Pathol 118:173–181
3. Astruc N, Marcos JF, Macquaire G, Candresse T, Pallás V (1996) Studies on the diagnosis of *Hop stunt viroid* in fruit trees: identification of new hosts and application of a nucleic acid extraction procedure based on non-organic solvents. Eur J Plant Pathol 102:837–846
4. Bouani A, Al Rwahnih M, Abou Ghanem-Sabanadzovic N, Alami I, Emzami M, Myrta A et al (2004) A preliminary account of the sanitary status of stone-fruit trees in Morocco. EPPO Bull 34:399–402
5. Cañizares MC, Aparicio F, Amari K, Pallás V (2001) Studies on the aetiology of apricot "viruela" disease. Acta Hortic 550:249–258
6. El-Dougdoug KA, Dawoud RA, Rezk AA, Sofy AR (2012) Incidence of fruit tree viroid diseases by tissue print hybridization in Egypt. Int J Virol 8:114–120
7. Gümüs M, Paylan IC, Matic S, Myrta A, Sipahioglu HM, Erkan S (2007) Occurrence and distribution of stone fruit viruses and viroids in commercial plantings of *Prunus* species in western Anatolia, Turkey. J Plant Pathol 89:265–268
8. Hooftman R, Arts MJ, Shamloul AM, Van Zaayen A, Hadidi A (1996) Detection of chrysanthemum stunt viroid by reverse transcription polymerase chain reaction and by tissue blot hybridization. Acta Hortic 432:120–128
9. Hurtt SS, Podlekis EV, Howell WE (1996) Integrated molecular and biological assays for rapid detection of *Apple scar skin viroid* in pear. Plant Dis 80:458–462
10. Kaponi MS, Sano T, Kyriakopoulou PE (2013) Natural infection of sweet cherry trees with *Apple scar skin viroid*. J Plant Pathol 95:429–433. https://doi.org/10.4454/JPP.V95I2.018
11. Lolic B, Afechtal M, Matic S, Myrta A, Di Serio F (2007) Detection by tissue-printing of pome fruit viroids and characterization of *Pear blister canker viroid* in Bosnia and Herzegovina. J Plant Pathol 89:369–375
12. Mandic B, Al Rwahnih M, Myrta A, Gómez G, Pallás V (2008) Incidence and genetic diversity of *Peach latent mosaic viroid* and *Hop stunt viroid* in stone fruits in Serbia. Eur J Plant Pathol 120:167–176
13. Matsushita Y, Yanagisawa H, Sano T (2018) Vertical and horizontal transmission of Pospiviroids. Viruses 10(12):706. https://doi.org/10.3390/v10120706
14. Matsushita Y, Osaka M (2019) Screening of *Chrysanthemum seticuspe* accessions reveals different degrees of resistance to chrysanthemum stunt viroid. Eur J Plant Pathol 154(4):1–8. https://doi.org/10.1007/s10658-019-01726-x
15. Michelutti R, Myrta A, Pallás V (2005) A preliminary account on the sanitary status of stone fruits at the clonal genebank in Harrow, Canada. Phytopathol Mediterr 44:71–74
16. Palacio-Bielsa A, Foissac X, Duran-Vila N (1999) Indexing of citrus viroids by imprint hybridization. Eur J Plant Pathol 105:897–903
17. Podleckis EV, Hammond RW, Hurtt SS, Hadidi A (1993) Chemiluminescent detection of potato and pome fruit viroids by digoxigenin-labeled dot blot tissue blot hybridization. J Virol Methods 43:147–158
18. Romero-Durbán J, Cambra M, Duran-Vila N (1995) A simple imprint hybridization method for detection of viroids. J Virol Methods 55:37–47
19. Stark-Lorenzen P, Guitton MC, Werner R, Mühlbach HP (1997) Detection and tissue distribution of *Potato spindle tuber viroid* in

infected tomato plants by tissue print hybridization. Arch Virol 142:1289–1296

20. Tomassoli L, Faggioli F, Zaccaria A, Caccia R, Albani M, Barba M (2004) Molecular diagnosis of *Chrysanthemum stunt viroid* for routine indexing. Phytopathol Mediterr 43:285–288
21. Torres H, Gómez G, Stamo B, Shalaby A, Aoune B, Gavriel I et al (2004) Detection by tissue printing of stone fruit viroids from Europe, the Mediterranean and North and South America. Acta Hortic 657:379–384
22. WenXing X, Ni H, QiuTing J, Farooq AB, ZeQiong W, YanSu S, ChengChun W, LiPing W, GuoPing W (2009) Probe binding to host proteins: a cause for false positive signals in viroid detection by tissue hybridization. Virus Res 145:2630. https://doi.org/10.1016/j.virusres.2009.06.011
23. Cañizares MC, Marcos JF, Pallás V (1999) Molecular characterization of an almond isolate of *Hop stunt viroid* (HSVd) and conditions for eliminating spurious hybridization in its diagnosis in almond samples. Eur J Plant Pathol 105:553–558
24. Di Serio F, Aparicio F, Alioto D, Ragozzino A, Flores R (1996) Identification and molecular properties of a 306 nucleotide viroid associated with apple dimple fruit disease. J Gen Virol 77:2833–2837
25. Di Serio F, Malfitano M, Alioto D, Ragozzino A, Flores R (2002) *Apple dimple fruit viroid*: sequence variability and its specific detection by multiplex fluorescent RT-PCR in the presence of *Apple scar skin viroid*. J Plant Pathol 84:27–34
26. Li SF, Onodera S, Sano T, Yoshida K, Wang GP, Shikata E (1995) Gene diagnosis of viroids: comparisons of return-PAGE and hybridization using DIG-labeled DNA and RNA probes for practical diagnosis of hop stunt, citrus exocortis and apple scar skin viroids in their natural host plants. Ann Phytopathol Soc Japan 61:381–390
27. Kaponi MS (2009) Pome and stone fruit viroids in Greece-molecular characterization and phylogenetic relationships. PhD Dissertation, Agricultural University of Athens, pp 234 (In Greek with English Summary)
28. Maxam AM, Gilbert W (1977) A new method for sequencing DNA. Proc Natl Acad Sci U S A 74:560–564
29. Maniatis T, Sambrook J, Russel DW (1982) Molecular cloning: a laboratory manual, 1st edn. Cold Spring Harbor Laboratory Press, New York, pp 151–152

Chapter 12

Fluorescein-Based Electrophoretic Mobility Shift Assay

Shachinthaka D. Dissanayaka Mudiyanselage and Ying Wang

Abstract

RNA-protein complexes are functionally important in biology. Electrophoretic mobility shift assays (EMSA) have been widely used to study the molecular basis of protein-RNA interactions. Previous methods for EMSA mostly relied on radioactive RNA substrates, raising health and environmental concerns. Here, we describe a method based on fluorescein-labeled RNA for EMSA. In addition, we simplified the protocol to efficiently purify RNA-binding proteins from bacterial expression systems.

Key words EMSA, Fluorescein, Viroid

1 Introduction

RNA-protein interactions are widely involved in transcriptional and post-transcriptional gene regulation [1, 2]. In general, RNA-binding proteins recognize specific RNA sequences or structures to form ribonucleoproteins (RNPs), although other mechanisms exist for the formation of RNPs [3]. Multiple tools have been developed to analyze the binding kinetics and molecular basis of interaction specificities and elucidate the functional mechanisms of various RNPs. Among those tools, electrophoretic mobility shift assays (EMSA) have been widely used [4, 5]. In brief, an RNA-binding protein and its RNA substrate are mixed in an optimized buffer to form a higher molecular weight complex, and the mixture is then subjected to agarose or polyacrylamide gel electrophoresis to separate the unbound from bound RNAs. However, most EMSA protocols rely on radioactive RNA substrates, which generates serious concerns regarding health and environmental safety. Therefore, developing alternative non-radioactive approaches is of great interest in the research community.

Viroids are single-stranded covalently closed circular RNAs that can infect plants [6, 7]. These infectious RNAs co-opt host factors to complete their life cycles and alter host gene expression [6, 8]. Viroids serve as an excellent model system to describe the

Ayala L. N. Rao et al. (eds.), *Viroids: Methods and Protocols*, Methods in Molecular Biology, vol. 2316,
https://doi.org/10.1007/978-1-0716-1464-8_12, © Springer Science+Business Media, LLC, part of Springer Nature 2022

RNA structure-based formation of RNPs and the underlying mechanisms of their functions. Transcription factor IIIA (TFIIIA) is a single-copy gene in eukaryotes [9]. In land plants, a deeply conserved alternative splicing pattern in TFIIIA transcripts leads to the production of two proteins (TFIIIA-7ZF and TFIIIA-9ZF) [10–12]. Both proteins bind potato spindle tuber viroid (PSTVd) in vitro and in vivo [13, 14], providing a valuable platform to understand plant-viroid interactions as well as the general principles in coordinating multiple RNA-binding domains to achieve binding specificity.

Here, we describe a simple method to purify TFIIIA-9ZF protein from *Escherichia coli*. TFIIIA-9ZF protein purification is technically challenging, possibly due to the formation of inclusion bodies [15]. Instead of simply repeating previously reported methodologies with tedious procedures, we modified a SDS/Triton X-100-based purification system [16, 17] and obtained desirable amounts of active proteins with a simplified procedure. In addition, we optimized the conditions to apply fluorescein-labeled RNA substrates for EMSA. This protocol can be adapted to other EMSA assays using various combinations of RNA substrates and RNA-binding proteins.

2 Materials

2.1 Reagents for TFIIIA Purification

1. LB solution: 25 g LB broth (Miller) dissolved in 900 mL ddH_2O. Bring up the volume to 1 L using ddH_2O. Autoclave the solution and store at 4 °C.
2. 100 mg/mL ampicillin: 1 g ampicillin powder mixed with 9 mL ddH_2O. Bring up the volume to 10 mL using ddH_2O. S-terilize using a 0.22μm syringe filter. Aliquot and store at −20 °C.
3. Filter units: 0.22μm syringe filters with 5 mL syringes.
4. 34 mg/mL chloramphenicol: 0.34 g chloramphenicol powder mixed with 9 mL 100% dimethyl sulfoxide (DMSO). Bring up the volume to 10 mL using 100% DMSO. Sterilize using a 0.22μm syringe filter. Aliquot and store at −20 °C.
5. Rosetta competent cells from EMD Millipore.
6. A lab shaker that can set temperatures at 20 and 37 °C.
7. 1 M IPTG stock: 2.38 g isopropyl β-D-1-thiogalactopyranoside (IPTG) dissolved in 8 mL distilled water. Bring up the volume to 10 mL using distilled water. Aliquot and store at −20 °C.
8. 1 M $ZnCl_2$: 6.81 g zinc chloride powder dissolved in 50 mL distilled water. Autoclave and store at 4 °C.

9. Centrifuges: Eppendorf 5810R (for 1.5 and 2 mL tubes) and MX301 high-speed refrigerated microcentrifuge (for 15 and 50 mL Falcon tubes) from Amuza Inc. Similar centrifuges that allow the same settings can be used.
10. Centrifuge tubes: standard 1.5 and 2.0 mL tubes, as well as 15 and 50 mL Falcon tubes.
11. Native lysis buffer: 6.9 g $NaH_2PO_4 \cdot H_2O$, 17.54 g NaCl, and 0.68 g imidazole, and dissolve in 850 mL ddH_2O. Adjust pH to 8.0 using NaOH solution (*see* **Note 1**). Store at room temperature.
12. Sodium dodecyl sulfate (SDS) solution: 10% (w/v) solution in water (*see* **Note 2**). Store at room temperature.
13. Sonicator: Bioruptor® Plus sonication device from Diagenode.
14. Triton X-100: 100% Triton X-100 from MilliporeSigma. Store at room temperature.
15. Ni-NTA resin: 50% slurry from Qiagen. Store at 4 °C.
16. Wash buffer: 6.9 g $NaH_2PO_4 \cdot H_2O$, 17.54 g NaCl, and 1.36 g imidazole dissolved in 850 mL ddH_2O. Adjust pH to 8.0 using NaOH solution (*see* **Note 3**). Store at room temperature.
17. Elution buffer: 6.9 g $NaH_2PO_4 \cdot H_2O$, 17.54 g NaCl, and 17.00 g imidazole mixed with 850 mL ddH_2O. Adjust pH to 8.0 using NaOH solution (*see* **Note 4**). Store at room temperature.
18. Sephadex G-25 resin: 50 g Sephadex G-25 washed three times using 500 mL 65 °C ddH_2O. The resin is then mixed with 225 mL ddH_2O and 25 mL 100% ethanol. The resin slurry is stored at 4 °C.
19. Concentration units: Amicon Ultra-15 centrifuge filter units (3MWCO) from EMD Millipore.
20. 3× SDS loading buffer: 150 mM Tris–HCl (pH 6.8), 300 mM DTT, 6% SDS, 0.3% (w/v) bromophenol blue, and 30% (v/v) glycerol. Store at room temperature.
21. BSA stock: 0.1 g BSA powder mixed with 9 mL distilled water. Bring up the volume to 10 mL by adding distilled water. The solution is sterilized using a 0.22μm syringe filter, aliquoted, and stored at −20 °C.
22. 3× Bis-Tris gel buffer: 209.24 g Bis-Tris mixed with 900 mL ddH_2O. Adjust pH to 6.6, and then bring up the volume to 1 L by adding ddH_2O. Autoclave the solution before storing at room temperature.
23. 6% Bis-Tris PAGE gel mixture: 1.8 mL 30% acrylamide/bis solution (29:1), 100μL ammonium persulfate solution, 5μL TEMED, 3 mL 3× Bis-Tris gel buffer, and 4.1 mL ddH_2O.

24. 1× Bis-Tris gel running buffer: 10.46 g MOPS, 6.04 g Tris base; mix them with 900 mL ddH_2O. Add 2 mL 0.5 M EDTA (pH 8.0) and 10 mL 10% (w/v) SDS solution (*see* **Note 5**).
25. Gel running units: PAGE gels (SDS or native) are made and run using the Bio-Rad Mini-PROTEAN Tetra Cell system. The Bio-Rad PowerPac Basic unit is used for power supply.
26. Coomassie blue staining solution: 0.1% (w/v) Coomassie Brilliant Blue R-250, 50% (v/v) methanol, and 10% (v/v) glacial acetic acid. Store at room temperature.
27. De-staining solution: 40% (v/v) methanol and 10% (v/v) glacial acetic acid. Store at room temperature.

2.2 Reagents for EMSA

1. *Sma*I: restriction enzyme *Sma*I from New England Biolabs. Reaction buffer is included. Store at −20 °C.
2. Thermal cycler: Bio-Rad T100. Regular 0.2 mL PCR tubes are needed as well.
3. T3 MEGAScript kit: T3 MEGAScript kit from Thermo Fisher Scientific. The kit includes reaction buffer, nuclease-free water, T3 polymerase, TURBO DNase, and 75 mM solutions of UTP, ATP, CTP, and GTP. Store at −20 °C.
4. Fluorescein-labeled UTP: Fluorescein-12-UTP from MilliporeSigma. Store at −20 °C.
5. 100% glycerol: store at room temperature.
6. RNaseOUT: RNaseOUT from Thermo Fisher Scientific. Store at −20 °C.
7. MEGAclear kit: MEGAclear kit from Thermo Fisher Scientific for purification of in vitro transcripts.
8. 10× EMSA binding buffer: 500 mM Tris (pH 8.0), 500 mM NaCl, 1 mM EDTA, 100 mM 2-mercaptoethanol (*see* **Note 6**).
9. 6× EMSA loading buffer: 30% glycerol, 0.3% bromophenol blue. Store at 4 °C.
10. Ammonium persulfate: 10% (w/v) solution in water. Store at 4 °C.
11. *N*,*N*,*N*,*N′*-Tetramethyl-ethylenediamine (TEMED): Store at 4 °C.
12. 10× Tris-borate-EDTA (TBE): 108 g Tris base and 55 g boric acid, and dissolve in 900 mL distilled water. Add 40 mL 0.5 M EDTA (pH 8.0), and adjust volume to 1 L by adding distilled water. Autoclave and store at room temperature.
13. 6% native PAGE gel mixture: 1.8 mL 30% acrylamide/bis solution (19:1), 60μL ammonium persulfate solution, 5μL TEMED, 450μL 10× TBE, and 6.7 mL ddH_2O.

14. 0.5× TBE running buffer: 20 mL 10× TBE buffer with 380 mL distilled water.
15. Imaging system: The Bio-Rad Gel Doc XR+ system with the XcitaBlue™ conversion screen.

3 Methods

3.1 TFIIIA Purification

1. pDEST-17 plasmid containing TFIIIA-9ZF cDNA [14] is transformed to Rosetta competent cells via a standard heat shock procedure. One bacterial colony is inoculated in 5 mL LB solution with 5μL ampicillin (100 mg/mL) and 5μL chloramphenicol (34 mg/mL), shaking overnight at 37 °C. The overnight culture is added to 250 mL fresh LB solution with 250μL ampicillin (100 mg/mL). Grow the bacteria to OD_{600} 0.4–0.5.
2. Add 1 M IPTG stock to a final concentration of 0.4 mg/mL. Add $ZnCl_2$ to a final concentration of 5μM (*see* **Note 7**). The bacterial culture is kept in a shaker at 20 °C overnight.
3. Pellet the bacteria at 8,000 × *g* at 4 °C for 20 min.
4. Resuspend the pellet in 2.5 mL native lysis buffer, and then add 10% SDS solution to a final concentration of 1%. The mixture is sonicated for 10 min with a 30 s on/30 s off cycle (*see* **Note 8**).
5. Add 100% Triton X-100 to a final concentration 2%, and shake the solution at 4 °C for 1 h.
6. Dilute with 25 mL native lysis buffer, and shake at 4 °C overnight.
7. Wash 5 mL 50% Ni-NTA resin with 10 mL native lysis buffer in a 50 mL Falcon tube. Spin down the resin at 400 × *g* to remove all liquid.
8. Centrifuge the bacterial lysate from **step 6** at 15,000 × *g* at 4 °C for 20 min, and take the supernatant after centrifugation to mix with the washed resin in **step 7** at 4 °C for 1 h. Do not disturb the pellet when taking the supernatant.
9. Apply Ni-NTA resin and the lysate to an empty gravity flow column to allow the lysate to pass through via gravity.
10. Wash the resin twice with 20 mL wash buffer (*see* **Note 9**).
11. Elute the protein eight times with 0.5 mL elution buffer for each elution. Combine the portions that contain the pure target protein.
12. Purified proteins are loaded to Sephadex G-25 resin to remove excessive amount of salt. The flow-through fraction is collected and concentrated using an Amicon Ultra-15 centrifugal filter unit by centrifuging at 1,000 × *g* at 4 °C for 1 h. The protein

solution is supplemented with 10% glycerol, aliquoted, and stored at −20 °C.

13. Take 5μL purified protein, and mix it with 2.5μL 3× SDS loading buffer to estimate concentration. Run the purified the protein sample in a 6% Bis-Tris gel with 1× Bis-Tris gel running buffer, together with 0.1, 0.5, 1, 5, and 10 mg/mL BSA standard diluted from 10 mg/mL BSA stock.
14. The 6% Bis-Tris PAGE gel mixture is poured into Bio-Rad Mini-PROTEAN Tetra Cell gel cassette. Let the gel stand at room temperature for 1 h before using.
15. Run the Bis-Tris PAGE gel at 180 V for 30 min.
16. After electrophoresis, the Bis-Tris PAGE gel is incubated with 20 mL Coomassie blue staining solution for 1 h followed by incubation with 100 mL de-staining solution overnight.

3.2 RNA Substrate Preparation

1. The pInt$^{95-94}$(+) plasmid [14] is linearized with *Sma*I. The digestion reaction is kept at 25 °C for 2 h, followed by 65 °C incubation for 20 min.
2. The linearized plasmid is used as the template for in vitro transcription using T3 MEGAscript kit (Invitrogen). The reaction mixture contains 1μL MEGAScript reaction buffer, 1μL MEGAScript T3 polymerase, 1μL ATP (75 mM), 1μL GTP (75 mM), 1μL CTP (75 mM), 0.67μL UTP (75 mM), 1μL fluorescein-labeled UTP (10 mM), 2μL linearized plasmids (250 ng/μL), and 1.33μL nuclease-free water. The reaction is kept in dark at 37 °C overnight.
3. The in vitro transcription reaction is stopped by adding 1μL TURBO DNase and incubating at 37 °C for 20 min. The RNA was purified using MEGAclear kit (Invitrogen) following user manual (*see* **Note 10**).

3.3 EMSA

1. 50 ng fluorescein-labeled RNA in 1μL nuclease-free water is heated at 70 °C for 10 min and kept at room temperature for 10 min.
2. The treated RNA is incubated in a 30μL reaction system with 3μL 10× EMSA binding buffer, 1.5μL 100% glycerol, 1μL RNaseOUT, appropriate amounts of proteins, and nuclease-free water. The binding reactions are kept at 28 °C for 30 min.
3. 6μL 6× EMSA loading buffer is added to each reaction, and the samples are kept on ice before gel loading.
4. The 6% native PAGE gel mixture is poured into Bio-Rad Mini-PROTEAN Tetra Cell gel cassette. Let the gel stand at room temperature for 1 h before using.

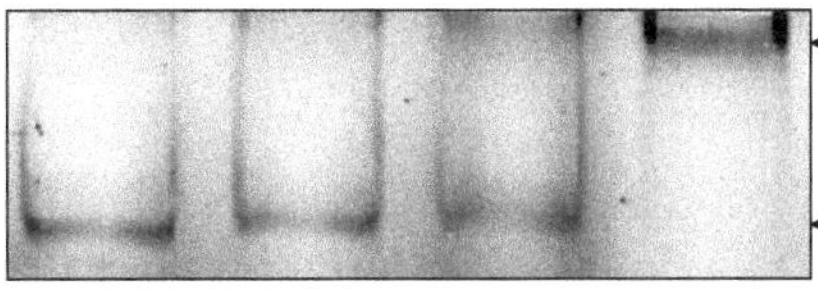

Fig. 1 A representative TFIIIA-9ZF/PSTVd EMSA result. In this EMSA assay, increasing amounts of TFIIIA-9ZF protein were used and mixed with 50 ng fluorescein-labeled PSTVd RNA. "Unbound" and "bound" indicate the free probe and the RNP, respectively

5. Samples are subjected to 6% native PAGE gel electrophoresis using 0.5× TBE running buffer. The gel is run on ice at 120 V for 50 min.
6. After electrophoresis, the native PAGE gel is visualized using a Bio-Rad Gel Doc XR+ gel documentation system (Fig. 1). It is visualized under SYBR safe mode with XcitaBlue™ conversion screen.

4 Notes

1. Do not autoclave native lysis buffer. We also added 1 M $ZnCl_2$ stock to a final concentration of 5μM to facilitate the formation of zinc finger domains. This is not necessary for the extraction of proteins without zinc finger domains.
2. Do not autoclave 10% SDS solution.
3. Do not autoclave wash buffer. We also added 1 M $ZnCl_2$ stock to a final concentration of 5μM to facilitate the formation of zinc finger domains. This is not necessary for the extraction of proteins without zinc finger domains.
4. Do not autoclave elution buffer. We also added 1 M $ZnCl_2$ stock to a final concentration of 5μM to facilitate the formation of zinc finger domains. This is not necessary for the extraction of proteins without zinc finger domains.
5. Do not autoclave 1× Bis-Tris gel running buffer. No need to adjust pH. It is optional to add 1 mM sodium bisulfite.
6. Based on our experience, TFIIIA-9ZF is not stable with buffer containing DTT. Thus, we used 2-mercaptoethanol instead. In addition, we found that dialysis may trigger the aggregation/precipitation of purified proteins, so we only use Sephadex G-25 resin to desalt the purified protein fraction.
7. $ZnCl_2$ is not required for the extraction of proteins without zinc finger domains.
8. The sonication step is optional.

9. The volume of wash buffer is proportional to the bed volume of Ni-NTA resin.
10. To measure the RNA concentration, perform Northern blot with a dilution series of un-labeled PSTVd in vitro transcripts. Fluorescein interferes with absorbance-based measurements.

References

1. Baltz AG, Munschauer M, Schwanhausser B et al (2012) The mRNA-bound proteome and its global occupancy profile on protein-coding transcripts. Mol Cell 46:674–690
2. Castello A, Fischer B, Eichelbaum K et al (2012) Insights into RNA biology from an atlas of mammalian mRNA-binding proteins. Cell 149:1393–1406
3. Hentze MW, Castello A, Schwarzl T et al (2018) A brave new world of RNA-binding proteins. Nat Rev Mol Cell Biol 19:327–341
4. Hellman LM, Fried MG (2007) Electrophoretic mobility shift assay (EMSA) for detecting protein-nucleic acid interactions. Nat Protoc 2:1849–1861
5. Ryder SP, Recht MI, Williamson JR (2008) Quantitative analysis of protein-RNA interactions by gel mobility shift. Methods Mol Biol 488:99–115
6. Ding B (2009) The biology of viroid-host interactions. Annu Rev Phytopathol 47:105–131
7. Flores R, Gago-Zachert S, Serra P et al (2014) Viroids: survivors from the RNA world? Annu Rev Microbiol 68:395–414
8. Flores R, Minoia S, Carbonell A et al (2015) Viroids, the simplest RNA replicons: how they manipulate their hosts for being propagated and how their hosts react for containing the infection. Virus Res 209:136–145
9. Layat E, Probst AV, Tourmente S (2013) Structure, function and regulation of transcription factor IIIA: from Xenopus to Arabidopsis. Biochim Biophys Acta 1829:274–282
10. Fu Y, Bannach O, Chen H et al (2009) Alternative splicing of anciently exonized 5S rRNA regulates plant transcription factor TFIIIA. Genome Res 19:913–921
11. Hammond MC, Wachter A, Breaker RR (2009) A plant 5S ribosomal RNA mimic regulates alternative splicing of transcription factor IIIA pre-mRNAs. Nat Struct Mol Biol 16:541–549
12. Layat E, Cotterell S, Vaillant I et al (2012) Transcript levels, alternative splicing and proteolytic cleavage of TFIIIA control 5S rRNA accumulation during Arabidopsis thaliana development. Plant J 71:35–44
13. Eiras M, Nohales MA, Kitajima EW et al (2011) Ribosomal protein L5 and transcription factor IIIA from Arabidopsis thaliana bind in vitro specifically Potato spindle tuber viroid RNA. Arch Virol 156:529–533
14. Wang Y, Qu J, Ji S et al (2016) A land plant-specific transcription factor directly enhances transcription of a pathogenic noncoding RNA template by DNA-dependent RNA polymerase II. Plant Cell 28:1094–1107
15. Ottonello S, Ballabeni A, Soncini C et al (1994) High level expression in E. coli and purification of yeast transcription factor IIIA. Biochem Biophys Res Commun 203:1217–1223
16. Clarke S (1981) Direct renaturation of the dodecyl sulfate complexes of proteins with Triton X-100. Biochim Biophys Acta 670:195–202
17. Tao H, Liu W, Simmons BN et al (2010) Purifying natively folded proteins from inclusion bodies using sarkosyl, Triton X-100, and CHAPS. BioTechniques 48:61–64

Part III

PCR Techniques

Chapter 13

Detection of Viroids by RT-PCR

Nerida J. Donovan, Grant A. Chambers, and Mengji Cao

Abstract

Reverse transcription-polymerase chain reaction (RT-PCR) is an effective method for detecting the presence of viroids in plant tissue. Viroid RNA is converted to cDNA and amplified to detectable levels, making it a fast and useful detection tool, even when the viroid is present at low levels. Methods of viroid detection using conventional RT-PCR are described in this chapter.

Key words Reverse transcription-polymerase chain reaction (RT-PCR), Viroid detection, Primers, Gel electrophoresis

1 Introduction

Reverse transcription-polymerase chain reaction (RT-PCR) was first reported for use in viroid detection in 1990 [1]. It is now a routine but valuable method for detecting the presence of viroids in plant tissue, even when present at low levels. In RT-PCR, a total nucleic acid extract from infected plant tissue is subject to reverse transcription, resulting in the conversion of viroid RNA to cDNA by reverse transcriptase enzymes. The cDNA is then exponentially amplified to detectable levels by PCR. The PCR reaction mixture contains the cDNA template, primers which are short lengths of DNA designed to bind to either side of the target section of DNA that you want to copy, nucleotide bases and polymerase enzyme to build the new strands of cDNA, and buffer chemicals to provide a suitable environment in which the reaction can happen. Primers typically target a conserved region of the viroid genome; a region that is common to all variants of a species containing essential genes that have remained unchanged through evolution. Primers may also be specific enough to discriminate between variants within a viroid species [2].

The process of PCR involves heating to separate the cDNA into two strands (denaturing), cooling where primers join onto complementary strands of the denatured cDNA (annealing), and heating

Ayala L. N. Rao et al. (eds.), *Viroids: Methods and Protocols*, Methods in Molecular Biology, vol. 2316,
https://doi.org/10.1007/978-1-0716-1464-8_13,

again when the new cDNA strand is made (extending). The process of heating and cooling is called thermocycling and is repeated 20–40 times to increase the number of copies of cDNA in the mixture. The quantity and size of the cDNA fragments amplified during PCR are then visualized using gel electrophoresis. The PCR product is loaded into a well in a gel, and a current pulls the negatively charged DNA fragments toward the positive charge at the end of the gel. Different-sized fragments move at different speeds through the gel, allowing different viroids to be identified.

RT-PCR for detection of viroids is fast (hours), sensitive, and specific. A disadvantage, common to all molecular detection methods, is the potential for false results if there is contamination in the reaction mixture, even at small levels: either a similar genetic sequence to the target fragment which results in a false detection or substances which inhibit the reaction and lead to a false negative result. However, detection of viroids using RT-PCR has been widely studied and successfully applied. This chapter describes straightforward user-friendly protocols to detect viroid infection in plant material using conventional RT-PCR.

2 Materials

1. Pipettes capable of dispensing volumes of <1 to 1000μL and appropriate filter tips.
2. 1.5 mL sterile nuclease-free tubes.
3. One-step RT-PCR kit such as Qiagen OneStep RT-PCR Kit or Invitrogen Superscript III RT-PCR with Platinum Taq kit, which includes enzymes, $MgSO_4$, and dNTPs.
4. 0.2 mL thin-walled PCR tubes.
5. Thermocycler.
6. Cartridge purified primers are usually delivered lyophilized and will require resuspension using TE buffer or nuclease-free water to a concentration of 100μM. Stock primer solutions (100μM) are kept at −20 °C for extended periods. Working solutions of 10μM are diluted in TE buffer or nuclease-free water from the stock primer and are also stored at −20 °C.
7. TE buffer (10 mM Tris–Cl pH 7.5, 1 mM EDTA).
8. Nuclease-free water.
9. Agarose gels 2% (1.6 g agarose melted in 80 mL 1× TBE).
10. 1× TBE, pH 8.3: 89 mM Tris base, 89 mM boric acid, 2 mM EDTA.
11. Electrophoresis equipment.
12. GelRed™ (Biotium) nucleic acid gel stain (10,000×).

13. Bromophenol blue gel loading dye (6×).
14. Gel marker.
15. UV transilluminator.
16. Gloves.

3 Methods

1. Prior to setting up a reaction, determine a protocol (Table 1) based on the number of samples, the primers (Table 2), and the protocol that is included with the commercial master mix.
2. Wear gloves to handle reagents, remove from the freezer, and thaw gradually on ice.
3. In the PCR setup area, prepare a master mix for your number of assay tubes by adding ingredients in the order outlined in the example below. Keep all components on the ice during the reaction setup. Refer to the relevant reference to determine the specific reaction components for each viroid.
4. Mix the master mix by pipetting gently up and down.
5. Dispense 18μL of master mix into each 0.2 mL PCR tube.
6. Move PCR tubes to a separate area for template addition.
7. Dispense 2μL of template total RNA to the appropriate tube. Total RNA extracted from plant tissue is the template used for RT-PCR amplification. Tubes should have a final reaction volume of 20μL.
8. For each set of reactions, ensure that you include a negative control (plant sample with no viroid present), positive control

Table 1
An example of reaction components for use in a conventional RT-PCR

Component	Each reaction tube (μL)	8 reaction tubes (μL)	Final concentration 20μL reaction
Nuclease-free water	5.6	44.8	
5× PCR reaction mix	10	80	1×
$MgSO_4$ (5 mM)	0.8	6.4	1.8 mM[a]
Sense primer (10 mM)	0.4	3.2	200μM
Antisense primer (10 mM)	0.4	3.2	200μM
RT/*Taq* enzyme mix	0.8	6.4	

[a]Commercial kit has $MgSO_4$ final concentration of 1.6 mM which works for most targets

Table 2
Primers for viroid detection by RT-PCR

Viroid	Primer name	Primer sequence (5′–3′)	Product size (bp)	References
Pospiviroidae				
Apscaviroid				
ADFVd	ADFVd-for	CCCCCCTGCGCTACTGAC TAAAAG	260	[3]
	ADFVd-rev	GTGTTTTACCCTGGAGGCTCCAC TC		
ASSVd	ADAS-36	GCCTTCGTCGACGACGACAG	330	[4]
	AS-37fl	CGGTGACAAAGGAGCTGCCAG		
AGVd	AGVd-For	GTCGACGAAGGGTCC TCAGCAGAG	375	[5]
	AGVd-Rev	CTCGACGACGAGTCGCCAGG TGAG		
CBLVd	CBLVd-R1	TTCGTCGACGACGACCAGTC	327	[2]
	CBLVd-F1	GGCTCGTCAGCTGCGGAGGT		
CBLVd/CVd-I-LSS	CBLVd-F	GGTCGTCGTCGACGAAGGCTC STCAGC	231–241	[6]
	CBLVd-R	AGGAGCCCTCAGGGGTTC		
CDVd	CVd-III-F1	GGCAGCTAAGTTGGTGACGC	294	[2]
	CVd-III-R1	TTCGTCGACGACGACAGGTA		
CVd-V	CVd-V F	GACGAAGGCCGGTGAGCAG TAAGCC	294	[7]
	CVd-V R	GACGACGACAGGTGAGTACTTTC		
CVd-VI	CVd-VI-F	CGTCGACGAAGGCATGTGAGCT	331	[8]
	CVd-VI-R	ACGACAGGTGAGTTCTCCTT		
CVd-VII	VIIF1	CTTGCCTTGAGAAGCGAAGC	279	[9]
	VIIR3	AAGCAGTTCCAGTTACAG		
GYSVd1	HI^{+}	GCTAAGAGGTCTCCGGA	223	[10]
	CI^{-}	GCGAACGTGAATAGGCTAGAA		
GYSVd2	GYSVd-2-P1	ACTAGTACTTTCTTCTATC TCCGAAGC	375	[11]
	GYSVd-2-P2	ACTAGTCCGAGGACCTTTTC TAGCGCTC		
PBCVd	cPBCVd	CCTTCGTCGACGACGAGCCGAG	315	[12]
	hPBCVd	GTCTAGAAGCCTGGGCGCTGG		
Cocadviroid				
CBCVd	CBCVd-For68-89	CCGGGGAAATCTCTTCAGACTC	284	[13]
	CBCVd-Rev71-48	CCGGGGATCCCTCTTCAGGTATG T		

(continued)

Table 2
(continued)

Viroid	Primer name	Primer sequence (5′–3′)	Product size (bp)	References
CCCVd	GV4 GVR4	ACTCACGCGGCTCTTACC TGTATCCACCGGGTAGTCTC	270–297	[14]
CTiVd	D4-2 D1-2	GTCGCCGATTCGTGCTGG TTGGGCTTCGTC ACCAGGTACGCTCCCTACCTCGG TTGCTAGG	254	[15]
HLVd	HLVd200-181 HLVd201-220	CCACCGGGTAGTTTCCAACT ATACAACTCTTGAGCGCCGA	256	[16]
Coleviroid				
CbVd1	CbVd-UCCR-M CbVd-UCCR-P	GTTCCCTGGCAGCGCTGCAA TTGCAGCGCTGCAACGGAAT	250	[17]
CbVd2	CbVd-2F CbVd-2R	AGCTTACCTGGGTTCCCT CTCTCCTCTATTTACTCTCTTC TC	301	[18]
CbVd3	CbVd-UCCR-M CbVd-UCCR-P	GTTCCCTGGCAGCGCTGCAA TTGCAGCGCTGCAACGGAAT	250	[17]
CbVd5	CbVd5R CbVd5F	AATTGAGGTCAAACCTCTTT GACTAGAACAGTAGTAAAGA	270	[19]
Hostuviroid				
DLVd	DLVd-P1 DLVd-P2	GGGGCAACTCCGAGAGTGCTG GGGGCTCCTCAGAGAGTCTC	342	[20]
HSVd	HSVd-F1 HSVd-R1	GGGGCAACTCTTCTCAGAATCC GGGGCTCCTTTCTCAGGTAAGTC	299	[2]
Pospiviroid				
Pospiviroid universal primers[a]	Pospi1-RE Pospi1-FW	AGCTTCAGTTGTWTCCACCGGGT GGGATCCCCGGGGAAAC	197	[21]
CSVd	CSV-2B CSV-1M	CTTAGGACCCCACTCCTGCG CCGCGATCTCGTCGGACTTC	348	[22]
CEVd	CEVd-F1 CEVd-R1	GGAAACCTGGAGGAAGTCG CCGGGGATCCCTGAAGGA	371	[2]
CLVd	pCLV4 pCLVR4	GGGGCTCCTGAGACCGCTCTTG GGGGCAACTCAGACCGAGC	370	[23]
IrVd-1	IrVd-FW1 IrVd-RE1	GCGGAAGAAACAGGAGCTCGWCT CGCGWGGAGTTCTCCGGTC TTTA	370	[24]

(continued)

Table 2
(continued)

Viroid	Primer name	Primer sequence (5′–3′)	Product size (bp)	References
PCFVd	AP-FW1 AP-RE2	ACCCTTCCTTTCTTCGGGTTTCC CAGCGGGGATTACTCCTGTCA	334	[25]
PSTVd	PSTVd H PSTVd C	GGGGAAACCTGGAGCGAAC CCTGAAGCGCTCCTCCGAGC	355	[26]
TASVd[b]	CEVd-FW2 CEVd-RE2	GTGCTCACCTGACCCTGCAGG ACCACAGGAACCTCAAGAAAG	364	[27]
TCDVd	TCDVd FP3 TCDVd RP4	ACTGGCAAAAGGCGGCAGGGAGC TTGTGGA AGGGCTAAACACCC TCGCCCCGGTAGCAGC	228	[28]
TPMVd	MPVd-F MPVd-R	GGGATCCCCGGGGAAACCT GGGATCCCTGAAGCGCTCC	360	[29]
MPVd	MPVd-F MPVd-R	GGGATCCCCGGGGAAACCT GGGATCCCTGAAGCGCTCC	360	[29]
Avsunviroidae				
Avsunviroid				
ASBVd	ASBVd-SB1-F1 ASBVd-SB1-R1	TGGGAAGAACACTGATGAG TCTTTCCCTGAAGAGACGA	247–250	[30]
Elaviroid				
ELVd	PIII PIV	AGGATGTGTTCCCTAGGAGG TGAAARRGGATAGTACCTGGG	332–335	[31]
Pelamoviroid				
CChMVd	CChMVd-F CChMVd-R	ACGTCGGTGTCCTGATGAAG CAGTGGCTGTGACTACGAGA	316	[32]
PLMVd	cPLMVd hPLMVd	AACTGCAGTGCTCCGAA TAGGGCAC CCCGATAGAAAGGCTAAGCACC TCG	337	[33]

[a] *Pospiviroidae* universal primers will detect CEVd, CSVd, IrVd-1, MPVd, PCFVd, PSTVd, TASVd, TCDVd, and TPMVd
[b] TASVd primers detect both TASVd and CEVd

Table 3
An example of thermocycling conditions

cDNA synthesis Pre-denaturation	55 °C for 30 min 94 °C for 2 min	1 cycle
Denaturation Annealing Extension	94 °C for 15 s 60 °C for 30 s 68 °C for 30 s	40 cycles
Final extension	68 °C for 5 min	1 cycle

(sample that is known to have viroid present), and a no template control (NTC) that contains water in the place of a template.

9. Place tubes in a pre-heated thermocycler with a preset program of cDNA synthesis and denaturation of RT enzyme, followed by PCR amplification cycles and a final extension step. Refer to the relevant reference for the specific reaction conditions to be programmed into the thermocycler for each viroid (Table 2; example provided in Table 3).
10. Remove PCR tubes from thermocycler following the completion of the run, and store in the refrigerator until the samples are subjected to agarose gel electrophoresis.
11. Dissolve 1.6 g agarose in 80 mL 1× TBE by slowly heating using a microwave oven. Allow the gel to cool so it is comfortable to touch and pour into a gel tray containing a comb with an appropriate number of teeth to form wells for your sample (s). Once the gel has set, remove the comb, and place the gel in an electrophoresis tank. Cover with 1× TBE buffer, pH 8.3, to 5 mm above the gel.
12. Load 2μL of PCR product mixed with 1μL loading dye into wells on the 2% agarose gel. Ensure that an appropriate molecular marker is also loaded. Gels are run at 90 V for 60 min and stained with GelRed™ according to the manufacturer's instructions.
13. Gels are observed and photographed under UV light using UV transilluminator equipment. A single band of the appropriate size, depending on the primer set used, acts as a confirmation of the presence of the viroid in question.
14. If available, sequence analysis of the PCR product would provide you with confirmatory sequence data.

4 Notes

1. One-step RT-PCR kits from commercial manufacturers are recommended to reduce the risk of contamination. There are many commercial manufacturers of one-step RT-PCR kits. For viroid detection, you require a high-quality, robust kit.
2. It is recommended that separate areas are designated in the laboratory for PCR setup, template addition, and PCR/gel analysis to reduce the risk of contamination of samples and reagents.
3. Magnesium concentration in the master mix can be optimized by adding extra $MgSO_4$.

References

1. Hadidi A, Yang X (1990) Detection of pome fruit viroids by enzymatic cDNA amplification. J Virol Methods 30:261–270
2. Bernad L, Duran-Vila N (2006) A novel RT-PCR approach for detection and characterization of citrus viroids. Mol Cell Probes 20:105–113
3. Chiumenti M, Torchetti EM, Di Serio F, Minafra A (2014) Identification and characterization of a viroid resembling apple dimple fruit viroid in fig (*Ficus carica* L.) by next generation sequencing of small RNAs. Virus Res 188:54–59
4. Serio FD, Malfitano M, Alioto D, Flores AR (2002) Apple dimple fruit viroid: sequence variability and its specific detection by multiplex fluorescent RT-PCR in the presence of apple scar skin viroid. J Plant Pathol 84:27–34
5. Navarro B, Pantaleo V, Gisel A, Moxon S, Dalmay T, Bisztray G, Di Serio F, Burgyan J (2009) Deep sequencing of viroid-derived small RNAs from grapevine provides new insights on the role of RNA silencing in plant-viroid interaction. PLoS One 4(11):e7686. https://doi.org/10.1371/journal.pone.0007686
6. Wildman O (2013) Developing molecular diagnostics for citrus viroids. Honours thesis, University of Western Sydney, Australia, p. 135
7. Serra P, Eiras M, Bani-Hashemian SM, Murcia N, Kitajima EW, Daròs JA, Flores R, Duran-Vila N (2008) Citrus viroid V: occurrence, host range, diagnosis, and identification of new variants. Phytopathology 98:1199–1204
8. Cao M, Wu Q, Yang F, Wang X, Li R, Zhou C, Li Z (2017) Molecular characterization and phylogenetic analysis of citrus viroid VI variants from citrus in China. Eur J Plant Pathol 149:885–893
9. Chambers GA, Donovan NJ, Bodaghi S, Jelinek SM, Vidalakis G (2018) A novel citrus viroid found in Australia, tentatively named citrus viroid VII. Arch Virol 163:215–218
10. Wah WCYF (1996) Viroids in grapevines: transmission via seeds and persistence in meristem regenerated vines. PhD thesis, The University of Adelaide, Australia, p. 128
11. Jiang D, Guo R, Wu Z, Wang H, Li S (2009) Molecular characterization of a member of a new species of grapevine viroid. Arch Virol 154:1563–1566
12. Fekih Hassen I, Roussel S, Kummert J, Fakhfakh H, Marrakchi M, Jijakli MH (2010) Development of a rapid RT-PCR test for the detection of peach latent mosaic viroid, pear blister canker viroid, hop stunt viroid and apple scar skin viroid in fruit trees from Tunisia. J Phytopathol 154:217–223
13. Wang J, Boubourakas IN, Voloudakis AE, Agorastou T, Magripis G, Rucker TL, Kyriakopoulou PE, Vidalakis G (2013) Identification and characterisation of known and novel viroid variants in the Greek national germplasm collection: threats to the industry. Eur J Plant Pathol 137:17–27
14. Vadamalai G (2005) An investigation of orange spotting disorder in oil palm. PhD thesis, The University of Adelaide, Australia p. 155
15. Hodgson RAJ, Wall GC, Randles JW (1998) Specific identification of coconut tinangaja viroid for differential field diagnosis of viroids in coconut palm. Phytopathology 88:774–781
16. Matoušek J, Patzak J (2000) A low transmissibility of hop latent viroid through a generative phase of *Humulus lupulus* L. J. Biologia

Plantarum 43:145. https://doi.org/10.1023/A:1026531819806
17. Tsushima T, Sano T (2018) A point-mutation of Coleus blumei viroid 1 switches the potential to transmit through seed. J Gen Virol 99:393–401
18. Fu FH, Li SF, Jiang DM, Wang HQ, Sang LW (2011) First report of Coleus blumei viroid 2 from commercial coleus in China. Plant Dis 95:494
19. Hou WY, Sano T, Li F, Wu Z-J, Li L, Li SF (2008) Identification and characterization of a new coleviroid (CbVd-5). Arch Virol 154:315–320. https://doi.org/10.1007/s00705-008-0276-6
20. Verhoeven JTJ, Meekes ET, Roenhorst JW, Flores R, Serra P (2013) Dahlia latent viroid: a recombinant new species of the family Pospiviroidae posing intriguing questions about its origin and classification. J Gen Virol 94:711–719. https://doi.org/10.1099/vir.0.048751-0
21. Verhoeven JTHJ, Jansen CCC, Willemen TM, Kox LFF, Owens RA, Roenhorst JW (2004) Natural infections of tomato by citrus exocortis viroid, Columnea latent viroid, potato spindle tuber viroid and tomato chlorotic dwarf viroid. Eur J Plant Pathol 110:823–831
22. Li S, Hataya T, Furuta K, Horita H, Sano T, Shikata E (1997) Occurrence of chrysanthemum stunt disease in Hokkaido and detection of chrysanthemum stunt viroid by electrophoresis and hybridization. Ann Rep Plant Prot North Japan 48:113–117. (in Japanese)
23. Spieker RL (1996) A viroid from *Brunfelsia undulata* closely related to the Columnea latent viroid. Arch Virol 141:1823–1832
24. Verhoeven JTJ, Jansen CCC, Botermans M, Roenhorst JW (2010) First report of Iresine viroid 1 in *Celosia plumose* in the Netherlands. Plant Dis 94:920
25. Verhoeven JTJ, Jansen CCC, Roenhorst JW, Flores R, de la Pena M (2009) Pepper chat fruit viroid: biological and molecular properties of a proposed new species of the genus Pospiviroid. Virus Res 144:209–214
26. Hailstones D, Tesoriero L, Terras M, Dephoff C (2013) Detection and eradication of potato spindle tuber viroid in tomatoes in commercial production in New South Wales, Australia. Australas Plant Path 32:317–318
27. Verhoeven JTJ, Jansen CCC, Roenhorst JW, Steyer S, Schwind N, Wassenegger M (2008) First report of *Solanum jasminoides* infected by citrus exocortis viroid in Germany and the Netherlands and tomato apical stunt viroid in Belgium and Germany. Plant Dis 92:973
28. Hammond RW, Zhang S (2016) Development of a rapid diagnostic assay for the detection of tomato chlorotic dwarf viroid based on isothermal reverse-transcription-recombinase polymerase amplification. J Virol Methods 236:62–67
29. Li R, Padmanabhan C, Ling KS (2017) A single base pair in the right terminal domain of tomato planta macho viroid is a virulence determinant factor on tomato. Virology 500:238–246
30. Semancik JS, Szychowski JA (1994) Avocado sunblotch disease: a persistent viroid infection in which variants are associated with differential symptoms. J Gen Virol 75:1543–1549
31. Fadda Z, Daròs JA, Fagoaga C, Flores R, Duran-Vila N (2003) Eggplant latent viroid, the candidate type species for a new genus within the family Avsunviroidae (hammerhead viroids). J Virol 77:6528–6532
32. Jeon SM, Naing AH, Kim HH, Chung MY, Lim KB, Kim CK (2016) Elimination of chrysanthemum stunt viroid and chrysanthemum chlorotic mottle viroid from infected chrysanthemum by cryopreservation. Protoplasma 253:1135–1144
33. Loreti S, Faggioli F, Cardoni M, Mordenti G, Babini AR, Poggi Pollini C, Barba M (1999) Comparison of different diagnostic methods for detection of peach latent mosaic viroid. Bull OEPP 29:433–438

Chapter 14

Detection of Viroids Using RT-qPCR

Subhas Hajeri, Georgios Vidalakis, and Raymond K. Yokomi

Abstract

Viroids are the smallest known infectious pathogens. They are nonprotein-encoding, single-stranded, circular, naked RNA molecules that can cause several diseases in economically important crops. With the advent of thermal cyclers incorporating fluorescent detection, reverse transcription coupled to the quantitative polymerase chain reaction (RT-qPCR) has transformed the way the viroids are detected. The method involves using sequence-specific primers that anneal to the viroid RNA of interest. The viroid RNA serves as a template during reverse transcription, in which the enzyme reverse transcriptase generates a cDNA copy of a portion of the target RNA molecule. After first-strand cDNA synthesis, RNA template from cDNA:RNA hybrid molecule is removed by digestion with RNase H to improve the sensitivity of PCR step. This cDNA is then be used as a template for amplification of viroid sequence in PCR.

Key words Viroid, RT-qPCR, CEVd, HSVd, One-step RT-qPCR

1 Introduction

Viroids are uncapsidated, covalently closed circular, non-encoding, single-stranded infectious RNA pathogens that can cause several diseases in economically important crops [1, 2]. Viroids with very simple structures consist of RNA genomes of only a few hundred nucleotides in length (250–440 nt), which is significantly smaller than the genome of the smallest virus. Since viroids lack coat protein encapsidation, serological assays cannot be employed, and assays have to be developed based on RNA detection methods such as polyacrylamide gel electrophoresis/silver staining, imprint/dot-blot hybridization, Northern blotting, reverse transcription-polymerase chain reaction (RT-PCR), and next-generation sequencing. The quantitative PCR (qPCR) approach combines the simplicity of PCR amplification with the sensitivity of fluorescence detection during amplification. Reverse transcription coupled with qPCR (RT-qPCR) has transformed the way the RNA pathogens are diagnosed. Over the decades, RT-qPCR analysis has

Ayala L. N. Rao et al. (eds.), *Viroids: Methods and Protocols*, Methods in Molecular Biology, vol. 2316, https://doi.org/10.1007/978-1-0716-1464-8_14,

become the method of choice not only for quantitative and accurate measurement of RNA pathogens but also for sensitive detection of samples with low pathogen titer [3]. It also gives results more quickly because it does not require post-PCR processing. The protocol discussed in this chapter is based on the detection of *Citrus exocortis viroid* (CEVd) [4] and *Hop stunt viroid* (HSVd) [5], but it can be extended to other viroids by designing viroid-specific primers and optimizing reaction conditions.

2 Materials

2.1 Plant Material

Harvest phloem-rich tissue such as leaf petiole and midrib, young bark tissue, or fruit peduncle from suspect plant material to isolate RNA and test by RT-qPCR. Alternatively, for CEVd, graft-inoculate suspect citrus plant material on to "Etrog" citron (*Citrus medica* L.) selection "Arizona 861-S1" propagated on rough lemon (*C. jambhiri* Lush.) rootstock seedlings [6]. Maintain greenhouse temperatures about 32–40 °C during the day and 27–30 °C during the night. Monitor for typical symptoms induced by CEVd such as leaf epinasty, vein and petiole necrosis, and stunting at around 2–6 months post-inoculation. HSVd induces necrotic spots on the stem just below insertion of the petiole and petiole wrinkling at around 6–9 months post-inoculation [7, 8]. Harvest phloem-rich tissues for RNA isolation and test by RT-qPCR.

2.2 RNA Isolation

2.2.1 TRIzol Method (Invitrogen Kit)

1. TRIzol™ Reagent (Invitrogen Cat#15596026).
2. Isopropanol.
3. Ethanol, 75%.
4. RNase-free water (with 0.1 mM EDTA) or TE buffer (10 mM Tris, 1 mM EDTA, pH 8).
5. DNase I.
6. Centrifuge and rotor capable of reaching 12,000 × *g* and 4 °C.
7. Water bath or heat block at 55–60 °C.
8. 2 mL microcentrifuge tubes.
9. Mortar and pestle.
10. Liquid nitrogen or dry ice.

2.2.2 Commercial Kit (Spectrum Plant RNA Isolation Kit by Sigma-Aldrich)

1. Spectrum Plant Total RNA Kit (Cat # STRN50).
2. 100% ethanol, molecular biology grade.
3. DNase I.
4. Mortar and pestle.
5. Liquid nitrogen or dry ice.

6. 2 mL microcentrifuge tubes.
7. Centrifuge and rotor capable of reaching 12,000 × *g* and 4 °C.
8. Water bath or heat block at 55–60 °C.

2.3 Reverse Transcriptase-Quantitative Polymerase Chain Reaction (RT-qPCR)

1. SuperScript® III First-Strand Synthesis System (Invitrogen #18080051), which contains:
   (a) 10× RT buffer.
   (b) 0.1 M DTT.
   (c) 10 mM dNTP mix (deoxyribonucleoside triphosphates).
   (d) 25 mM $MgCl_2$.
   (e) SuperScript™ III RT (200 U/μL).
   (f) RNaseOUT™ (40 U/μL).
   (g) RNase H ((2 U/μL).
   (h) DEPC-treated water.
2. PerfeCTa® SYBR® Green SuperMix (QuantaBio. Cat# 95053 to 95056, depending on qPCR instrument).
3. CEVd- or HSVd-specific primers (Table 1).
4. Heating blocks or water bath, capable of 55–60 °C.
5. Ice.
6. qPCR machine (e.g., Bio-Rad CFX96).
7. Chloroform, molecular biology grade.
8. Microcentrifuge capable of 12,000 × *g* at 4 °C.
9. Sterile filter (barrier) pipette tips.
10. Dedicated, annually calibrated micropipettes.
11. Spectrophotometer.

Table 1
Primers for reverse transcription-quantitative polymerase chain reaction (RT-qPCR) detection of *Citrus exocortis viroid* and *Hop stunt viroid*

Viroid	Primers	Sequence 5′ to 3′	Reference
Citrus exocortis viroid	Forward	GAAGCTTCAACCCCAAACC	[4]
	Reverse	CTTTTTTCTTTTCCTGCCTGCA	
Hop stunt viroid	Forward	CCCTCTCTCCACGCCCCCG	[5]
	Reverse	CCCTCTCTCCTACGCCTCTCG	

12. Vortex.
13. EDTA, molecular biology grade.

3 Methods

3.1 Infectivity Assay

1. For infectivity or biological indexing assay, use at least three citrus plants (citron bud-grafted on rough lemon rootstock) per treatment, and treatments include unknown, known healthy (negative), and known positive controls.
2. Graft inoculate the unknown, negative, and positive controls by chip- or bud- or side-graft method.
3. Maintain all plants at elevated greenhouse temperatures (32–40 °C) optimum for citrus viroid replication and symptom expression.

3.2 RNA Isolation (See Notes 1–4)

Two to four weeks after inoculation, collect the symptomatic leaf and/or bark tissue for viroid RNA isolation. Four weeks after inoculation, symptoms in citron test plants should begin to appear, but some mild strains may take up to 10 weeks to induce symptoms.

3.2.1 TRIzol Method (Invitrogen Kit Manual)

1. Grind 100 mg of leaf petiole or bark tissue in pestle and mortar with liquid nitrogen.
2. Homogenize the ground tissue powder in 1 mL of TRIzol™ Reagent, and transfer the homogenate to microcentrifuge tubes.
3. Incubate for 5 min to permit complete dissociation of the nucleoprotein complex.
4. Add 0.2 mL of chloroform, vortex briefly, and incubate for 2–3 min.
5. Centrifuge the sample for 15 min at 12,000 × *g* at 4 °C.
6. The mixture separates into a lower red phenol-chloroform, and interphase, and a colorless upper aqueous phase.
7. Transfer the upper aqueous phase containing the RNA to a new tube by angling the tube at 45° and pipetting the solution out while avoid transferring any of the interphase or organic layer into the pipette.
8. Add 0.5 mL of isopropanol to the aqueous phase. Vortex briefly and incubate for 10 min.
9. Centrifuge for 10 min at 12,000 × *g* at 4 °C.
10. Total RNA precipitate forms a white gel-like pellet at the bottom of the tube. Discard the supernatant with a micropipette.

Table 2
DNase I digestion to remove genomic DNA from RNA sample

Component	Concentration
Total nucleic acid	10μg
10× incubation buffer	5μL
DNase I enzyme, RNase-free (1 U)	10μL
Water, RNase-free	25μL
Final volume	**50μL**

11. Resuspend the pellet in 1 mL of 75% ethanol.
12. Vortex the sample briefly, and then centrifuge for 5 min at 7500 × *g* at 4 °C.
13. Discard the supernatant with a micropipette.
14. Air-dry the RNA pellet for 5–10 min.
15. Resuspend the pellet in 50μL of RNase-free water (with 0.1 mM EDTA) or TE buffer (10 mM Tris, 1 mM EDTA) solution by pipetting up and down.
16. Incubate in a water bath or heat block set at 55–60 °C for 10–15 min.
17. Determine the RNA yield by spectrophotometer absorbance reading at $A_{260}/A_{280\text{nm}}$.
18. Proceed to downstream applications, proceed to DNase digestion, or store the RNA at −70 °C.
19. Perform digestion of DNA by preparing mixture as shown in Table 2.
20. Incubate the DNA digestion mixture at 25–37 °C for 15–20 min.
21. Stop the reaction by adding 2μL of 0.2 M EDTA, pH 8.0, to a final concentration of 8 mM and heating to 75 °C for 10 min. Alternatively, DNase I can be inactivated and removed by phenol extraction.

3.2.2 Commercial Kit (Spectrum Plant RNA Isolation Kit by Sigma-Aldrich)

Note: Materials in this section are included in the kit.

1. Grind 100 mg of petiole or bark tissue in pestle and mortar with liquid nitrogen.
2. Homogenize the ground tissue powder in 500μL of lysis solution/2-mercaptoethanol mixture, and transfer the homogenate to microcentrifuge tubes.
3. Vortex immediately and vigorously for at least 30 s.
4. Incubate the sample at 56 °C for 3–5 min.

5. Centrifuge the sample at maximum speed for 3 min to pellet cellular debris.
6. Transfer the lysate supernatant into a filtration column (blue retainer ring) seated in a 2 mL collection tube.
7. Centrifuge at maximum speed for 1 min to remove residual debris.
8. To the clarified flow-through lysate, add 500μL of binding solution, and vortex briefly.
9. Pipette 700μL of the mixture into a binding column (red retainer ring) seated in a 2 mL collection tube.
10. Centrifuge at maximum speed for 1 min to bind RNA.
11. Decant the flow-through liquid, and tap the collection tube (upside down) briefly on a clean absorbent paper to drain the residual liquid.
12. Return the column to the collection tube and pipette the remaining mixture to the column and repeat the centrifugation and decanting steps.
13. Pipette 500μL of wash solution 1 into the column, and centrifuge at maximum speed for 1 min.
14. Decant the flow-through liquid, and tap the collection tube (upside down) briefly on a clean absorbent paper to drain the residual liquid. Return the column to the collection tube.
15. Prepare DNase digestion solution by adding 10μL of DNase I enzyme with 70μL of DNase digestion buffer, and mix gently by pipetting. Do not vortex the DNase I vial or the mixture of DNase I and DNase I digestion buffer because DNase I is sensitive to physical denaturation.
16. Pipette 80μL of the mixture directly onto the center of the filter inside the binding column.
17. Close the cap and incubate the sample at room temperature for 15 min.
18. Pipette 500μL of wash solution 1 into the binding column. Close the cap, and centrifuge at maximum speed for 1 min to remove the digested DNA. Decant the flow-through liquid, and tap the collection tube briefly on a clean absorbent paper to drain the residual liquid.
19. Pipette 500μL of wash solution 2 into the column, and centrifuge at maximum speed for 30 s.
20. Decant the flow-through liquid, and tap the collection tube (upside down) briefly on a clean absorbent paper to drain the residual liquid. Return the column to the collection tube.
21. Repeat the third wash with wash solution 2 as described in **steps 19** and **20**.

22. Centrifuge the column at maximum speed for 1 min to dry.
23. Transfer the column to a new, clean 2 mL collection tube.
24. Pipette 50μL of elution solution directly onto the center of the binding matrix inside the column.
25. Close the cap and let the tube sit for 1 min.
26. Centrifuge at maximum speed for 1 min to elute.
27. Purified RNA is now in the flow-through elute and ready for immediate use or storage at −20 °C (short term) or −70 °C (long term).

3.3 Reverse Transcriptase-Quantitative Polymerase Chain Reaction (RT-qPCR)

Reverse transcription (first-strand cDNA synthesis) followed by SYBR Green I-based quantitative PCR assay [9].

3.3.1 First-Strand cDNA Synthesis

1. Mix components of the first-strand cDNA synthesis as shown in Table 3.
2. Incubate the mixture at 65 °C for 5 min for denaturing RNA template, and immediately place it on ice for at least 1 min.
3. To this RNA/reverse primer mixture, add reverse transcription reaction components as shown in Table 4.
4. Vortex the mixture for 10 s and spin down for 30 s.
5. Incubate the final mixture (volume of 20μL) at 50 °C for 50 min.
6. Stop the reaction by incubating at 80 °C for 10 min.
7. Use this first-strand cDNA in qPCR immediately or store at −20 °C for later use.

3.3.2 SYBR Green-Based qPCR

Different real-time PCR systems employ different strategies for the normalization of fluorescent signals and correction of well-to-well optical variations. It is critical to match the appropriate qPCR reagent to your specific instrument. Choose the appropriate PerfeCTa® SYBR® Green SuperMix based on the available qPCR machine because some instruments need ROX as a reference and others do not. Among those who need ROX, some need low ROX and other regular strength ROX. PerfeCTa SYBR Green SuperMix (2×) contains optimized concentrations of $MgCl_2$, dNTPs (dATP, dCTP, dGTP, dTTP), AccuStart Taq DNA polymerase, SYBR Green I, ROX, and stabilizers. AccuStart Taq DNA polymerase contains monoclonal antibodies that bind to the polymerase and keep it inactive prior to the initial PCR denaturation step. Upon heat activation (2 min at 95 °C), the antibodies denature irreversibly, releasing fully active, unmodified Taq DNA polymerase. This

Table 3
First-strand cDNA synthesis reaction components

Component	Volume (μL)
RNA	2.0
Reverse primer (10μM)	0.5
dNTPs (10 mM)	1.0
DEPC-treated water/nuclease-free water	6.5
Final volume	**10μL**

Table 4
Reverse transcription reaction components

Component	Volume (μL)
RNA/reverse primer mixture	10.0
RT buffer (10×)	2.0
$MgCl_2$ (25 mM)	4.0
DTT (0.1 M)	2.0
RNaseOUT	1.0
SuperScript™ III reverse transcriptase	1.0
Final volume	**20μL**

Table 5
qPCR reaction components

Component	Volume (μL)
PerfeCTa SYBR green SuperMix (2×)	12.5
Forward primer (10μM)	0.5
Reverse primer (10μM)	0.5
cDNA template	2.0
DEPC-treated water/nuclease-free water	9.5
Final volume	**25μL**

enables specific and efficient primer extension with the convenience of room temperature reaction assembly.

1. In a PCR tube, assemble the qPCR components as shown in Table 5.
2. Incubate the qPCR reaction mixture in a real-time thermal machine as shown in Table 6.

Table 6
qPCR reaction protocol

Initial denaturation	95 °C for 3 min
PCR cycling	40 cycles of 95 °C, 15 s and 60 °C, 60 s
Melt curve	60 °C to 95 °C with measurement of fluorescence taken at every 0.5 °C

3.3.3 Analysis of RT-qPCR Data

Analyze the real-time PCR results with CFX96 data analysis software (Bio-Rad). A data file is automatically generated after a run with the gene expression module. The specificity of a real-time PCR assay is determined by the melting curve analysis.

4 Notes

1. RNases are ubiquitous and very stable enzymes and generally do not need cofactors for enzymatic activity. While endogenous RNases are denatured during the lysis step and removed subsequently during RNA purification, care must be taken to avoid introducing exogenous RNases during RNA preparation steps.
2. Use RNaseZap™ RNase Decontamination Solution (Cat. no. AM9780) or similar products to remove RNase contamination from work surfaces and non-disposable items such as centrifuges and pipettes used during purification.
3. Use disposable, individually wrapped, sterile plastic ware and sterile, disposable RNase-free pipettes, pipette tips, and tubes. Use RNase-free pipette tips, preferably those with an aerosol barrier.
4. Wear disposable gloves while handling reagents and RNA samples to prevent RNase contamination from the surface of the skin; change gloves frequently, particularly as the protocol progresses from crude extracts to more purified materials.

Acknowledgments

This work was supported in part by USDA Agricultural Research Service, In-House Appropriated Project Number 2034-22000-013-00-D, California Department of Food and Agriculture Specialty Crops Block Grant SCB10005, California Citrus Nursery Board Yok-13, and the California Citrus Research Board. Mention of trade names or commercial products in this publication is solely for the purpose of providing specific information and does not imply recommendation or endorsement by the US Department of Agriculture.

References

1. Semancik JS, Weathers LG (1972) Exocortis virus: an infectious free-nucleic acid plant virus with unusual properties. Virology 46:456–466
2. Semancik JS, Morris TJ, Weathers LG, Rordorf GF, Kearns DR (1975) Physical properties of a minimal infectious RNA (viroid) associated with the exocortis disease. Virology 63(1):160–167
3. Mackay IM, Arden KE, Nitsche A (2002) Real-time PCR in virology. Nucleic Acids Res 30 (6):1292–1305
4. Hajeri S, Ramadugu C, Manjunath K, Ng J, Lee R, Vidalakis G (2011) In vivo generated citrus exocortis viroid progeny variants display a range of phenotypes with altered levels of replication, systemic accumulation and pathogenicity. Virology 417:400–409
5. Loconsole G, Önelge N, Yokomi RK, Kubaa RA, Savino V, Saponari M (2013) Rapid differentiation of citrus hop stunt viroid variants by real-time RT-PCR and high resolution melting analysis. Mol Cell Probes 27(5–6):221–229
6. Roistacher CN, Calavan EC, Blue RL, Navarro L, Gonzales R (1977) A new more sensitive citron indicator for detection of mild isolates of citrus exocortis viroid (CEV). Plant Dis Rep 61:135–139
7. Semancik JS, Roistacher CN, Rivera-Bustamante R, Duran-Vila N (1988) Citrus cachexia viroid, a new viroid of citrus: relationship to viroids of the exocortis disease complex. J Gen Virol 69:3059–3068
8. Roistacher CN (1991) Techniques for biological detection of specific citrus graft-transmissible diseases. In: Roistacher CN (ed) In: graft-transmissible diseases of citrus: handbook for detection and diagnosis. International Organization of Citrus Virologists and Food and Agricultural Organization of the United Nations, Rome, pp 13–151
9. Hajeri (2010) 'Study of molecular and biological properties of citrus exocortis viroid and Dweet mottle virus' Doctoral dissertation, University of California, Riverside

Chapter 15

Multiplex RT-PCR

Francesco Faggioli and Marta Luigi

Abstract

Amplification of different nucleic acid targets in the same reaction (multiplex polymerase chain reaction) is challenging but an extremely useful tool especially for viroid diagnosis. In the amplification mixtures, several pairs of primers work together in the same conditions to detect different targets. Here, we describe the development and use of a multiplex reverse transcription polymerase chain reaction protocol highlighting the most crucial factors that can significantly affect the quality of the method. First, particular attention must be paid to primer design. Then, the amplification mixture and temperature conditions must be calibrated precisely to avoid cross reactivity or loss in sensitivity. Finally, the detection system of the amplification results must allow a specific identification of the amplified target(s).

Key words Two-step mRT-PCR, One-step mRT-PCR, Viroids, Primer design, Amplicon size, Agarose gel

1 Introduction

Simultaneous amplification of different nucleic acid targets has the potential to produce considerable savings in time and effort within the laboratory without compromising test validity. Since its introduction, multiplex polymerase chain reaction (mPCR) or multiplex reverse transcription-PCR (mRT-PCR) has been successfully applied in many areas of nucleic acid diagnostics; this technique has been shown to be a valuable method for the detection and identification of viruses, bacteria, fungi, and/or parasites in plants [1]. mRT-PCR can reduce costs, time, and labor particularly for pathogen detection where large-scale diagnostic tools such as enzyme-linked immunosorbent assay (ELISA) are not available, as is the case for viroids [2] due to their lack of protein coats.

Viroids are composed of small naked circular single-strand RNA molecules without a protein capsid. They do not code for any protein and can infect plant species of high economical value [3]. Different species of viroids can affect the same plant species as single or mixed infections [4]; moreover, multiple viroids infection

Ayala L. N. Rao et al. (eds.), *Viroids: Methods and Protocols*, Methods in Molecular Biology, vol. 2316,
https://doi.org/10.1007/978-1-0716-1464-8_15,

can attenuate or enhance symptoms of each viroid comparing to a single infection [5]. The availability of reliable and economic detection methods that can detect multiple viroids at once would be most valuable in certification and quarantine programs. Several mRT-PCR protocols have been developed for the simultaneous detection of viroids infecting chrysanthemum, tomato, citrus, fruit trees, grapevine, and hop [2, 5–13]. Many parameters must be considered in setting up a mRT-PCR protocol; special attention must be paid to primer design, which is critical for the success of mRT-PCR; and optimized reaction conditions are also more critical in mRT-PCR than in single RT-PCR [14, 15].

2 Materials

Prepare all solutions using nuclease-free water, which can be bought or prepared using double distilled water, incubated overnight under agitation with 0.1% diethyl pyrocarbonate (DEPC) as an RNase inhibitor, and autoclaved to remove DEPC. Store all reagents at −20 °C. Primer preparation is as follows:

1. Primers for mRT-PCR are bought as desalted and lyophilized. Centrifuge primer tubes prior to opening them, and then resuspended them to make a stock solution of 100μM (or 100 pmol/μL) with DEPC-water using the nanomoles reported on the certificate of analysis released by the producer.
2. Dilute the primer stock solution to a working solution of 10μM using DEPC-water. Aliquot the working solution in several tubes (50–100μL per tube) to limit freeze/thaw events.
3. Calculate a primer mixture combining all the primers involved in mRT-PCR, and prepare it for 50 or 100 mRT-PCR reactions to avoid mistakes in pipetting small volumes of many different solutions.

2.1 Reverse Transcription Reaction

1. Reverse transcriptase buffer: 20 mM Tris–HCl, pH 7.5, 100 mM NaCl, 0.1 mM EDTA, 1 mM DTT, 0.01% (v/v) NP-40, and 50% (v/v) glycerol.
2. 1 mM deoxyribonucleoside triphosphates (dNTPs): dATP, dGTP, dCTP, and dTTP, 250μM each.
3. 0.1 M dithiothreitol (DTT).
4. Random primer: random hexamers 5′-NNNNNN-3′ or random nonamers 5′-NNNNNNNNN-3′.
5. RNase inhibitor.
6. Reverse transcriptase: avian myeloblastosis virus (AMV), reverse transcriptase or Moloney murine leukemia virus (MMLV), reverse transcriptase or recombinant MMLV.

2.2 Polymerase Chain Reaction

1. Taq buffer: 500 mM KCl, 100 mM Tris–HCl, pH 9.0, at 25 °C, 15 mM $MgCl_2$, and 1% Triton X-100.
2. 12.5 mM $MgCl_2$ or 5 mM $MgSO_4$.
3. 1 mM dNTPs: dATP, dGTP, dCTP, and dTTP, 250μM each.
4. Primer mixture: premixed work solution of all the primers used in the mRT-PCR.
5. Taq polymerase: standard thermostable DNA polymerases or hot-start (HS) polymerases or high-fidelity polymerases (Hi-Fi).

2.3 One-Step-One-Tube Kit

1. One-step RT-PCR buffer: Tris–HCl, KCl, $(NH_4)_2SO_4$, 12.5 mM $MgCl_2$, DTT; pH 8.7 with or without 0.4 mM dNTPs.
2. 12.5 mM $MgCl_2$ or 5 mM $MgSO_4$.
3. 1 mM dNTPs: dATP, dGTP, dCTP, and dTTP, 250μM each.
4. Primer mixture: premixed work solution of all the primers used in the mRT-PCR.
5. Enzymes mixture: premixed reverse transcriptase and Taq polymerase (preferably hot-start).

2.4 Detection

1. TAE 50×, pH 8.5: 2 M Tris base; 0.05 M EDTA; 1 M acetic acid or TBE 10×, pH 8.3: 1 M Tris base; 1 M boric acid; 0.02 M EDTA.
2. Agarose.
3. Fluorescent stains (i.e., ethidium bromide, SYBER safe, GelRed according to the manufacturer's instructions).
4. 6× loading buffer: 10 mM Tris–HCl, pH 7.6, 0.03% bromophenol blue, 0.03% xylene cyanol FF, 60% glycerol 60 mM EDTA.
5. Molecular weight ladder diluted, adding 6× loading dye and water, or undiluted according to the manufacturer's instruction.

2.5 Laboratory Equipment

1. Centrifuge.
2. Ice maker.
3. UV sterilization chamber.
4. Thermocycler.
5. Equipment for horizontal gel electrophoresis.
6. Gel imaging system.
7. Volume variable pipettes ranging from 0.2 to 1000μL.
8. Tubes.
9. Sterile filter tips.

3 Methods

Amplification success depends on the capacity of the primers to anneal to the target(s) and on the ability of the mixture to support the extension during all the amplification steps.

The simultaneous presence of many primers working together poses several difficulties in the optimization of the protocol, including lack in sensitivity, specificity, and/or different primer efficiency [1]. Primers must first be tested in silico using specific bioinformatics software to verify the general parameters are satisfactory. Primers must then be validated in vitro in single RT-PCR prior to being used in the multiplex reactions. Generally, adding a primer pair that is able to amplify an endogenous, stably expressed plant mRNA (i.e., ubiquitin, 18S) is useful to verify the efficiency of the reaction and confirm the validity of negative results for pathogen detection purposes (Fig. 1).

Some biochemical properties of viroids can further complicate the development of the mRT-PCR protocol. Viroids have a very small genome size ranging from 246 to 401 nt [16], and often the obtained amplicons are not size compatible with the requirements of the mRT-PCR. Moreover, the two viroid families (*Pospiviroidae* and *Avsunviroidae*) differ in their mutation rates; most *Pospiviroidae* have a central conserved region and a low mutation rate, while *Avsunviroidae* results in a complex spectrum of quasi-species isolates with a higher mutation rate [16]. These characteristics make choosing the regions in which primers must be designed particularly challenging.

3.1 Primer Design: Parameter for Multiplex RT-PCR

In general, the primer parameters needed for mRT-PCR are a stricter version of the normal primer parameters, that is, when many primers are required, short length, between 18 and 22, nt is preferred. Primers must have a similar melting temperature (T_m), with a difference of more than 5 °C not being recommended;

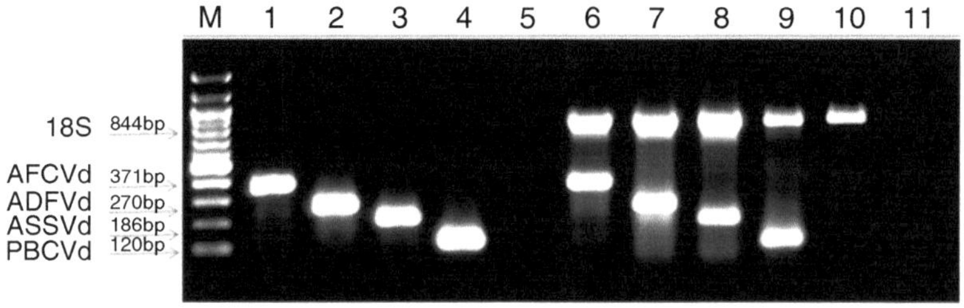

Fig. 1 Amplification of the target RNA (apple fruit crinkle viroid (AFCVd); apple dimple fruit viroid (ADFVd); apple scar skin viroid (ASSVd); pear blister canker viroid (PBCVd)) by single RT-PCR (lanes 1–5) and co-amplification of the endogenous control 18S (lanes 6–11). Lane M marker; (lanes 5 and 10) healthy pear; (lane 11) water [15]

moreover, it is very important to verify the specificity of all the primers to avoid competition when more targets are in the same sample. Checking dimers formation for all the primers present in a reaction mixture is another crucial control point of the primer design. Dimerization of the primers can produce either unspecific amplifications that consume reaction components or false-negative results.

For primer design follow the following steps:

1. Retrieve all the genome sequences of the targets from databases.
2. Multiple-align all sequences belonging to the same viroid species highlighting the more conserved regions.
3. Use bioinformatics software or manually design primers for the selected genome segments.
4. Evaluate if the length of the amplicons can be easily resolved during electrophoresis.
5. Test the primers in silico for:
   (a) Specificity: compare the nucleotide sequence of the primers and all sequences of the species present in database.
   (b) Selectivity: compare the nucleotide sequence of the primers and all the non-target sequences (from related viroids, hosts, other pathogens, etc.).
   (c) Hairpin formation.
   (d) Self-dimerization.
   (e) Hetero-dimerization.
6. Test and validate in vitro the efficiency of each couple of primers using a single RT-PCR protocol with serial dilutions of the target, and compare the efficiency of each primer pair (Fig. 2).

3.2 Sample Preparation (Tissue Selection, RNA Extraction, DNase)

Sample preparation and quality of the extracted total RNA can significantly affect any amplification method, especially in multiplex analysis. Viroids can affect many different plant species including monocots and dicots, vegetable crops, ornamentals, and woody perennials [3, 16, 17]. Furthermore, viroid distribution is not always even, and their titers can vary among species and environmental condition [18, 19]. To optimize the nucleic acids extraction step, the following must be considered: the season of sampling, the type of tissue, and the method of extraction. Once the proper time of sampling and the most suitable tissue are selected, it is important to choose the most reliable extraction method considering the possibility of secondary metabolites interfering in the amplification reactions (Fig. 3). In addition, RNA extraction for each target pathogen must be validated before the setup of the protocol.

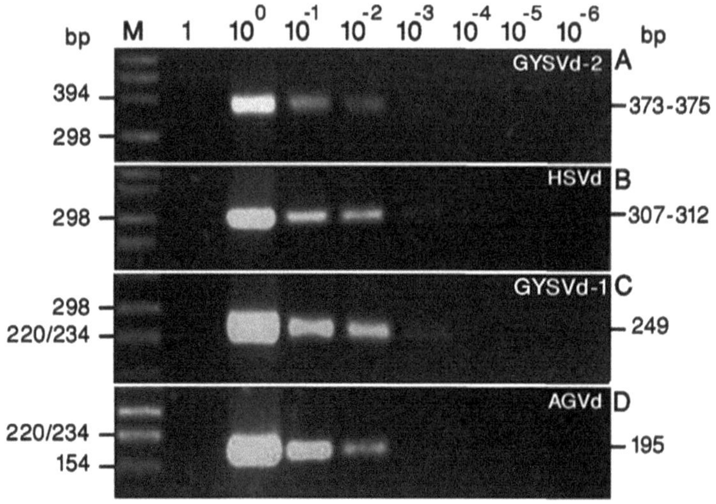

Fig. 2 Agarose gel electrophoresis analysis of DNA fragments amplified by preliminary single RT-PCR (psRT-PCR). Total nucleic acid (TNA) preparations, from a positive control naturally infected by Australian grapevine viroid (AGVd), grapevine yellow speckle viroid-1 (GYSVd-1), grapevine yellow speckle viroid-2 (GYSVd-2), and hop stunt viroid (HSVd), were diluted serially in TNA extracted from healthy plants. Panels A, B, C, and D psRT-PCR performed using GYSVd-2-, HSVd-, GYSVd-1-, and AGVd-specific primer pairs, respectively; Lane M, DNA marker with fragment sizes reported on the left; lane 1, healthy control; lanes from 0 to 10—6 indicate tenfold serial dilutions of the positive control. Expected viroid cDNAs are indicated on the right [13]

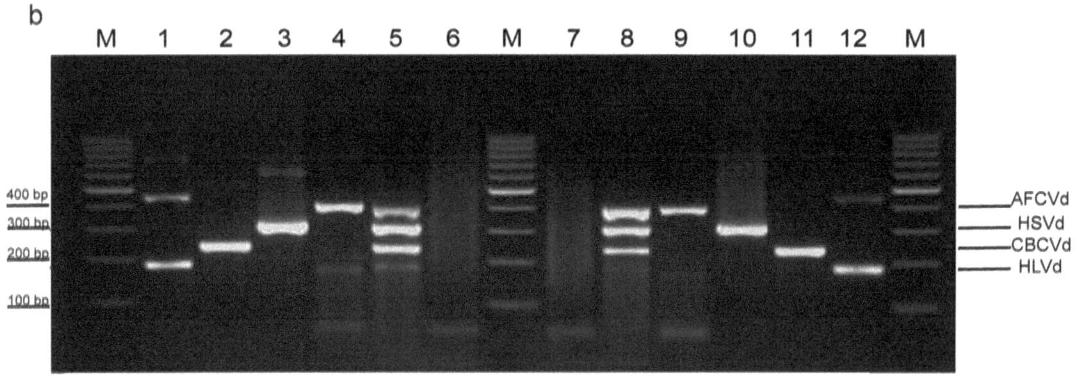

Fig. 3 Comparison of single and multiple viroid samples (AFCVd; HSVd; citrus bark cracking viroid (CBCVd); hop latent viroid (HLVd)) extracted with a commercially available kit (lanes 1–6) and common CTAB method (lanes 7–12). Lanes from 1 to 4 and 9 to 12 are single viroids. Lanes 5 and 8 are multiplex reactions, and lanes 6 and 7 are healthy controls. Lane M, 100 bp nucleotide size marker [5]

3.3 Mixture Setup and Development of the Method

Once the design of the primers is finalized, many other factors can prevent optimal amplification rates, namely, suboptimal buffer constituents, enzymatic activity, and optimal availability of components. Often the use of a pre-constituted master mix RT-PCR can overcome many problems, though expensive; on the other hand, if a homemade mixture is chosen, it is necessary to consider different factors listed below.

3.3.1 Mixture Setup

It is essential to choose between a one-step or a two-step assay. One-step assay combines reverse transcription and amplification in a single reaction, using reverse transcriptase together with DNA polymerase. In a two-step assay, the RNA reverse transcription and cDNA amplification reactions are performed in separate tubes, with different optimized buffers, reaction conditions, and priming strategies. The one-step procedure is the fastest and highly reproducible and reduces the risk of contamination, though it can be more difficult to optimize and can lose sensitivity. When the two-step strategy is chosen, cDNA is produced using random primers able to anneal at multiple points along the RNA; the most used are six to nine bases long (dN6 or dN9). It is then necessary to identify the optimal reverse transcriptase enzyme. The most commonly used reverse transcriptase enzymes are derived from the *pol* gene of AMV or MMLV. The AMV is a 170 kDa heterodimer with an optimal reaction temperature range of 42–48 °C. The AMV reverse transcriptase possesses strong RNase H activity. The MMLV reverse transcriptase is a 75 kDa enzyme with an optimal reaction temperature around 37 °C. MMLV reverse transcriptase is capable of synthesizing longer cDNA (<7 kb) at a higher efficiency, due to its lower RNase H activity. To further improve cDNA synthesis, MMLV reverse transcriptase has been engineered for even lower RNase H activity (i.e., mutated RNase H domain, or RNase H^-), higher thermostability (up to 55 °C), and enhanced processivity (65 times higher).

For the amplification step, a wide range of polymerase enzymes is available; most are robust enough to be used in many different PCR assays. The original Taq polymerase was also modified to become more suitable to different downstream applications: hot-start and high-fidelity. The hot-start DNA polymerase is chemically modified so that it is completely inactive until the covalent bonds are broken during the initial heat activation step. High-fidelity PCR enzymes generally provide a 3′ to 5′ exonuclease activity for removing incorrectly incorporated bases.

An equimolar solution of dNTPs (dATP, dCTP, dGTP, and dTTP) is generally used in the PCR steps. Normal PCR requires a dNTPs final concentration of 0.2 mM, but some adjustment in multiplex development can occur especially in the presence of high levels of magnesium ion (Mg^{2+}). Attention must be paid in not exceeding optimal concentrations of dNPTs as they can inhibit PCR.

Many other components can be added to the mixture to improve the amplification conditions. The supplied reaction buffer is generally composed to maintain favorable pH and ionic strength for the reaction, but in mRT-PCR, where many amplifications work together, some minor adjustments can be helpful. The most common components added are (a) DTT, a potent reducing agent widely used in molecular biology as an enzyme stabilizing agent to keep proteins in a reduced state; (b) an RNase inhibitor is usually included in the reaction buffer or added to the reverse transcription reaction to prevent RNA degradation; and (c) Mg^{2+} a cofactor of polymerase. It is commonly added as $MgCl_2$ or $MgSO_4$ to help the incorporation of dNTPs and facilitate formation of the complex between the primers and DNA templates by stabilizing negative charges on their phosphate backbones [20]. Low Mg^{2+} concentrations result in little or no PCR product, due to reduced activity of the polymerase; however, high Mg^{2+} concentrations can produce nonspecific PCR products from enhanced stability of primer-template complexes.

3.3.2 Cycling Condition

Setting up cycling condition depends on the type of enzyme, primers, buffer (especially saline conditions), and the complexity of the target. The optimal temperature range for reverse transcriptase is 37–50 °C depending on the thermostability of the enzyme. The time of incubation depends on the transcriptase processivity; wild-type MMLV requires almost 60 min (min) to synthetize cDNA, while AMV or engineered MMLV shows a higher processivity.

For amplification, an initial denaturation step is generally performed at 94–98 °C for 1–3 min. Subsequent PCR cycles begin with a separate denaturation step that lasts 0.5–2 min at 94–98 °C; if there are some additives (i.e., glycerol, DMSO) in the mixture, the incubation can be shorter and at lower temperatures. Annealing can be performed in 0.5–2 min; in a mRT-PCR protocol, the annealing temperature is very critical and must be optimized with great precision (Fig. 4). The next step is the synthesis of the daughter strand with the incorporation of dNTPs; the temperature and the time of each cycle depend on the intrinsic characteristic of the polymerase; typically, 1 min at 70–75 °C is necessary for the synthesis of 1 kb products. The number of cycles (from 25 to 40) depends on the amount of input target and the final extension on the amplicons' length.

3.4 Preparation of the Assay

1. Prepare a detailed list of the samples to be tested.
2. Prepare a suitable schematic in which all the data of the experiment are reported. For each extraction and amplification event, enter a series of controls: include a positive control, a negative control for each species tested, and a water control (*see* **Note 2**).

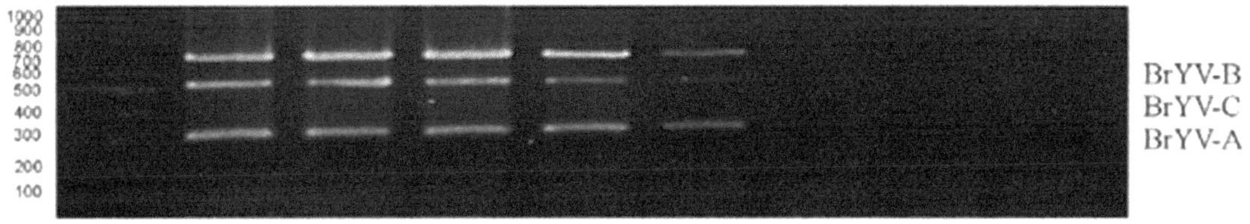

Fig. 4 Optimization of the annealing temperature for multiplex RT-PCR (Brassica yellows virus (BrYV)). First lane: 100 bp DNA marker; second to ninth lanes 2–9, 60.0 °C, 60.6 °C, 61.4 °C, 62.4 °C, 63.8 °C, 64.9 °C, 65.5 °C, 66.0 °C [21]

3. Use pipettes, sterile filter tips, and tubes and, if you have a working hood for PCR, keep them under UV light for 10 min before use.
4. Always wear gloves to avoid RNase contamination. Change gloves if you suspect they are contaminated.
5. Put all the reagents on ice until thawed; prepare and label sterile PCR tubes (*see* **Note 1**).
6. Open and close all sample tubes and reaction plates carefully. Do not splash or spray PCR samples.

3.4.1 Two Steps-Two Tubes

1. Prepare reverse transcriptase mixture as reported in Table 1.
2. Divide 18μL of the mixture in each tube, and add 2μL of total RNA target with a concentration of 10 pg to 5μg (use the same amount of water for the negative amplification control).
3. Incubate at least for 1 h at the temperature required by the reverse transcriptase enzyme (42 °C for the MMLV; 46 °C for the AMV; and >50 °C for the recombinant MMLV). Then put the cDNA on ice or freeze it (−20 °C or −80 °C) until the PCR mixture is ready.
4. Prepare the polymerase chain reaction mixture as reported in Table 2.
5. Divide 18–23μL of the mixture in each tube and add 2μL of cDNA.
6. Load the samples in the thermocycler and start the amplification program (Table 3).

3.4.2 One Step-One Tube

1. Prepare the mixture for one-step RT-PCR according to the kit instructions (Table 4).

   Or prepare everything to make a *homemade* RT-PCR mixture (Table 5).
2. Divide 18–48μL of the mixture in each tube, and add 2μL of total RNA target with a concentration of 1 pg to 2μg.
3. Load the samples in the thermocycler, and start the amplification program (Table 6).

Table 1
List of material needed for reverse transcriptase (RT) reaction

Reagent	Final concentration	Description
RT buffer	1×	Enzyme buffer (20 mM Tris–HCl (pH 7.5), 100 mM NaCl, 0.1 mM EDTA, 1 mM DTT, 0.01% (v/v) NP-40, 50% (v/v) glycerol)
dNTPs	400μM	10 mM each dATP, dGTP, dCTP and dTTP at neutral pH
DTT (if needed)	100μM	
Random primers	50–250 ng	Random hexamers dN_6 or Random nonamers dN_9
RT enzyme	According to the user's manual (0.5–1μL/20μL reaction)	AMV reverse transcriptase MMLV reverse transcriptase Recombinant MMLV
RNase inhibitor	According to the user's manual	If the reverse transcriptase has an intrinsic RNase H activity

Table 2
List of material needed for amplification (PCR) reaction

Reagent	Final concentration	Description
Taq buffer	1×	Enzyme buffer (500 mM KCl, 100 mM Tris–HCl (pH 9.0 at 25 °C), 15 mM $MgCl_2$, and 1% Triton X-100)
dNTPs	200–300μM	10 mM each dATP, dGTP, dCTP, and dTTP
Primer mixture	0.15–0.6μM each	Diluted in nuclease-free water at the work solution of 10μM, and premixed
$MgSO_4$ or $MgCl_2$ (if needed)	1.4 and 2.5 mM	
Taq polymerase	According to the user's manual (0.25–0.5μL/25μL reaction)	Standard thermostable DNA polymerase Hot-start (HS) polymerase High-fidelity polymerase (Hi-Fi)

3.5 Discrimination Methods

The most common discrimination method for the viroids mRT-PCR protocol is resolving the amplicons on agarose gel. The choice of appropriate agarose and its concentration depend mostly on the length of the amplified products, as well as the buffer for the gel running: TAE buffer is more suitable for large DNA (>1 kb); TBE buffer is a better choice for shorter fragments (<1 kb) (the case of viroid amplicons).

Table 3
PCR amplification conditions

	Temperature	Time	Additional comments
Initial activation step	95 °C	2–5 min or 15 min	If a Hot-Start Taq DNA polymerase is used, a longer activated step is needed
Denaturation	95 °C	30–60 s	25–40 cycles
Annealing	50–62 °C	30–60 s	
Extension	68–72 °C	60 s/kb	
Final extension	68–72 °C	10 min	

Table 4
List of material needed for a one-step one-tube RT-PCR reaction using a kit

Reagent	Final concentration	Description
One-step RT-PCR buffer	1×	Kit buffer (contains Tris–HCl, KCl, $(NH_4)_2SO_4$, 12.5 mM $MgCl_2$, DTT; pH 8.7 (20 °C))
dNTPs	200–400μM	10 mM each of dATP, dCTP, dGTP, and dTTP
$MgSO_4$ or $MgCl_2$ (if needed)	1.4 and 2.5 mM	
Primer mixture	0.15–0.6μM each	Diluted in nuclease-free water at the work solution of 10μM, and premixed
Enzyme mixture	According to the user's manual (1μL/25μL reaction	Premixed reverse transcriptase and Taq polymerase (generally Hot-start)

1. Prepare a stock solution for TAE or TBE buffer. Verify the pH (8.5 for TAE and 8.3 for TBE), and store at room temperature in a glass bottle.
2. Dilute the stock solution to a working solution (1×).
3. Weigh the agarose powder to obtain the desired percentage of w/v; typically, the reported protocol for mRT-PCR of viroids range from 1% [22] to 2.5% w/v [23].
4. In a beaker (with size two to four times the volume of the solution), add 1× TAE or TBE buffer, the pre-measured agarose powder, and the florescent stains. To avoid the evaporation, cover the beaker with plastic wrap, or weigh the solution before and after heating.
5. Heat the beaker in a microwave oven until bubbles appear, and then remove the beaker, and gently swirl it to resuspend any settled powder and gel pieces.

Table 5
List of material for a one-step one-tube RT-PCR reaction using a *homemade* mixture

Reagent	Final concentration	Description
Reaction buffer	1×	Taq polymerase buffer (500 mM KCl, 100 mM Tris HCl (pH 9.0 at 25 °C), 15 mM $MgCl_2$, and 1% Triton X-100)
dNTPs	200–400μM	10 mM each of dATP, dCTP, dGTP, and dTTP
DTT (if needed)	100μM	It inhibits RNase activity and preserves mRNA for the RT reaction
$MgSO_4$ or $MgCl_2$ (if needed)	1.4 and 2.5 mM	
Primer mixture	0.05–1.0μM each	Diluted in nuclease-free water at the work solution of 10μM, and premixed
RT enzyme	According to the user's manual (0.5–1μL/20μL reaction)	AMV reverse transcriptase MMLV reverse transcriptase Recombinant MMLV
Taq polymerase	According to the user's manual (0.25–0.5μL/25μL reaction)	Standard thermostable DNA polymerase Hot-start (HS) polymerase High-fidelity polymerase (Hi-Fi)
RNase inhibitor	According to the user's manual	If the reverse transcriptase has an intrinsic RNase H activity

Table 6
RT-PCR amplification conditions

	Temperature	Time	Additional comments
Reverse transcriptase	42–50 °C	15–60 min	
Initial activation step	95 °C	2–5 min or 15 min	If Hot-Start Taq DNA polymerase is used, a longer activated step is needed
Denaturation	95 °C	30–60 s	25–40 cycles
Annealing	50–62 °C	30–60 s	
Extension	68–72 °C	60 s/kb	
Final extension	68–72 °C	10 min	

6. Reheat until completely dissolved. Mix thoroughly and cool the solution to 60 °C prior to casting.
7. While the agarose solution is cooling, assemble the gel casting tray, and level it prior to pouring the agarose solution. Ensure that any residuals of dried agarose from the comb teeth were removed.

8. Pour the agarose solution into the gel tray, and let it solidify at room temperature for 30 min.
9. Once the gel is ready, place the gel casting tray into the electrophoresis chamber. Fill the chamber with running buffer and slowly remove the comb. Load the DNA amplicons with 6× loading buffer and an appropriate molecular weight ladder.
10. Run the gel at 80–150 V until the dye line is approximately 75–80% of the way down the gel (to track the running use as reference the line of the dye, bromophenol blue, and xylene cyanol FF). A typical run time is about 1–1.5 h, depending on the gel concentration, its size, and voltage.
11. Visualize the band under the appropriate lights (generally UV).

To distinguish the amplicons, two other methods can be used: a multiplex RT-PCR probe capture hybridization and a fluorescent RT-PCR [6, 7]. In the first method, also called RT-PCR ELISA, after amplification with digoxigenin tagged primers, discrimination is performed with biotinylated cDNA capture probe, then the reaction is incubated with polyclonal anti-DIG Fab fragments conjugated to peroxidase, and hybridization is revealed using a substrate solution. In fluorescent RT-PCR, primers tagged with different fluorescent molecules (i.e., fluorescein, rhodamine) are used. In this method the fluorescent signal allows to recognize the different amplicons rapidly without any staining of the gel.

3.6 Evaluation of the Performance

Before adopting the protocol, the mRT-PCR must be validated.

According to the European and Mediterranean Plant Protection Organization (EPPO PM 7/98 (3)) [24], the criteria that must be evaluated are analytical sensitivity, analytical specificity (inclusivity and exclusivity), repeatability, and reproducibility. Moreover, the method must be evaluated in routine testing of plant material. All the samples must be analyzed either in single or in multiplex RT-PCR to verify the reliability, robustness, and flexibility of the method. No discrepancy between results must be revealed. If all the values are acceptable, the protocol is now validated. For a complete overview of the decisions to be made in development of a multiplex RT-PCR protocol, check the flow chart in Fig. 5.

4 Notes

1. *Master mix preparation.* Amplification reactions should always be prepared on ice. When making amplifications of several samples at the same time, a master solution (master mix) containing all the necessary reagents except for the target nucleic acids must be prepared in a single 1.5 mL or 2 mL tube. The

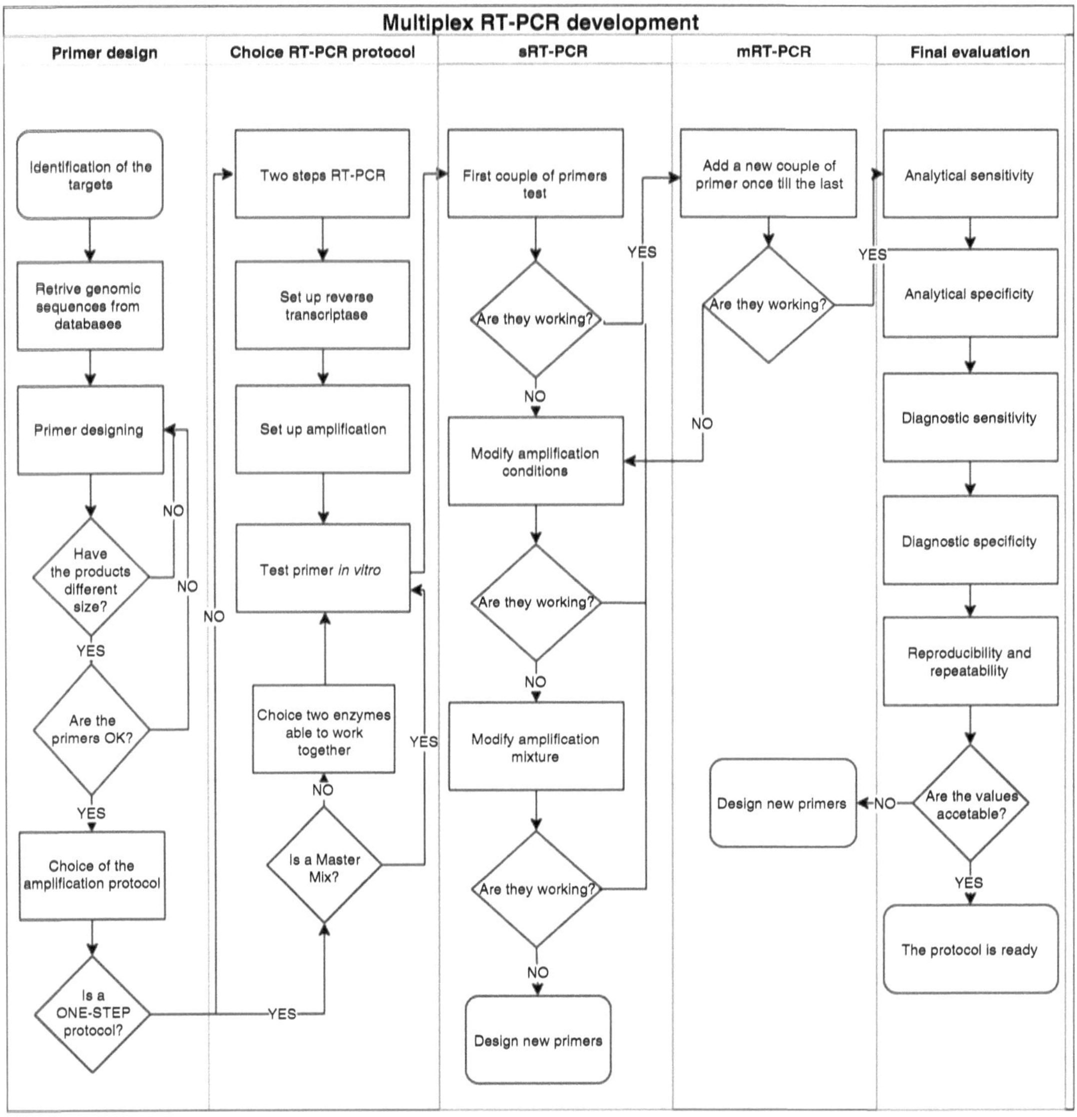

Fig. 5 Flow chart for the development of a multiplex RT-PCR assay

master mix will then be aliquoted into as many test tubes as samples to be amplified. It is essential to mark each individual tube with a number or letter that will correspond to each specific sample. The nucleic acid samples are then added individually into the marked tubes.

2. *Controls.* For each mRT-PCR method described above, it is always necessary to include the following controls:
   (a) Positive amplification control (PAC): to monitor the efficiency of the amplification. It consists of an already extracted nucleic acid of the target organism (viroid RNA).

(b) Negative amplification control or no template control (NAC): to rule out false positives due to contamination during the preparation of the reaction mix. It consists of amplification of molecular grade water used to prepare the reaction mix.

(c) Negative isolation control (NIC): to monitor contamination with target viroid during nucleic acid extraction. It consists of a nucleic acid extraction and subsequent amplification of a sample from uninfected tissue.

(d) Positive isolation control (PIC): to ensure that nucleic acids of sufficient concentrations and quality are isolated. It consists of a nucleic acid extraction and subsequent amplification of target infected sample tissue(s).

(e) Internal control (IC): to eliminate the possibility of PCR false negatives due to nucleic acid extraction failure or degradation or the presence of PCR inhibitors. It consists in the use of a plant gene as endogenous control.

3. *Critical points.* The weak point of the mRT-PCR is the high sensitivity of the reaction, which on one hand allows the detection of a high number of positive samples but on the other hand, if not used correctly, can give false-positive samples or cause contamination in the laboratory (reagents, thermocycler block, micropipettes, etc.). It is therefore recommended to work with extreme care and always use negative controls (NAC and NIC) to verify the validity of each amplification event. To avoid contamination, it is recommended to:

(a) Organize the molecular diagnostics laboratory, if possible, with separate rooms (a laboratory for extraction, a laboratory for amplification, and laboratory for electrophoresis). If this is not possible, the use of separate work benches should be implemented to include physical separation of the three phases (extraction, amplification, electrophoresis) and sets of micropipettes dedicated to each phase. Pay close attention to the electrophoresis workbench, where amplified products are handled.

(b) It is advisable to aliquot all reagents, and at the first suspicion of contamination, eliminate them all, and start again with new aliquots (in the event of contamination, it is very difficult to trace the contaminated reagent or sample).

(c) Use only sterile RNase-free water.

(d) Change gloves frequently during preparation operations of the reaction mixture and loading of the targets (RNA extracts).

(e) Use only sterile tips with filter.

(f) When loading the thermal cycler block, make sure the tubes are tightly closed; if at the end of a PCR test open tubes are found inside the thermocycler (due to a non-hermetic closure of the lid or to failed tubes), treat the block with commercially available DNase solutions.

4. *Additional additive.* Occasionally, some genomic DNA can be retrieved in RNA extractions, and it may lead to false-positive results, higher background, or false negatives (i.e., if the endogenous internal control is included in the primer mixture). Thus, a DNase I treatment step to completely remove the undesired DNA should be included. Moreover, DMSO, glycerol, formamide, and BSA may be added to the reaction mixture to enhance the DNA polymerase activity; they are generally used for G-C-rich templates and can affect primer annealing, template denaturation, and enzyme activity.

Acknowledgments

The authors would like to thank Dr. Ahmed Hadidi for the critical reading and review of the chapter.

References

1. Elnifro EM, Ashshi AM, Cooper RJ, Klapper PE (2000) Multiplex PCR: optimization and application in diagnostic virology. Clin Micro Rev 13:559–570. https://doi.org/10.1128/CMR.13.4.559
2. Ito T, Ieki H, Ozaki K (2002) Simultaneous detection of six citrus viroids and apple stem grooving virus from citrus plants by multiplex reverse transcription polymerase chain reaction. J Virol Methods 106:235–239. https://doi.org/10.1016/S0166-0934(02)00147-7
3. Hadidi A, Flores R, Randles JW, Palukaitis P (eds) (2017) Viroids and satellites. Academic Press/Elsevier, United Kingdom and United States
4. Luigi M, Luison D, Tomassoli L, Faggioli F (2011) Natural spread and molecular analysis of Pospiviroids infecting ornamentals in Italy. J Plant Pathol 93:1–5
5. Guček T, Jakše J, Matoušek J, Radišek S (2019) One-step multiplex RT-PCR for simultaneous detection of four viroids from hop (Humulus lupulus L.). Eur J Plant Pathol 154:273–286. https://doi.org/10.1007/s10658-018-01654-2
6. Di Serio F, Malfitano M, Alioto D, Ragozzino A, Flores R (2002) Apple dimple fruit viroid: sequence variability and its specific detection by multiplex fluorescent RT-PCR in the presence of apple scar skin viroid. J Plant Pathol 84:27–34
7. Shamloul AM, Faggioli F, Keith JM, Hadidi A (2002) A novel multiplex RT-PCR probe capture hybridization (RT-PCR-ELISA) for simultaneous detection of six viroids in four genera: Apscaviroid, Hostuviroid, Pelamoviroid, and Pospiviroid. J Virol Methods 105:115–121. https://doi.org/10.1016/S0166-0934(02)00090-3
8. Ragozzino E, Faggioli F, Barba M (2004) Development of a one tube-one step RT-PCR protocol for the detection of seven viroids in four genera: Apscaviroid, Hostuviroid, Pelamoviroid and Pospiviroid. J Virol Methods 121:25–29. https://doi.org/10.1016/j.jviromet.2004.05.012
9. Hosokawa M, Matsushita Y, Uchida H, Yazawa S (2006) Direct RT-PCR method for detecting two chrysanthemum viroids using minimal amounts of plant tissue. J Virol Methods 131:28–33. https://doi.org/10.1016/j.jviromet.2005.07.001
10. Levy L, Hadidi A, Garnsey SM (1992) Reverse-transcription polymerase chain reaction assay for the rapid detection of citrus viroids using multiplex primer sets. Proc Int Soc Citriculture 2:800–803

11. Wang X, Zhou C, Tang K, Zhou Y, Li Z (2009) A rapid one-step multiplex RT-PCR assay for the simultaneous detection of five citrus viroids in China. Eur J Plant Pathol 124:175–180. https://doi.org/10.1007/s10658-008-9386-y
12. Matsushita Y, Usugi T, Tsuda S (2010) Development of a multiplex RTPCR detection and identification system for potato spindle tuber viroid and tomato chlorotic dwarf viroid. Eu. J Plant Pathol 128:165–170. https://doi.org/10.1007/s10658-010-9672-3
13. Hajizadeh M, Navarro B, Bashir NS, Torchetti EM, Di Serio F (2012) Development and validation of a multiplex RT-PCR method for the simultaneous detection of five grapevine viroids. J Virol Methods 179:62–69. https://doi.org/10.1016/j.jviromet.2011.09.022
14. Ito T, Ieki H, Ozaki K, Iwanami T, Nakahara K, Hataya T, Ito T, Isaka M, Kano T (2002) Multiple citrus viroids in citrus from Japan and their ability to produce exocortis-like symptoms in citron. Phytopathology 92:542–547. https://doi.org/10.1094/PHYTO.2002.92.5.542
15 Lin L, Li R, Mock R, Kinard G (2012) One-step multiplex RT-PCR for simultaneous detection of four pome tree viroids. Eu. J Plant Pathol 133:765–772
16. Flores R, Gago-Zachert S, Serra P, Sanjuán R, Elena SF (2014) Viroids: survivors from the RNA world? Annu Rev Microbiol 68:395–414. https://doi.org/10.1146/annurev-micro-091313-103416
17. Hadidi A, Flores R, Randles JW, Semancik JS (eds) (2003) Viroids. CSIRO Publishing, Collingwood, Australia
18. Palacio-Bielsa A, Foissac X, Duran-Vila N (1999) Indexing of citrus viroids by imprint hybridization. Eur J Plant Pathol 105:897–903. https://doi.org/10.1023/A:1008711402856
19. Malfitano M, Barone M, Alioto D, Duran-Vila N (2005) A survey of citrus viroids in Campania (southern Italy). Plant Dis 89:434. https://doi.org/10.1094/PD-89-0434C
20. Steitz TA (1998) A mechanism for all polymerases. Nature 391(6664):231–232
21. Zhang X, Peng Y, Wang Y, Zhang Z, Li D, Yu J, Han C (2016) Simultaneous detection and differentiation of three genotypes of brassica yellows virus by multiplex reverse transcription-polymerase chain reaction. Virol J 13:189. https://doi.org/10.1186/s12985-016-0647-7
22. Nguyen PT, Chen YC (2017) Simultaneous detection of two Viroids infecting grapevines in Taiwan by multiplex RT-PCR. Chiang Mai J Sci 44(3):721–730
23. Zhao X, Liu X, Ge B, Li M, Hong B (2015) A multiplex RT-PCR for simultaneous detection and identification of five viruses and two viroids infecting chrysanthemum. Arch Virol 160(5):1145–1152. https://doi.org/10.1007/s00705-015-2360-z
24. Bulletin OEPP/EPPO Bulletin (2018) PM 7/98 (3) Specific requirements for laboratories preparing accreditation for a plant pest diagnostic activity. EPPO Bulletin 48(3):387–404

Chapter 16

Real-Time Detection of Viroids Using Singleplex and Multiplex Quantitative Polymerase Chain Reaction

Fatima Osman and Georgios Vidalakis

Abstract

Multiplex quantitative polymerase chain reaction (multiplex qPCR) enables the amplification of more than one target in a single reaction using different reporter dyes with distinct fluorescent spectra. The number of reporter fluorophores is typically restricted to three or four, depending upon the capability of the real-time PCR platform and software used. Each target is amplified by a different set of primers and a uniquely labeled probe that distinguishes each PCR amplicon. Thus, the levels of several targets of interest can be quantified in real time. By combining several reactions in a single tube, multiplex qPCR reduces the quantity, and cost of reagents needed to screen a sample for multiple targets. Specificity and efficiency are not affected by the inclusion of the three assays in a multiplex reaction.

Key words Fluorescence, dsDNA intercalating molecules, Hydrolysis probes, Minor groove binding, MIQE

1 Introduction

Real-time quantitative polymerase chain reaction (qPCR) has significantly accelerated the detection of nucleic acid targets during the past decade, and guidelines for qPCR evaluation parameters (e.g., MIQE) have standardized the process of development and validation for qPCR-based detection assays [1–3]. Reverse transcription (RT) followed by qPCR, i.e., RT-qPCR assays, have been extensively used for the detection of one or more citrus viroid species [4–13]. The development of various qPCR chemistries such as dsDNA intercalating molecules (e.g., SYBR Green I) and fluorophore-labeled oligonucleotides (e.g., hydrolysis probes), as well as novel instrumentation platforms, has led to improved diagnostic speed, sensitivity, reproducibility, and reduced risk of carryover contaminations (for a review see [14, 15]). The hydrolysis qPCR probes are designed to bind to a specific region of the targeted DNA, thus eliminating the need for melting curve analysis. The reporter dye on the qPCR probe can fluoresce only after

Ayala L. N. Rao et al. (eds.), *Viroids: Methods and Protocols*, Methods in Molecular Biology, vol. 2316, https://doi.org/10.1007/978-1-0716-1464-8_16, © Springer Science+Business Media, LLC, part of Springer Nature 2022

amplification of the targeted DNA [15–17]. Hydrolysis probes are conjugated at their 3′ end to minor groove binding (MGB) ligands, which are small tripeptides forming non-covalent bonds with dsDNA. As a result, the "target DNA-MGB probe" structure is greatly stabilized (increased melting temperature) providing higher target DNA-binding specificity and sensitivity [15].

The use of differently labeled fluorescent MGB probes made the simultaneous detection of more than one pathogen possible in multiplex qPCR assays, thus substantially decreasing labor time and cost of diagnostics [18]. Optimization and validation of multiplex qPCR assays is essential since interactions between the qPCR assays can interfere with the sensitivity of detection. Typical problems include primer dimers, competition between assays for dNTP's and polymerase, and possible overlap in emissions spectra between reporter fluorophores. In addition, the effects of competition among qPCR assays included in the multiplex are reduced if all assays have similar efficiency. This is especially important for assays which use a housekeeping gene to normalize data; similar efficiencies ensure that the housekeeping gene and target gene are amplified at proportional rates.

Proper design of the qPCR primers/probes and careful selection of the fluorophores to minimize overlap in spectra while remaining compatible with the equipment and filters available guarantee an optimum multiplex qPCR reaction; however, they do not guarantee the viroid target detection. Careful sample preparation and strict laboratory protocols for overall workflow as well as sample, reagent, and equipment handlings are equally critical components for accurate viroid detection and avoiding cross contamination issues.

2 Materials

1. Access to GenBank sequences (nucleotide) https://www.ncbi.nlm.nih.gov/nuccore (*see* **Notes 1** and **2**).
2. Sequence Analysis and Molecular Biology Data Management software Vector NTI Advance™ 11 (Invitrogen; Life Technologies, Grand Island, NY) to align sequences used for designing qPCR assays (*see* **Note 2**).
3. Primer Express™ software (Thermo Fisher Scientific, Carlsbad, California, USA) to design primers and qPCR probes.
4. Vortex.
5. PDI® Sani-Cloth® Plus Germicidal Disposable Wipes (for general cleaning) and PDI® Sani-Cloth® Plus Surface Disinfectant Cleaner with nuclease action.

6. Two dead air box PCR workstations placed in different rooms (*see* **Note 3**).
7. RNAse-free water provided in kits (*see* **Note 4**).
8. Microfuge tubes.
9. Mini Benchtop Centrifuge for microfuge tubes and PCR plates, Grainger.
10. Ice bucket with ice.
11. Flat bottom 96 well polystyrene microtiter plate or 384 well plate (Thermo Fisher Scientific, Carlsbad, California, USA).
12. MicroAmp Optical Adhesive Film (Thermo Fisher Scientific, Carlsbad, California, USA).
13. Single- and multichannel filter tipped pipettes (2μL, 100μL, 200μL).
14. Purified RNA extracts (*see* **Note 5**).
15. Primers and labeled qPCR probes from various vendors; primers, Eurofins genomics (Louisville, KY), qPCR probes, Thermo Fisher (Thermo Fisher Scientific, Carlsbad, California, USA).
16. Designed plasmids (Eurofins MWG Operon, Huntsville, AL) for assay validation.
17. QuantStudio™ 12K Flex qPCR instrument (Thermo Fisher Scientific, Carlsbad, California, USA) (*see* **Note 6**).
18. AgPath-ID™ One-Step RT-PCR Kit, Part Number 4387424. (Thermo Fisher Scientific, Carlsbad, California, USA), for singleplex RT-qPCR.
19. QuantiFast Multiplex RT-PCR Kit, Cat No./ID: 204852 (Qiagen, Valencia, CA), for multiplex RT-qPCR.
20. PASW Statistics 18 software (SPSS Inc., Chicago, IL).

3 Methods

3.1 RT-qPCR Assay Design: Primers and qPCR Probes

1. Retrieve all sequences in the GenBank of your target viroids (as well as locally generated sequences), and save each viroid sequence separately in FASTA format in a text file (*see* **Note 2**).
2. In Vector NTI Advance software, go to Align tab and open New Alignment window. In Align X, go to Project tab, and click Add files to import the viroid sequence files (*see* **step 1** in Subheading 3.1). Then click Align selected sequences.
3. Create a separate alignment for each targeted viroid species using Sequence Analysis and Molecular Biology Data Management Software Vector NTI Advance™ 11.0.

4. Determine the region in the aligned viroid genomes with as close as possible to 100% conserved sequence (*see* **Note 7**).
5. Use this conserved region to design the primers/qPCR probes of the singleplex qPCR assay using the guidelines set by the Primer Express™ software (*see* **Notes 1**, **2**, and 7 and **steps 6–8** in Subheading 3.1).
6. The optimum primers' melting temperature (T_m) should be 58 °C with the T_m of the qPCR probe(s) 10 °C higher to prevent the formation of primer dimers.
7. For multiplexing purposes, the amplicon length of all designed singleplex assays should be similar and should not to exceed 100 nucleotides.
8. Confirm the homology of the designed qPCR assay (primers/qPCR probes) by blasting against the nonredundant database of GenBank (NCBI) using the Basic Local Alignment Search Tool (BLAST) of the GenBank (https://blast.ncbi.nlm.nih.gov/Blast.cgi).
9. Use fluorophore dyes with different, nonoverlapping, wavelengths when designing multiplex qPCR assays. For example, the fluorophore used for one viroid can be 6-carboxyfluorescein FAM, another can be Vic or Ned, and the 3′ quencher could be the Black Hole Quencher (BHQ).

3.2 Resuspension of Primers and qPCR Probes for RT-qPCR Assay Preparation

1. In the clean room, sanitize working surfaces using the PDI Sani-Cloth disposable wipes (alternatively use 10% household bleach solution; *see* Chapter 5, Subheading 3.1), prepare the ice bucket with fresh ice, thaw any frozen materials, and place all working microfuge tubes on ice (*see* also Subheading 3.7 and **Note 3**).
2. Determine how many nanomoles (nmol) of primer and probes are present in the microfuge tubes. This is typically in the documentation accompanying the ordered materials or is listed on the side of the tubes.
3. On ice, add ten times the number of nmols of microliters (μL) of RNAse-free water to the primer tube. For example, if you have 20.5 nmol of primer, add 205μL of RNAse-free water. Leave primers and probes on the bench (room temperature) for 30 min for complete resuspension. Vortex each tube briefly, and then spin down for a few seconds with the Mini Benchtop Centrifuge. Primers and probes are now100 picomole (pmol)/μL concentration and ready to use. Prepare aliquots of your primer and probes. It is good practice to prepare many aliquots of your 100 pmol/μL working primer/probe solution (*see* **Note 8**).

4. Prepare the qPCR primer/probes mixes for both singleplex and multiplex by mixing 20μL each of the 100 pmol/μL forward primers, 20μL each of the 100 pmol/μL reverse primers, and 4μL each of the 100 pmol/μL probes per viroid in a final volume of 240μL of RNAse-free water, and label the mix as primer probe mix (*see* **Note 9**).

3.3 Singleplex RT-qPCR Procedure

1. Prepare the AgPath Master Mix solution for singleplex RT-qPCR reaction using the AgPath-ID™ One-Step RT-PCR Kit starting with 2.65μL RNAse-free water, 6.25μL 2× RT buffer, 0.6μL primer/probe mix (100 pmol/μL) (*see* **steps 1–4** in Subheading 3.2), and 0.5μL 25× RT mix.
2. For a total of 12μL final reaction mix, dispense 10μL of the AgPath Master Mix solution (*see* **step 2** in Subheading 3.3) into the 96 well microtiter plate, using a multichannel pipette.
3. Dispense 2μL of non-template (water), negative, and positive controls (preferably in this order to avoid cross contamination) into the assigned microtiter plate wells (*see* **Notes 4**, **5**, and **10**).
4. Dispense 2μL of the RNA template into the assigned microtiter plate wells (*see* **Notes 5** and **10**).
5. Tightly seal the microtiter plate with MicroAmp Optical Adhesive Film, and pulse down the plate for ~10 s using a mini benchtop centrifuge (*see* **Note 11**).
6. Singleplex cycling conditions is 45 °C for 10 min, 95 °C for 10 min for the first cycle, then 40 cycles at 95 °C for 15 s, and 60 °C for 45 s as per AgPath-ID™ One-Step RT-PCR Kit.
7. Analyze all samples using QuantStudio™ 12K Flex qPCR instrument. Fluorescent is collected during the annealing temperature and the quantitative cycle (C_q), calculated, and exported. The C_q is calculated by the qPCR instrument using an algorithm with a set range of cycles at which the first detectable significant increase in fluorescence occurs (*see* **Note 6**).

3.4 Multiplex RT-qPCR Procedure

1. Use the same primer/probe mix concentrations used for singleplex (*see* **Steps 1–4** in Subheading 3.2).
2. Prepare the QuantiFast Master Mix solution for multiplex RT-qPCR reaction using the QuantiFast Multiplex RT-PCR Kit starting with 0.045μL RNAse-free water, 6.25μL QuantiFast RT Mix, 0.58μL primer/probe mix (100 pmol/μL) (*see* **step 1** in Subheading 3.4), and 0.125μL 25× RT-PCR mix.
3. For a total of 12μL final reaction mix, dispense 7μL of the QuantiFast Master Mix solution (*see* **step 2** in Subheading 3.4) into the 96 well microtiter plate, using a multichannel pipette.

4. Dispense 5μL of positive, negative, and non-template control (water), into the assigned microtiter plate wells (*see* **Notes 4**, **5**, and **10**).
5. Dispense 5μL of the RNA template into the assigned microtiter plate wells (*see* **Notes 5** and **10**).
6. Tightly seal the microtiter plate with MicroAmp Optical Adhesive Film, and pulse down the plate for ~10 s using a mini benchtop centrifuge (*see* **Note 11**).
7. Multiplex cycling conditions is 50 °C for 20 min for the first cycle, 95 °C for 5 min for the second cycle, then 40 cycles at 95 °C for 15 s, and 60 °C for 1 min as per QuantiFast Multiplex RT-PCR Kit.
8. Analyze all samples using QuantStudio™ 12K Flex qPCR instrument (*see* **step** 7 in Subheading 3.3).
9. C_q value for each fluorophore representing each multiplexed is analyzed separately. For example, viroid X has Fam fluorophore, and viroid Y has Vic fluorophore (*see* **Note 12**).

3.5 Validation of Singleplex and Multiplex RT-qPCR Assays

1. Validate the designed singleplex and multiplex RT-qPCR assays using applicable parameters that are proposed in the "Guidelines for validation of qualitative real-time PCR methods" (*see* **Note 13**).
2. Critical parameters include specificity and cross-reactivity (*see* Subheading 3.5.1), efficiency (*see* Subheading 3.5.2), sensitivity, and viroid load (*see* Subheading 3.5.3), inter- and intra-assay variation (*see* Subheading 3.5.4).
3. To perform validation experiments, absolute standard curves must be generated to determine the starting number of copies. Standard curves must be created by running a serial tenfold dilution of plasmids carrying a known copy number of the viroid inserts (amplicon sequence spanning the forward and reverse primers) (*see* **Note 14**).

3.5.1 Specificity and Cross-Reactivity

1. Evaluate the specificity of the assay both in silico and experimentally.
2. For in silico evaluation, compare the sequence of each viroid amplicon (sequence spanning the forward and reverse primers) against the NCBI GenBank sequences using BLASTn.
3. The results of the in silico specificity test should show 100% homology with the targeted viroid and 0% homology with any other GenBank deposited sequence.
4. For experimental evolution, run the singleplex (*see* Subheading 3.3) or multiplex (*see* Subheading 3.4) qPCR assays using a variety of biological samples with known viroid infection status from broad geographical origins whenever possible.

5. The results of the specificity experimental test can be expressed as a percentage of false positives or negatives using the formulae given below. No false-positive or false-negative results should be seen. If unexpected signals are observed, they should be further investigated.

$$\text{False neg rate} = \frac{100 \times \text{number misclassified known pos samples}}{\text{to number known pos samples}}$$

$$\text{False pos rate} = \frac{100 \times \text{number misclassified known neg samples}}{\text{to number known neg samples}}$$

6. To assess cross-reactivity, test RNA derived from different non-inoculated host species and varieties as well as RNA from hosts inoculated with nontargeted viroid and other viral pathogens of the host.
7. All the results of the cross-reactivity test should be negative to confirm no cross-reactivity occurred.

3.5.2 Efficiency

1. To construct the absolute standard curves, obtain viroid amplicons sequences spanning the forward and reverse primers used to create the qPCR assay and order-linearized standard plasmids from a biotech company such as Eurofins MWG Operon, Huntsville, Alabama, USA (*see* **Notes 13–17**).
2. Create a tenfold serial dilution of the linearized plasmids that harbor the individual viroid amplicons.
3. To validate singleplex RT-qPCR assays, use the tenfold serial dilutions of the linearized plasmid corresponding to the targeted viroid. In multiplex RT-qPCR, combine in a single amplification tube, the serial dilutions of all linearized plasmids of the targeted viroids.
4. Run in triplicate the singleplex (*see* Subheading 3.3) or multiplex (*see* Subheading 3.4) qPCR assays of the tenfold serial dilutions.
5. To generate the standard curves, plot the C_q values against the logarithm of the concentration for the exponential phase of the reaction, and fit a straight line to plotted data by simple linear regression (*see* **Note 14**).
6. Calculate the slope of the standard curve and the coefficient of determination (R^2) using linear regression. The R^2 for each viroid target should be ≥ 0.98 (*see* **Note 14**).
7. Calculate the efficiency (E) with the formula $E = 10(-1/\text{slope}) - 1$ (*see* **Note 14**).

3.5.3 Sensitivity and Viroid Load

1. To calculate the sensitivity (or absolute limit of detection, LOD) and quantify the viral load or the starting number of copies of your target viroid in the tested samples, you will need

the values generated from the absolute standard curves (*see* Subheading 3.5.1).

2. The sensitivity (LOD) of the RT-qPCR assay is calculated using the equation LOD = 3.33 × standard deviation of the regression line (σ)/slope (S) of the absolute standard curves.
3. The viral load or copy numbers for each sample are calculated using the equation: viroid load = (sample C_q − Y intercept)/− slope (*see* **Note 18**).

3.5.4 Inter- and Intra-Assay Variation

1. Access the inter-assay and intra-assay variation of the developed singleplex and multiplex RT-qPCR assays within one plate and in different plates, respectively.
2. Extract total RNAs from three randomly selected viroid-infected samples (*see* **Note 5**).
3. Prepare tenfold serial dilutions for each sample and test by singleplex (*see* Subheading 3.3) and multiplex RT-qPCR (*see* Subheading 3.4) in triplicates, and calculate the C_q values.
4. Calculate the intra-assay and inter-assay variations of each tested sample, by determining the percentage of coefficient of variation (CV%), as follows:

$$CV\% = \frac{\text{Mean of the standard deviation of the duplicates}}{\text{Grand mean of the duplicates}} \times 100$$

5. Test the reproducibility of the assay by analysis of variance (ANOVA) comparing the C_q values of the samples using PASW Statistics 18 (SPSS Inc., Chicago, IL) or other available statistics software. P values < 0.05 are considered significant.

3.6 Evaluation of Singleplex and Multiplex RT-qPCR Assays

1. After RT-qPCR assays have successfully completed the validation process (*see* Subheading 3.5), they should be evaluated with a large number of biological samples under different laboratory conditions.
2. Critical tests include test with a large number of diverse host and viroid titer samples (*see* Subheading 3.6.1), robustness (*see* Subheading 3.6.2), and evaluation of applicability, practicability, and transferability (*see* Subheading 3.6.3) (*see* **Note 19**).

3.6.1 Use with a Large Number of Diverse Host and Viroid Titer Samples

1. Evaluate the newly validated singleplex and multiplex RT-qPCR assays by testing RNA extracts (*see* **Notes 5** and **10**) derived from:
   (a) Host plants inoculated singly or with mixtures of targeted viroids (positive controls).
   (b) Non-inoculated plants (negative controls).
   (c) Plants inoculated with nontargeted pathogens (cross-reactivity controls).

2. You should get all expected results with correct sample classification:
   (a) Positive, for 1.1 (positive controls).
   (b) Negative, for 1.2 (negative controls).
   (c) Negative, for 1.3 (cross-reactivity controls).

3.6.2 Robustness

1. To evaluate the robustness of the RT-qPCR assays, different experimental conditions can be slightly changed and their impact on the results evaluated (e.g., a slight deviation in annealing temperature, small pipetting errors, different qPCR instruments and kits, etc.).
2. Test the RT-qPCR assays using:
   (a) A deviation in annealing temperatures (±2 °C).
   (b) A deviation in reaction volumes (±2μL).
   (c) Different manufactures qPCR instruments (e.g., Bio-Rad, Thermo-Fisher, Agilent, Roche Diagnostics, Eppendorf, Qiagen, etc.)
   (d) Different RT-qPCR kits for singleplex and multiplex.
3. All tests should provide comparable results, and no test should result in the wrong classification of a sample as false positive or false negative (*see* **Note 19**).

3.6.3 Applicability, Practicability, and Transferability

1. After an RT-qPCR assay has successfully completed the in-house validation (*see* Subheading 3.5) and evaluation (*see* Subheading 3.6), it should be tested by different laboratories. This will evaluate the assay's performance (applicability, practicability, and transferability) under different laboratory conditions (e.g. different operators, equipment, etc.).
2. Provide detailed assay protocols to the participating laboratories, and allow them to run assay independently using their own biological positive and negative controls and standard operating procedures.
3. Compare the results with those obtained by the in-house evaluation. All tests should provide comparable results, and no test should result in the wrong classification of a sample as false positive or false negative.

3.7 RT-qPCR Swab Test and Decontamination Practices

1. For successful validation, evaluation, and use of RT-qPCR in any laboratory setting, it is extremely important to perform swap tests on all instruments and lab workstations that will be used for your experiment including primer and probe resuspension and RT-qPCR testing, to detect any possible contamination that may show up as false positives in viroid tests.
2. Swab testing is performed using qPCR to test all pathogens that are routinely used in the lab.

3. Select ten common areas to test. The best common areas to test are those used for sample processing, nucleic acid extraction, primer probe reconstitution, and RT-qPCR setup.
4. Aliquot 200μL of nuclease-free water into labeled tubes for the ten selected areas.
5. Swipe locations using cotton Q-tips swaps, break the tip of the Q-tip, and place it upside down in each of the ten nuclease-free water tubes. Make sure the swab head is submerged in the water.
6. Run the ten samples against qPCR designed to test all pathogens that are routinely used in your lab.
7. When and if contamination is detected, lab decontamination SOP should be implemented by sanitizing the surfaces and equipment with bleach and PDI Sani-Cloth Plus Wipes as follows:
   (a) Clean all benches/tables with 10% bleach (to remove contaminants).
   (b) Wipe down with PDI® Sani-Cloth® Plus Germicidal Disposable Wipes (for general cleaning), if plant material/nucleic acids have been processed in that area, also wipe down with PDI® Sani-Cloth® Plus Surface Disinfectant Cleaner with nuclease action (to break down nucleic acids).
   (c) Wipe down with 70% ethanol (to remove residue from wipes and bleach and prevent corrosion), and check for contamination with swab tests (*see* **steps 1–5** in Subheading 3.7).
8. Keep decontaminating and swabbing until no false-positive reading will show up on any qPCR test from tested locations.

4 Notes

1. It is very important to thoroughly review the assay specific design and validation criteria for quantitative PCR set by the MIQE guidelines [1] as well as for qualitative PCR [2] before you start designing your assay.
2. To design a robust and accurate qPCR assay, use all GenBank available sequence from which your qPCR assays are to be designed. If you have access to local positive controls specific to your targeted pathogen, design forward and reverse primers spanning your qPCR template, sequence this amplicon, and align with all corresponding sequences in the GenBank. When aligning the sequences from the NCBI GenBank, make sure to include all sequences in the GenBank for the specific viroid sequence as well as locally sequenced amplicons.

3. Separate the lab workspace in three rooms. A dedicated clean room with a −20 °C freezer for RT-qPCR reagents storage and a dead air box PCR workstation, equipped with UV light, for nontarget reaction mix preparation. Use the UV light for approximately 5 min and thoroughly clean the hood space for nucleic acid transfer to a microplate using the PDI® Sani-Cloth® Plus Germicidal Disposable Wipes. A dedicated set of pipettes should be used for preparing primers and PCR reagent master mixes. Always use filtered tips. This will help prevent RNA template contamination of stock PCR reagents and primers. A second room with UV light-equipped dead air box PCR workstation should also be established. This room will be used strictly to dispense the master mixes, add the RNA templates, and finalize qPCR reaction setup. Use the sanitation process as described above before using this room as well. If possible the PCR/qPCR thermocycler instruments should be located in an isolated dedicated third room.
4. Prepare all solutions using ultrapure water either provided in various kits or autoclaved (prepared by purifying deionized water).
5. Purified RNA extract of infected samples and positive control are derived from MagMAXTM Express-96 Magnetic Particle Processor with magnetic head (Thermo Fisher Scientific, Carlsbad, California, USA) using a modified 5× MagMaxTM-96 Viral RNA Isolation Kit optimized for citrus tissues [19] (catalog number AMB1836-5) (*see* Chapter 22). Total RNA was eluted in 100μL elution buffer and used as template for RT-qPCR.
6. The presented protocol has been optimized for the QuantStudio™ 12K Flex qPCR instrument; however, any qPCR instrument can be used as long as all validation tests have been performed per published guidelines [1, 2].
7. It is preferable that the qPCR probe is designed from 100% conserved sequence location. If there is sequence variability, it recommended designing couple of primers to include the mismatches instead of including a degenerate nucleotide in one primer. Multiple primers and qPCR primers are encouraged to design to cover all diversity of sequences in your target template.
8. It is good practice to prepare many aliquots of your 100 picomole (pmol)/μL working primer/probe solution to prevent repeated freeze and thaw of the main qPCR primers and probe stock.
9. If the qPCR assay (singleplex or multiplex) has multiple primers and qPCR probes, add appropriate volume; for primers add 20μL each of the 100 pmol/μL, and for probes add 4μL each of the 100 pmol/μL probes, and adjust the water volume to 240μL final volume.

10. It is very important to assess the integrity of the RNA of every sample tested using the cytochrome oxidase gene (COX) RT-qPCR assay designed as an internal control [19], using the same singleplex reaction conditions (*see* Subheading 3.3).
11. For pulsing down the qPCR plate, we recommend the Eppendorf 5430 centrifuge with the swing-bucket rotor (A-2 MTP), but any other centrifuge that can accommodate PCR plates is sufficient.
12. It is important when designing singleplex qPCR assays that will be multiplexed to choose fluorophore with nonoverlapping wavelengths.
13. It is important to review the principals of standard curves calculations and optimal amplification efficiency before you start validating your assay [20–22].
14. Create standard curves generated from the linearized plasmids that harbor the individual viroid amplicons by creating serial dilutions of the plasmid. Calculate the slope of the standard curve and the coefficient of determination (R^2) using linear regression (the R^2 for each target should be $\geq$0.98) [20] and the formula $E = 10(-1/\text{slope}) - 1$ [21, 22]. A plot of the C_qs versus the logarithm of the target concentrations is constructed and is expected to be linear with a negative slope. For a tenfold dilution series, the slope is -3.33 when $E = 100\%$.
15. The DNA standard curve to assess PCR efficiency must be based on a clean matrix (i.e., plasmid and not nucleic acids from plant extracts) [21] to assess the performance of the PCR assay in absence of interference. Therefore, standard curve is designed using a serial dilution of a plasmid carrying the targeted qPCR template to avoid accounting for inhibition and interference in the field samples.
16. The PCR efficiency depends on many factors including (1) the assay performance, which depends on the primers' and template sequences and structures. Secondary structure and opportunity for undesired intramolecular interactions reduce PCR efficiency; (2) the sample matrix, which may contain inhibitors and other interfering substances from the sample or carry-overs agents from upstream processing steps; (3) reagents used and their concentrations [21].
17. You may choose to construct the linearized standard plasmids in your lab; however, extreme caution must be applied when handling plasmids because based on our experience the risk of cross- and carry-over contamination is extremely high. In addition, biotech companies can guarantee the absolute purity (i.e., clean matrix [21]) of the plasmid and inserted viroid amplicon. Linearize the standard plasmid with a known restriction enzyme that cuts outside the intended viroid amplicon. Usually Hind III enzyme is used, to increase the efficiency of dilutions.

18. Calculate the copy numbers for each sample, using this formula viroid load = (sample C_q − Υ intercept)/−slope as described by [23]. All values (Υ intercept and slope) are derived from the standard curves described in Subheading 3.5.1.
19. Evaluation parameters of RT-qPCR assays may vary among different laboratories. In this chapter we followed well-established protocols from the Citrus Clonal Protection Program (CCPP) Citrus Diagnostic and Research Laboratory for the detection of graft-transmissible pathogens of citrus. For more details on the use of diverse biological samples and evaluation of robustness, applicability, practicability, and transferability, *see* [24].

Acknowledgments

This work was supported by the California Citrus Nursery Board (CCNB) project "Implementation and streamlining of the newly developed high throughput diagnostic system for citrus nurseries registration" and by the Citrus Research Board (CRB) project "Citrus Clonal Protection Program" (6100) awarded to G. Vidalakis. Additional support was provided by the US Department of Agriculture (USDA) National Institute of Food and Agriculture (NIFA), Hatch (project 1020106), and the National Clean Plant Network (NCPN) which operates under the auspices of USDA Animal and Plant Health Inspection Service (APHIS) (12-8100-1544-CA; 14-, 15-, 16- 8130-0419-CA; AP17PPQS&T00C118; AP18PPQS&T00C107). The findings and conclusions in this publication are those of the author(s) and should not be construed to represent any official USDA or US Government determination or policy. URLs to sponsors' websites listed below.

1. CRB: https://www.citrusresearch.org
2. CCNB: http://www.ccnb.info/
3. USDA NIFA: https://nifa.usda.gov
4. NCPN: http://nationalcleanplantnetwork.org.

References

1. Bustin SA, Benes V, Garson JA, Hellemans J, Huggett J, Kubista M, Mueller R, Nolan T, Pfaffl MW, Shipley GL (2009) The MIQE guidelines: minimum information for publication of quantitative real-time PCR experiments. Clin Chem 55:611–622
2. Broeders S, Huber I, Grohmann L, Berben G, Taverniers I, Mazzara M, Roosens N, Morisset D (2014) Guidelines for validation of qualitative real-time PCR methods. Trends Food Sci Technol 37:115–126
3. Boonham N, Kreuze J, Winter S, Van der Vlugt R, Bergervoet J, Tomlinson J, Mumford R (2014) Methods in virus diagnostics: from ELISA to next generation sequencing. Virus Res 186:20–31
4. Osman F, Dang T, Bodaghi S, Vidalakis G (2017) One-step multiplex RT-qPCR detects three citrus viroids from different genera in a wide range of hosts. J Virol Methods 245:40–52

5. Lin C-Y, Wu M-L, Shen T-L, Yeh H-H, Hung T-H (2015) Multiplex detection, distribution, and genetic diversity of hop stunt viroid and citrus exocortis viroid infecting citrus in Taiwan. Virol J 12:11
6. Vidalakis G, Wang J (2013) Molecular method for universal detection of citrus viroids. US Patent Publication number 20130115591
7. Loconsole G, Onelge N, Yokomi RK, Kubaa RA, Savino V, Saponari M (2013) Rapid differentiation of citrus Hop stunt viroid variants by real-time RT-PCR and high-resolution melting analysis. Mol Cell Probes 27:221–229
8. Monger W, Tomlinson J, Booonham N, Marn MV, Plesko IM, Molinero-Demilly V, Tassus X, Meekes E, Toonen M, Papayiannis L, Perez-Egusquiza Z, Mehle N, Jansen C, Nielsen SL (2010) Development and inter-laboratory evaluation of real-time PCR assays for the detection of pospiviroids. J Virol Methods 169:207–210
9. Papayiannis LC (2014) Diagnostic real-time RT-PCR for the simultaneous detection of citrus exocortis viroid and Hop stunt viroid. J Virol Methods 196:99
10. Rizza S, Nobile G, Tessitori M, Catara A, Conte E (2009) Real time RT-PCR assay for quantitative detection of citrus viroid III in plant tissues. Plant Pathol 58:181–185
11. Saponari M, Loconsole G, Liao HH, Jiang B, Savino V, Yokomi RK (2013) Validation of high-throughput real time polymerase chain reaction assays for simultaneous detection of invasive citrus pathogens. J Virol Methods 193:478–486
12. Sun N, Deng C, Zhao X, Zhou Q, Ge G, Liu Y, Yan W, Xia Q (2014) Extraction of total nucleic acid based on silica-coated magnetic particles for RT-qPCR detection of plant RNA virus/viroid. J Virol Methods 196:204–211
13. Tessitori M, Rizza S, Reina A, Catara V (2005) Real-time RT-PCR based on Sybr-green I for the detection of citrus exocortis and citrus cachexia disease. In: Hilf ME, Duran-Vila N, Rocha-Peña MA (eds) Proceedings of the 16th conference of the International Organization of Citrus Virologists, 2004 Riverside. IOCV, CA, USA, pp 456–459
14. Josefsen MH, Löfström C, Hansen T, Reynisson E, Hoorfar J (2012) Instrumentation and fluorescent chemistries used in qPCR. In: Filion M (ed) Quantitative real-time PCR in applied microbiology Caister. Academic Press, Norfolk, UK, pp 27–52
15. Navarro E, Serrano-Heras G, Casatano Aroca MJ, Solera J (2015) Real-time PCR detection chemistry. Clin Chim Acta 439:231–250
16. Mackay IM, Arden KE, Nitsche A (2002) Real-time PCR in virology. Nucleic Acids Res 30:1292–1305
17. Keer JT, Birch L (2008) Quantitative real-time PCR analysis essentials of nucleic acid analysis: a robust approach. Royal Soc Chem 2008:132–155
18. Pallas V, Sanchez-Navarro J, Varga A, Aparicio F, James D (2009) Multiplex polymerase chain reaction (PCR) and real-time multiplex PCR for the simultaneous detection of plant viruses. Methods Mol Biol 508:193–208
19. Osman F, Hodzic E, Kon S-J, Wang J, Vidalakis G (2015) Development and validation of a multiplex reverse transcription quantitative PCR (RT-qPCR) assay for the rapid detection of citrus tristeza virus, citrus psorosis virus and citrus leaf blotch virus. J Viorl Methods 220:64–75
20. Rasmussen R (2001) In: Meuer S, Wittwer C, Nakagawara K-I (eds) Quantification on the LightCycler pages 21-34 in: rapid cycle real-time PCR: methods and applications. Springer Berlin Heidelberg, Berlin, Heidelberg
21. Svec D, Tichopad A, Novosadova V, Pfaffl MW, Kubista M (2015) How good is a PCR efficiency estimate: recommendations for precise and robust qPCR efficiency assessments. Biomol Detect Quantif 3:9–16
22. Pfaffl M (2004) Quantification strategies in real-time PCR A-Z of Quantitative PCR. International University Line (IUL), La Jolla, CA, USA, pp 89–113
23. Leutenegger CM (2001) The real-time TaqMan PCR and applications in veterinary medicine. Vet Sci Tomorrow 1:1–15
24. Tan S-h, Osman F, Bodaghi S, Dang T, Greer G, Huang A, Hammado S, Abu-Hajar S, Campos R, Vidalakis G (2019) Full genome characterization of 12 citrus tatter leaf virus isolates for the development of a detection assay. PLoS One 14(10):e0223958. https://doi.org/10.1371/journal.pone.0223958

Chapter 17

Reverse Transcription Loop-Mediated Isothermal Amplification (RT-LAMP) at the Service of Viroid Detection

Iraklis N. Boubourakas and Panayota E. Kyriakopoulou

Abstract

A wide range of nucleic acid-based detection methods is available for viroid diagnosis. In this chapter, we describe a novel reverse transcription loop-mediated isothermal amplification (RT-LAMP). The technique involves the use of nucleic acids, plant sap, or even traces of plant tissue. Adding a set of four to six primers, *Bst* DNA polymerase, AMV, or another thermostable reverse transcriptase, allows the fast detection of RNA targets like viroids after a short (20–40 min) incubation at 63 °C either in an open or close-tube system.

Key words PLMVd, Metal ion detectors, Filed detection, RT-LAMP, Specificity test, HNB

1 Introduction

LAMP was first described and evaluated for the detection of *hepatitis B virus* (HBV) [1] and became popular after 2003 following the emergence of *West Nile Virus* (WNV) and *severe acute respiratory syndrome viruses* (SARS) [2, 3]. Because of its speed, robustness, and simplicity of use, RT-LAMP became popular among researchers in human medicine diagnostics [4] and, more recently, in plant health, including virus and viroid detection. The first step was done by [5] with *chrysanthemum stunt viroid* (CSVd) [5], and 4 years later based on the protocol proposed by [6], *peach latent mosaic viroid* could be detected only within 32 min and with a sensitivity that exceeds at 100 times that of RT-PCR. Since then, many RT-LAMP protocols have been established for the detection of viroids like *potato spindle tuber viroid* (PSTVd) [7, 8], *chrysanthemum chlorotic mottle viroid* (CChMV) [9], CSVd [10], and *coconut cadang-cadang viroid* (CCCVd) [11]. In the case of the protocol proposed by [10], the simultaneous detection of CSVd and the *chrysanthemum virus* B (CVb) was achieved.

Ayala L. N. Rao et al. (eds.), *Viroids: Methods and Protocols*, Methods in Molecular Biology, vol. 2316,
https://doi.org/10.1007/978-1-0716-1464-8_17,

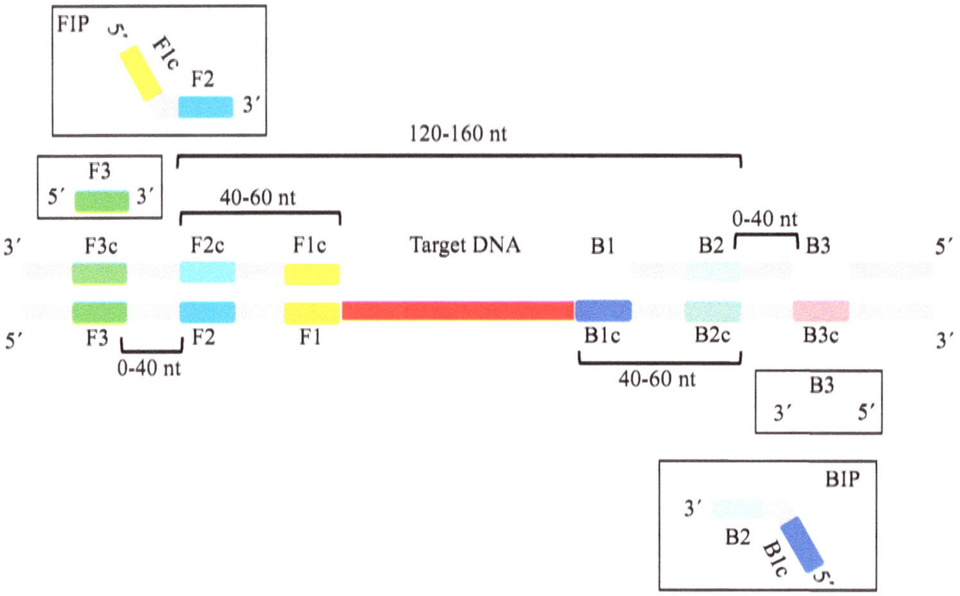

Fig. 1 Information about RT-LAMP primers used for the detection of PLMVd

LAMP was invented by Notomy [1]; it is a novel nucleic acid amplification method performed in the presence of a polymerase with strand displacement activity, like *Bst* polymerase isolated from ***Bacillus stearothermophilus*** and a set of at least four different primers: F3 (forward outer), B3 (backward outer), FIP (forward inner primer), BIP (backward inner primer) (Fig. 1), specifically designed to recognize six distinct regions on the target, under isothermal conditions (60–65 °C), [1, 12, 13]. Optionally two additional primers, the forward loop and backward loop can be added to the reaction mixture offering enhanced sensitivity and specificity [6].

Single-stranded loop regions serve as primer binding sites, allowing amplification to proceed without thermal denaturation of the template [14]. For the amplification of RNA targets, a reverse transcription step can be performed in a separate tube, before LAMP [9, 11], or simultaneously in the same tube with LAMP [6, 7]. A reverse transcriptase with high-temperature stability (60–65 °C) like *avian myeloblastosis virus reverse polymerase* (AMV) is added to the reaction mixture. In this case, the technique is called reverse transcription loop-mediated isothermal amplification (RT-LAMP).

For the detection of an RNA target with LAMP, the template could be either a cDNA of a previously reverse-transcribed RNA or a total RNA (t-RNA) in the case of one-tube, one-step RT-LAMP. Methods used to extract RNA template for the LAMP vary. Commercially available RNA extraction kits (TRIzol, RNeasy Plant Mini Kit, RNAiso) and laboratory-based protocols have been used for the RT-LAMP detection of viroids [6, 8, 9]. Considering that LAMP is less sensitive to inhibitors than PCR is, plant sap was used to detect CSVd and PSTVd [5, 7]. As per previous references [7, 15], traces of plant tissue obtained by a sterilized toothpick or an insect pin serve as a template. Final products are double-stranded stem-loop DNAs with several inverted repeats of the target and cauliflower-like structures with multiple loops.

RT-LAMP reaction produces a high DNA yield (>10 μg) and generation of large quantities of magnesium pyrophosphate as byproducts allowing direct visual observation of a white precipitate in positive reactions [1, 7, 13]. Endpoint or even real-time turbidity measurement at 400 nm is also applicable using simple and relatively inexpensive instruments [6, 11]. LAMP products can be analyzed as PCR products by gel electrophoresis. Staining reveals a ladder-like pattern in contrast to a single band in the case of PCR due to the cauliflower-like structures obtained [4]. Color change methods can also be used for the identification of positive and negative RT-LAMP reactions. For example, the addition of SYBR Green or other intercalating dyes at the end of the reaction and a color change from orange to green is detected under UV light [9]. Unlike SYBR Green and the other intercalating dyes, metal ion indicators like calcein (Loopamp Fluorescent Detection Reagent) and hydroxynaphthol blue (HNB) are added directly to the initial reaction mixture rendering the whole system a close-tube one, minimizing this way the carry-over contaminations [6, 7]. In this chapter, we present a step-by-step protocol for RT-LAMP for the sensitive, reliable, effective, and fast detection of PLMVd in the lab and the field.

2 Materials

All buffers and solutions are prepared using sterilized DEPC-treated or RNase-free water and molecular grade reagents. All procedures except those of detection must be carried out (conducted) in a UV sterilization working station using UV sterilized pipettes, tips, and PCR tubes (0.2 ml).t

2.1 RT-LAMP Reaction Components

2.1.1 2× LAMP Reaction Buffer

A 2× buffer can be prepared either from 10× ThermoPol Buffer provided with *Bst* 2.0 polymerase (New England Biolabs) or from 10× Isothermal Amplification Buffer consisting of individual components.

10× ThermoPol Buffer: 20 mM Tris–HCl, 10 mM $(NH_4)_2SO_4$, 10 mM KCl, 2 mM $MgSO_4$, 0.1% Triton X-100, pH 8.8. Store at −20 °C.

10× Isothermal Amplification Buffer (100 ml):

20 ml Tris–HCl 1 M pH 8.8 (sterilized), 10 ml KCl 1 M (sterilized), 10 ml $(NH_4)_2SO_4$ 1 M (sterilized), 1 ml Tween 20 (filter sterilized) (*see* **Note 1**), and 49 ml DEPC-treated RNase-free water. Pass this buffer through 0.22 μm filter and store at 4 °C.

2× LAMP reaction buffer (1000 μl):

200 μl 10× ThermoPol Buffer or 10× Isothermal Amplification Buffer, 320 μl betaine 5 M (*see* **Note 2**), 160 μl $MgSO_4$ 100 mM, (*see* **Note 3**), 28 μl AdTP 100 mM, 28 μl TdTP 100 mM, 28 μl CdTP 100 mM, 28 μl GdTP 100 mM, and 208 μl DEPC-treated RNase-free water. Store in aliquots of 250 μl each at −20 °C (*see* **Note 4**).

2.1.2 RT-LAMP Primers (Table 1)

The working solution of each LAMP primer category is as follows: 5 μM for the outer primers (F3, B3), 40 μM for the inner primers (FIP, BIP), and 20 μM for the (Loop F, Loop B).

From the primer stock solutions (100 μM in TE buffer pH 8.0) (*see* **Note 5**), prepare two 25× mixtures, one of the core primers (F3, B3 and FIP, BIP), and one of the Loop primers (Loop F, Loop B).

25× core primer solution (50 μl): 20 μl FIP 100 μM, 20 μl BIP 100 μM, 2.5 μl F3 100 μM, 2.5 μl B3 100 μM, and 5 μl DEPC-treated RNase-free water.

25× Loop primer solution (50 μl): 5 μl Loop F 100 μM, 5 μl Loop B 100 μM, and 40 μl DEPC-treated RNase-free water.

2.1.3 Template

t-RNA.

Plant sap (*see* **Note 6**).

Traces of plant tissues taken by a sterilized toothpick: leaves from infected plants and sterilized toothpicks are needed.

Control (negative and positive) samples must be also included in the tests.

Table 1
Primers used for the detection of PLMVd

Primer name	Primer length (nt)	Position	Sequence
F3	18 nt	32–49	TCAGTGGGCTAAGCCCAG
B3	19 nt	232–249	TGAATCCCGGGTAGACGTC
FIP	36 nt	123–141 71–88	GCGAGGTGCTTAGCCTTTC TCTCAGCCCCTCCACCT
BIP	38 nt	153–173 201–217	GTGGGACTTTTCCTTCTGGAA ACGGCGGTACCTGGATC
Loop F	21 nt	89–109	GTGCYYCGAATAGGGCACYCC
Loop B	21 nt	176–196	AAGCGGTTGGTTCCGAGGGGG

*2.1.4 Enzymes (See **Note 7**)*

Reverse transcriptase: *avian myeloblastosis virus* reverse transcriptase (10 U/μl).

DNA polymerase with strand displacement activity: *Bst* 2.0 DNA polymerase (10 U/μl).

2.1.5 Other Reagents

Dithiothreitol 1 M.

2.2 Detection

2.2.1 Endpoint Detection of RT-LAMP Products by Gel Electrophoresis

1. Agarose powder.
2. 10× TBE buffer (890 mM Tris, 890 mM boric acid, 20 mM EDTA pH 8.0.
3. 6× loading dye: xylene cyanol 0.2% w/v, 0.32% w/v bromophenol blue, and 60% v/v glycerol.
4. 100 bp molecular weight marker.
5. Horizontal gel electrophoresis apparatus set (comb, cassette, tank, power pack) UV transilluminator.

*2.2.2 Endpoint Detection of RT-LAMP Products Using Fluorescent Metal Indicators (See **Note 8**)*

1. Hydroxynaphthol blue (HNB) is dissolved in distilled water at 20 mM to prepare a stock solution. Then a working solution of 3.125 mM in water is prepared.
2. Loopamp Fluorescent Detection Reagent and UV lamp (15 W).

2.2.3 Real-Time Detection

1. Real-time turbidity meter LA 200 Teramecs.

2.3 Specificity Test of RT-LAMP Products

1. NEBCutter V2.0: http://toolsneb.com/NEBCutter2 software (*see* **Note 9**).
2. *Rsa*I restriction enzyme.
3. *10*× *Rsa*I reaction buffer.
4. RT-LAMP product.
5. Thermocycler PTC-200 DNA Engine Cycler (MJ Research, Waltham, Massachusetts, USA).
6. PCR tubes 0.2 ml.

2.4 Use of RT-LAMP in the Field (Fig. 2)

1. Microcentrifuge tubes 1.5 ml.
2. 1× LAMP reaction buffer (*see* **Note 10**).
3. Heat block.
4. Timer.
5. Sterilized toothpicks.
6. UV lamp (15 W).
7. Voltage converter 12 V DC-150 W.

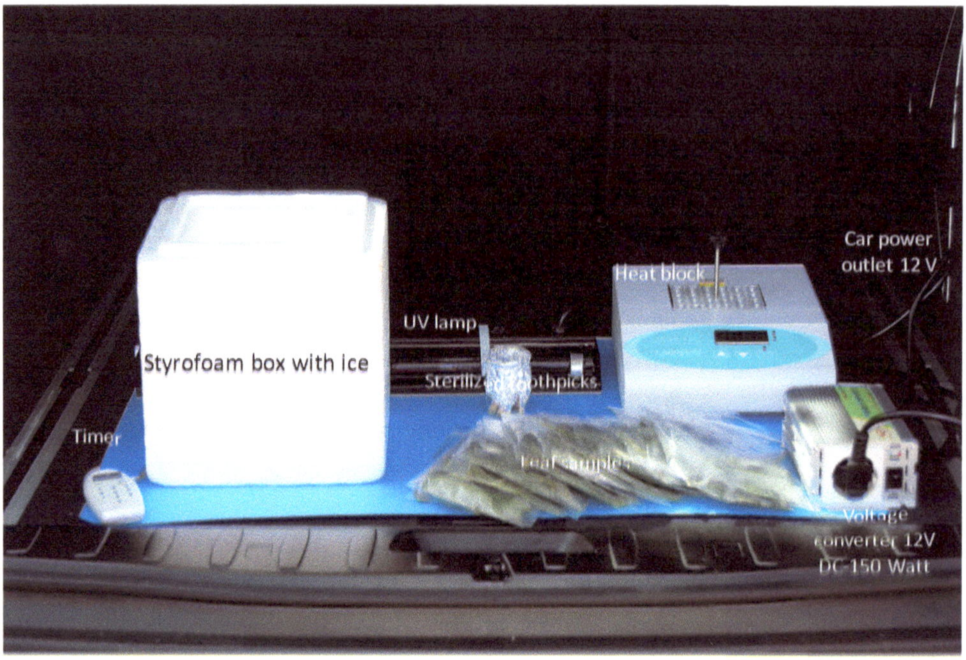

Fig. 2 RT-LAMP field lab

8. Car power outlet 12 V.
9. Leaf samples.
10. Styrofoam box with ice.

3 Methods

Prepare RT-LAMP reactions on ice, and carry out all other procedures at room temperature.

3.1 Primer Design

Primers can be designed manually and exercise special attention on the following key factors:

1. The melting temperature (T_m) has to be estimated using the nearest-neighbor method and is preferred to be calculated under fixed experimental conditions. F3, B3, F2, B2, and loop primer's T_m should be approximately 60 °C while F1 and B1 around 65 °C (Fig. 1).
2. Sequence areas with substantial secondary structure should be avoided. Ideally, primer sequences that do not form either hairpins or dimers (self-dimers or heterodimers), especially at the 3′ ends [14], are preferred.
3. The GC content of the primers should be about 50–60% in the case of GC-rich and about 40–50% for AT-rich targets.

4. A distance of 120–160 bases between the 5′ end of F2 and B2 (the region amplified by the LAMP method) and 40–60 bases between F2 and F1 must be maintained. The primers are also designed so that the distance between the 5′ end of F2 and the 3′ end of F3 to be 0 to 60 bases (Fig. 1).

   Primer can be designed using sophisticated design software programs like PrimerExplorer V4 (Eiken Chemical Company Ltd., Japan), LAMP Designer (Premier Biosoft, USA), LAMP Designer (OptiGene, UK) and various recommended settings (*see* **Note 11**) (Table 1).

3.2 RT-LAMP

1. Set the RT-LAMP reaction mixture in final volume of 25 μl or greater (*see* **Note 12**).
2. Prepare the reaction mix using the components shown in Table 2.
3. Calculate the volume of each reagent and primer needed for the number of reactions being performed (including a positive and negative control), plus one additional reaction per ten reactions to account for pipetting error.
4. Add the required amount of each reagent into a 1.5 ml micro-centrifuge tube, and mix thoroughly. Master-mix should be prepared on ice.
5. Pipette 24 μl of reaction mixture into the required number of 200 μl PCR tubes.

Table 2
Composition of RT-LAMP master-mix reaction buffer

Reagents	Volume (μl) per reaction	Final concentration of each component
2× LAMP reaction buffer	12.5	20 mM Tris–HCl pH 8.8 10 mM KCl 10 mM $(NH_4)_2SO_4$ 0.8 M betaine 8 mM $MgSO_4$ 1.4 mM each dNTP
25× core primer mixture	1	1.6 μM FIP, 1.6 μM BIP, 0.2 μM F3, 0.2 μM B3
25× loop primer mix	1	0.8 μM Loop F, 0.8 μM Loop B
DTT 1 M	0.125	5 mM
AMV reverse transcriptase (10 U/μl)	0.4	4 U
Bst 2.0 DNA polymerase (8 U/μl)	1	8 U
Template (t-RNA or plant sap)	1	
DEPC-treated RNase-free water	7.975	
Total volume	25	

6. Add 1 μl template (t-RNA or crude plant sap) into each PCR tube mix reaction by pipetting, and cap the tubes.
7. Then, briefly pulse-spin in a microcentrifuge (*see* **Note 13**).
8. Place tubes in a thermocycler (*see* **Note 14**), and incubate at 63 °C for 60 min, followed by a 2 min enzyme heat inactivation step at 80 °C (*see* **Note 15**).

3.3 Detection

3.3.1 Endpoint Detection of RT-LAMP Products by Gel Electrophoresis

1. Add 2 gr agar powder in a 250 ml conical flask containing 100 ml 1× TAE buffer, heat in a microwave for approximately 2.5 min until agar is completely dissolved. Swirl the agar in the flask to cool down, then pour it into the cassette gel, and immediately insert the comb for the appropriate number of samples. Leave the gel to set for approximately 30 min.
2. Place the gel in the tank containing enough 1× TAE buffer to cover the gel by a depth of 0.5 cm.
3. For each reaction, mix in a new microcentrifuge 1.5 ml tube 10 μl of RT-LAMP product with 2 μl of 6× loading dye, and load the sample.
4. Load 5 μl of the 100 bp molecular weight marker.
5. Run the gel at 100 V for 40 min.
6. Place the gel in a tank containing 0.5 mg/ml ethidium bromide (*see* **Note 16**) for 30 min and visualize products by placing the gel on a UV transilluminator.

   Gel electrophoresis analysis reveals a ladder-like pattern of RT-LAMP products in contrast to a single band in the case of PCR (Fig. 3) due to the cauliflower-like structures with multiple loops formed by annealing between alternately inverted repeats of the target in the same strand [4] (*see* **Note 17**).

3.3.2 Endpoint Detection of RT-LAMP Products Using Fluorescent Metal Indicators

1. Prepare the reaction mixture using the components shown in Table 2. However, add 1 μl 3.125 mM solution of HNB per 25 μl reaction to give a final concentration of 0.125 μM, or add 1 μl Loopamp Fluorescent Detection Reagent, and reduce the volume of water in the reaction mixture accordingly.
2. Follow **steps 3–8** of the Subheading 3.2 to set up the RT-LAMP reaction.
3. Observe reaction products under ambient light in the case of HNB or UV light for the Loopamp Fluorescent Detection. Observe for a color change from violet to blue or for a bright green, respectively, an indication of a positive reaction (Fig. 3) (*see* **Note 18**).

3.3.3 Real-Time Detection

1. Prepare the reaction mix using the components shown in Table 2, and follow Subheading 3.2 until **step 7**.

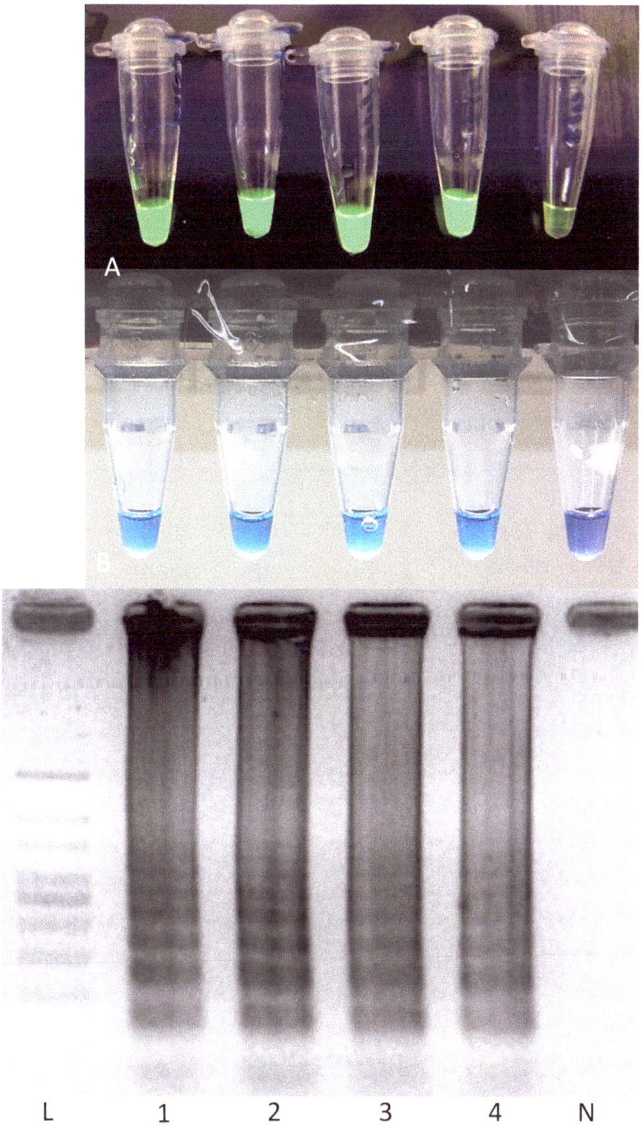

Fig. 3 Endpoint detection of positive RT-LAMP products (lanes 1–4) by naked eye inspection of (**a**) fluorescence under UV light after the addition of Loopamp Fluorescent Reagent, (**b**) blue sky color under ambient light, and (**c**) by agarose gel electrophoresis and ethidium bromide staining. N, negative sample, and L, 100 bp ladder

2. Place reaction tubes in the hot plate of real-time turbidity meter LA-200, and incubate them at 63 °C for 90 min, followed by a 2 min enzyme heat inactivation step at 80 °C (*see* **Note 19**).

Table 3
Composition of restriction analysis master-mix reaction buffer

Reagents	Volume (μl) per reaction	Final concentration of each component
10× *Rsa*I reaction buffer	1	20 mM Tris–HCl pH 9.0 50 mM KCl 0.12% triton X-100
*Rsa*I restriction enzyme (10 U/μl)	1	10 U
RT-LAMP product	3	
DEPC-treated RNase-free water	5	
Total volume	10	

3.4 Specificity Test of RT-LAMP Products

1. The restriction analysis is performed in a final volume of 10 μl.
2. Prepare the reaction mix using the components shown in Table 3, and follow instructions as in Subheading 3.2.
3. Master-mix should be prepared on ice.
4. Pipette 7 μl of reaction mix into the required number of 200 μl PCR tubes.
5. Add 3 μl of RT-LAMP product into each PCR tube mix reaction by pipetting and cap the tubes.
6. Then, briefly pulse-spin in a microcentrifuge.
7. Place tubes in a thermocycler, and incubate at 37 °C for 60 min, followed by a 20 min enzyme heat inactivation step at 65 °C.
8. Analyzed products by gel electrophoresis at 2–2.5% agarose gel analysis followed by EtBr staining (*see* Subheading 3.3.1) (*see* **Note 20**) (Fig. 4).

3.5 RT-LAMP in the Field

1. While in the lab, prepare the reaction mixture using the components shown in Table 2, follow the instruction of Subheading 3.3.2 using Loopamp Fluorescent reagent, and add water to a final volume of 25 μl instead of 24 μl (*see* **Note 14**).
2. Place reaction tubes in ice until the arrival of the field.
3. Collect leaf samples and place them on plastic bags.
4. Connect the voltage converter 12 V DC-150 W to car power outlet 12 V.
5. Turn the heat block device on after plugging it to the voltage converter, and set it to 63 °C.
6. With a sterilized toothpick, penetrate a leaf sample, then open a microcentrifuge tube, and immerse the toothpick to the reaction mixture. Mix reaction with toothpick and cap the tube.

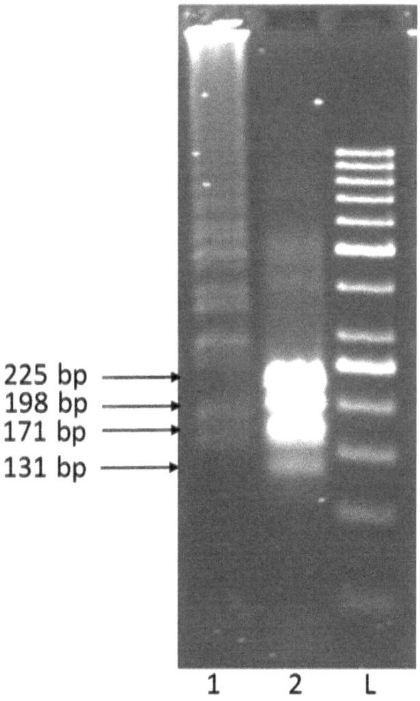

Fig. 4 Restriction enzyme digestion of RT-LAMP products. RT-LAMP products (lane 1) were digested by *Rsa*I (lane 2). L: molecular weight marker

7. Place the microcentrifuge tubes in the heat block, and start the countdown from 60 min.
8. Visualize products by placing tubes in front of the UV LAMP. A bright fluorescence indicates a positive reaction (Fig. 5).

4 Notes

1. For real-time turbidity detection, a ThermoPol Buffer should not be used for LAMP buffer due to Triton X-100, which interferes with turbidimetric detection. Instead, use Isothermal Amplification Buffer and Tween 20.
2. Sometimes for the optimization of the protocol, betaine concentration could be tested [6]. In this case, betaine is added directly at the reaction buffer. Usually, the highest rate is measured in a buffer solution with 0.8 M betaine. When betaine solution is made in house by dissolving solid form in water, do not forget to pass the solution through a 0.22 μm filter. Store the solution up to 2 years at 4 °C. Betaine is highly recommended to achieve consistent amplification and to improve the sensitivity and specificity of low copy detection. Betaine also helps to reduce the formation of primer dimers.

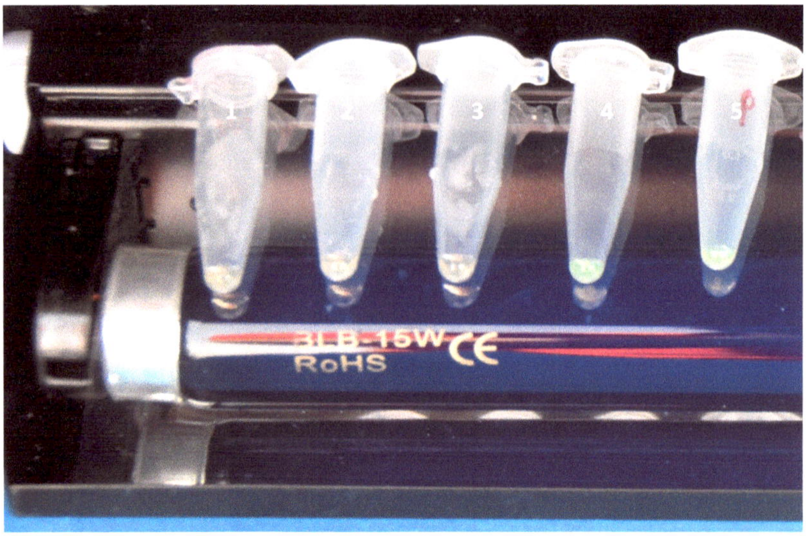

Fig. 5 Field detection of RT-LAMP products using the Loopamp Fluorescent Reagent. Lanes 1–3, negative samples; lanes 4–5, positive samples

3. Mg^{2+}, along with other parameters like dNTPs, plays a significant role in the sensitivity of the method. Researches show that Mg^{2+} concentration from 4 to 8 mM is optimal, while higher levels drop efficiency dramatically [16]. If using commercial polymerase buffers (2 mM Mg^{2+}), additional $MgSO_4$ is necessary to achieve 8 mM final concentration in RT-LAMP reaction buffer. A stock usually 100 mM solution can be purchased or made in house. In the last case, do not forget to pass the solution through a 0.22 μm filter.
4. Buffers and some components may precipitate during thawing but will readily suspend with vortexing. Buffers may be stored for many months and can undergo at least ten freeze/thaw cycles without losing activity.
5. TE buffer: (10 mM Tris–HCl pH 8.0, 1 mM EDTA pH 8.0). Primer purity and quality are important factors for successful LAMP amplification. It is suggested that LAMP primers and especially FIP and BIP have to be HPLC purified.
6. Homogenization buffer: 500 mM Tris–HCl pH 8.3, 2% PVP-40, 1% PEG-6000, 1.4 mM NaCl and 0.05% Tween-20.
7. *Bst* and other DNA polymerases have been demonstrated to have a reverse transcription (RT) activity, although it is variable and much lower than that of native reverse transcriptase. The combination of those two enzymes will provide faster amplification. In the case of one-tube-one-step RT-LAMP, a reverse transcriptase with high-temperature stability has to be added to the reaction mixture. *Avian myeloblastosis virus* reverse polymerase (AMV) is usually preferred on one-tube RT-LAMP

reactions instead of ThermoScript (Invitrogen). A comparison between these two reverse transcriptases with t-RNA or traces of plant tissue as PLMVd templates revealed a higher intensity in the case of AMV [15]. Recently, a new *Bst* enzyme with a reversibly bound aptamer, which inhibits polymerase activity at temperatures below 45 °C, is released.

This hot-start polymerase is called *Bst* 2.0 WarmStart. It resulted in a 50% faster amplification [17] and is considered appropriate for templates with high GC content like viroids.

8. The LAMP reaction results in large amounts of pyrophosphate ion byproduct. These ions react with Mg^{2+} ions to form the insoluble product magnesium pyrophosphate. Since Mg^{2+} ion concentration decreases as the LAMP reaction progresses, the LAMP reaction can be quantified by measuring the Mg^{2+} ion concentration in the reaction solution using fluorescent metal ion indicators like calcein (Loopamp Fluorescent Detection Reagent). HNB is added directly to the initial reaction mixture rendering the whole system a close-tube one, minimizing this way the carry-over contaminations. For calcein, the color is changed from orange to green and for HNB from violet to sky blue in the positive samples [11, 15].
9. To confirm the specificity of RT-LAMP products, a portion of the amplified products is subjected to restriction enzyme analysis [18]. Based on the sequence of the expected amplified products and using the NEBcutter V2.0 software (New England BioLabs, Hertfordshire, England, UK, http://tools.neb.com/NEBcutter2/), the appropriate restriction enzyme is selected.
10. 1 μl Loopamp Fluorescent Detection Reagent per reaction has already been added to the reaction mixture.
11. Loop primers reduced the reaction time (32 min) to less than half compared to that of the original method, while for PSTVd only 13–19 min was enough for the detection of positive samples [6, 8]. Moreover, the addition of the supplemented loop primer pair leads to increased specificity.

    LAMP is prone to non-template amplification due to the amount and number of primers, so care should be taken to avoid primer dimers or other effects. Software solutions will automatically avoid distinct problem sequences and are highly recommended [19].
12. A volume of 20 μl or higher is recommended, particularly if using a color change dye as this gives sufficient volume for the color change to be visible. Besides, a 20 μl or higher volume is needed to allow all the reagents to be added at the appropriate concentration [20].

13. When traces of plant tissue are used as template pipette 25 μl instead of 24 μl of master-mix per reaction tube.
14. If required, for example, in the case of a field version of RT-LAMP, the reaction may be incubated in a heat block or a water bath.
15. The priority in the development of a new RT-LAMP protocol is the selection of the most appropriate primer set, including loop primers (Fig. 6). As noticed by some researchers, more than one set of primers are designed for a single target, but only one finally is selected as the most effective [6, 8, 9, 19]. Then optimization is recommended for the following parameters: (1) $MgSO_4$ (4–8 mM), (2) betaine (0.2–1 M), dNTPs (0.8–2 mM), (3) primer concentration (F3, B3, 0.2–0.5 μM; FIP, BIP, 1–2 μM; Loop F, Loop B, 0.5–1 μM). However, to obtain the optimal reaction for a newly designed RT-LAMP protocol, several incubation temperatures and time must also be tested. Optimization tests should be performed over a range of temperatures between 60 and 65 °C for 15–90 min depending on the nucleic acid template and the primers. Prolonged incubation beyond 90 min is unlikely to increase sensitivity and may result in the eventual generation of nonspecific amplification artifacts [14].
16. Ethidium bromide is mutagenic and can be absorbed through skin and by inhalation. Always wear suitable gloves, and wear a face mask when making up solutions from powder. Alternatively, SYBR Safe can be used which is regarded as having a deficient mutagenic activity.
17. Endpoint detection methods which require opening the PCR tubes following the LAMP reaction like gel electrophoresis analysis increase the risk of contaminating other reaction tubes, work surfaces, and other laboratory equipment due to the large amount of target product produced during the LAMP reaction. On the contrary, close-tube systems like those where metal ion indicators like calcein (Loopamp Fluorescent Detection Reagent) and hydroxynaphthol blue (HNB) are added directly to the initial reaction mixture minimize the carry-over contaminations.
18. Endpoint RT-LAMP results can be easily and quickly be interpreted by naked eye inspection of the reaction mixture under natural or UV light depending on the color indicator used.
19. With the use of real-time turbidity meter LA-200, each sample's turbidity is measured every 6 s until the 90th minute. A turbidity equal to or higher than 0.1 is considered positive (Fig. 6).

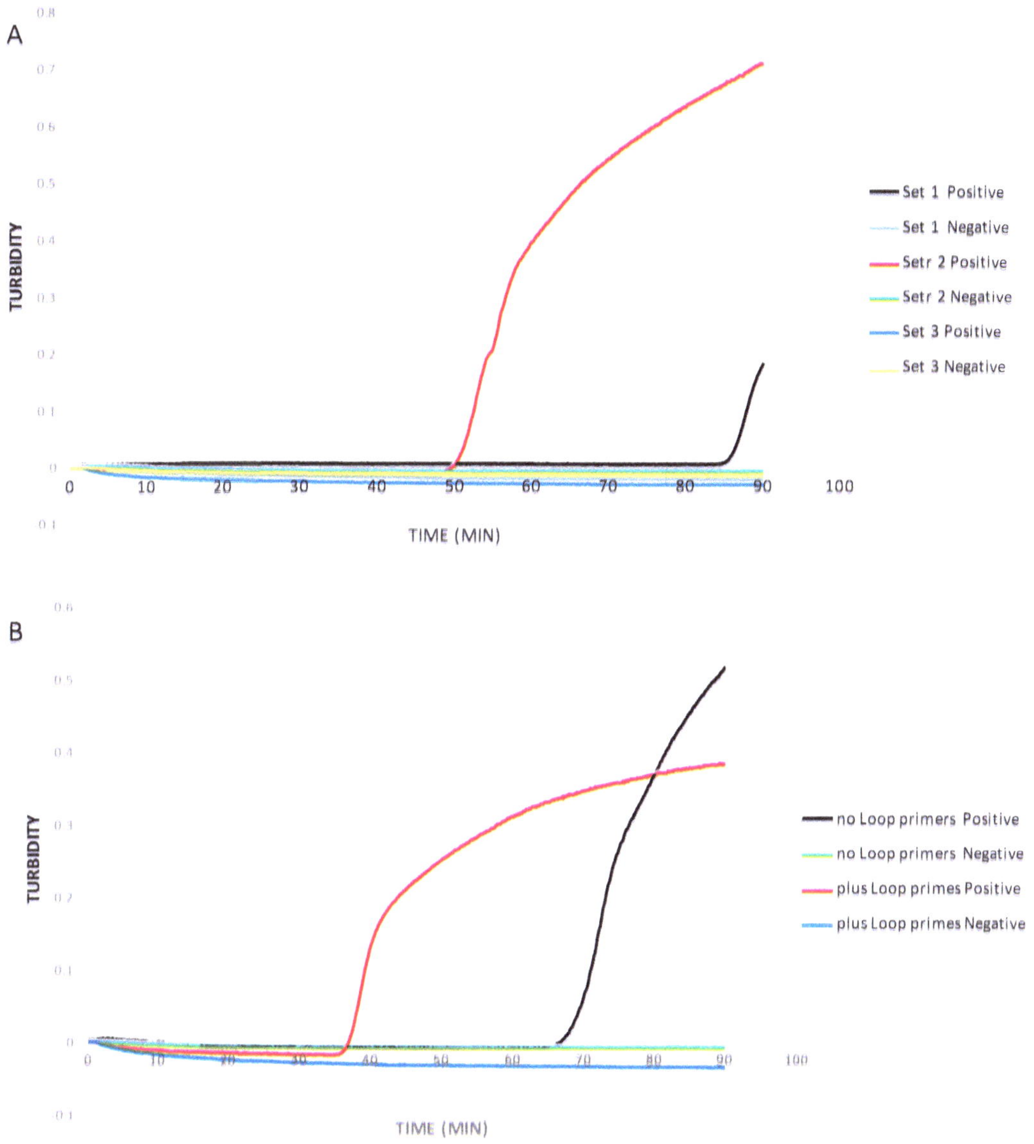

Fig. 6 Detection of RT-LAMP products in real-time using three different sets of RT-LAMP primes by measuring the turbidity using the real-time turbidity meter LA200 (Teramecs) (**a**). In (**b**), the Tn value drops to half when loop primers are used together with primer set 2

20. Based on the sequence of the RT-LAMP expected amplified products and using the NEBcutter V2.0 software (http://tools.neb.com/NEBcutter2/), the *Rsa*I restriction enzyme was found to cut the PLMVd RT-LAMP products in four fragments, namely, 131 bp, 171 bp, 198 bp and 225 bp (Fig. 4).

References

1. Notomi T, Okayama H, Masubucki H, Yonekawa T, Watanabe K, Amino N, Hase T (2000) Loop-mediated isothermal amplification of DNA. Nucleic Acids Res 28:E63
2. Parida M, Inoue GPS, Hasebe F, Morita K (2004) Real-time reverse transcription loop-mediated isothermal amplification for rapid detection of West Nile virus. J Clin Microbiol 42:257–263
3. Hong TC, Mai QL, Cuong DV, Parida M, Minekawa H, Notomi T, Hasebe F, Morita K (2004) Development and evaluation of a novel loop-mediated isothermal amplification method for rapid detection of severe acute respiratory syndrome coronavirus. J Clin Microbiol 42:1956–1961
4. Parida M, Sannarangaiah S, Dash PK, Rao PVL, Morita K (2008) Loop mediated isothermal amplification (LAMP): a new generation of innovative gene amplification technique; perspectives in clinical diagnosis of infectious diseases. Rev Med Virol 18:407–421
5. Fukuta S, Niimi Y, Oishi Y, Yoshimura Y, Anai N, Hotta M, Fukaya M, Kato T, Oya T, Kambe M (2005) Development of reverse transcription loop-mediated isothermal amplification (RT-LAMP) method for detection of two viruses and chrysanthemum stunt viroid. Annual report of the Kansai plant protection. Society 47:31–36
6. Boubourakas IN, Fukuta S, Kyriakopoulou P (2009) Sensitive and rapid detection of peach latent mosaic viroid by the reverse transcription loop-mediated isothermal amplification. J Virol Methods 160:63–68
7. Tsutsumi N, Yanagisawa H, Fujiwara Y, Ohara T (2010) Detection of potato spindle tuber viroid by reverse transcription loop mediated isothermal amplification. Res Bull Plant Prot Serv Jpn 46:61–67
8. Lenarčič R, Morisset D, Mehle N, Ravnikar M (2012) Fast real-time detection of potato spindle tuber viroid by RT-LAMP. Plant Pathol 62:1147–1156
9. Park J, Jung Y, Kil EJ, Kim J, Thi Tran D, Choi SK, Yoon JY, Cho WK, Lee S (2013) Loop-mediated isothermal amplification for the rapid detection of chrysanthemum chlorotic mottle viroid (CChMVd). J Virol Methods 193:232–237
10. Liu XL, Zhao XT, Muhammad I, Ge BB, Hong B (2014) Multiplex reverse transcription loop-mediated isothermal amplification for the simultaneous detection of CBV and CSVd in chrysanthemum. J Virol Methods 210:26–31
11. Thanarajoo SS, Kong LL, Kadir J, Lau WH, Vadamalai G (2014) Detection of coconut cadang-cadang viroid (CCCVd) in oil palm by reverse transcription loop-mediated isothermal amplification (RT-LAMP). J Virol Methods 202:19–23
12. Mori Y, Nagamine K, Tomita N, Notomi T (2001) Detection of loop-mediated isothermal amplification reaction by turbidity derived from magnesium pyrophosphate formation. Biochem Biophys Res Commun 289:150–154
13. Nagamine K, Kuzihara Y, Notomi T (2002) Isolation of single-stranded DNA from loop-mediated isothermal amplification products. Biochem Biophys Res Commun 290:1195–1198
14. Tomlinson J (2013) In-field diagnostics using loop-mediated isothermal amplification. In: Dickinson M, Hodgetts J (eds) Phytoplasma: methods and protocols, Methods in Molecular Biology. Springer, Berlin, pp 291–300
15. Boubourakas IN, Fucuta S, Luigi M, Faggioli F, Barba M, Kyriakopoulou PE (2010) Improvement of the reverse transcription loop mediated isothermal amplification (RT- LAMP) method for the detection of peach latent mosaic viroid (PLMVd). Julius-Kühn-Archiv 427:65–69
16. Nie X (2005) Reverse transcription loop-mediated isothermal amplification of DNA for detection of potato virus Y. Plant Dis 89:605–610
17. Tanner NA, Zhang Y, Evans TC Jr (2012) Simultaneous multiple target detection in real-time loop-mediated isothermal amplification. BioTechniques 53:81–89
18. Kubota R, Vine BG, Alvarez AM, Jenkins DM (2008) Detection of Ralstonia solanacearum by loop-mediated isothermal amplification. Phytopathology 98:1045–1051
19. Tanner NA, Evans TC (2014) Loop-mediated isothermal amplification for detection of nucleic acids. Curr Protoc Mol Biol 105:15.14.1–15.14.14
20. Ward LI, Harper SJ (2012) Loop-mediated isothermal amplification for the detection of plant pathogens. In: Sucher N (ed) Plant DNA fingerprinting and barcoding: methods and protocols, methods in molecular biology, vol 862. Springer, Berlin, pp 161–170

Chapter 18

SYBR® Green RT-qPCR for the Universal Detection of Citrus Viroids

Georgios Vidalakis, Jinbo Wang, Tyler Dang, Tavia Rucker, Sohrab Bodaghi, Shi-hua Tan, Irene Lavagi-Craddock, Alexandra Syed, Gerardo Uribe, Yi Hsieh, and Joana Carbajal-Moreno

Abstract

Quantitative polymerase chain reaction (qPCR) and reverse transcription (RT)-qPCR have now become the gold standard for molecular diagnostics because of its sensitivity, specificity, and reproducibility. In addition, qPCR diagnostics are flexible because they can be scaled for high- or low-throughput applications. Here we describe an optimized assay and workflow for the universal detection of eight citrus viroid species and their variants by RT-qPCR. The assay allows for quick and efficient molecular detection of viroids without the need to run RT-qPCR for each individual viroid species.

Keywords Degenerate primers, High-throughput screening, Melting temperature, CCPP, CDFA

1 Introduction

Molecular detection of citrus viroids has complemented traditional methods such as biological indexing [1], imprint hybridization [2, 3], and sequential polyacrylamide gel electrophoresis (sPAGE) [4]. Quantitative polymerase chain reaction (qPCR) and reverse transcription (RT)-qPCR have now become the gold standard for molecular detection. With the addition of high-throughput nucleic acid extractions (for an example *see* Chapter 5 in this book), an effective diagnostic procedure will allow regulatory programs, diagnostic laboratories, and research programs to test high volumes of samples in a shorter time.

Two SYBR® Green RT-qPCR assays were developed to detect eight different viroid species and their variants. The "apsca" set of primers were developed to detect viroid species of the *Apscaviroid* genus which includes citrus bent leaf viroid (CBLVd), citrus dwarfing viroid (CDVd), citrus viroid V (CVd V), citrus viroid VI (CVd VI), and citrus viroid VII (CVd VII). The "non-apsca" set of

Ayala L. N. Rao et al. (eds.), *Viroids: Methods and Protocols*, Methods in Molecular Biology, vol. 2316, https://doi.org/10.1007/978-1-0716-1464-8_18,

primers were developed to detect citrus exocortis viroid (CEVd), citrus bark cracking viroid (CBCVd), and hop stunt viroid (HSVd) of the *Pospiviroid*, *Cocadviroid*, and *Hostuviroid* genus, respectively [5]. This design allows for efficient screening for viroids without the need to run individual RT-qPCR assays for each viroid species. The assay has been used successfully with a variety of different extraction methods that include traditional phenol chloroform separation, spin column-based, and high-throughput magnetic-based RNA purification methods [6, 7]. The protocol and workflow presented here also help minimize the risk of cross contamination of RT-qPCR reactions allowing for more reliable results. These assays have been approved by California Department of Food and Agriculture (CDFA) for use in the mandatory "Citrus Nursery Stock Pest Cleanliness Program-Senate Bill 140" and have been used by the Citrus Clonal Protection Program (CCPP) for routine diagnostics of over 27,000 samples to date [8–10].

2 Materials

1. Bio-Rad CFX96™ Real-Time PCR Detection System.
2. "Apsca" primer set for the detection of CBLVd, CDVd, CVd V, CVd VI, and CVd VII: forward, GAR MMW YCK TGT GGT TCC TGT GG; reverse, HYV DWH GTC CGC TCG ACT AGC.
3. "Non-apsca" primer set for the detection of CEVd, CBCVd, and HSVd: forward, ARG GAK CCC CGG GGM AA; reverse, CTS KAC KCC AGW GMW CCG CGG C.
4. Plant internal control primers targeting nad5: forward, GAT GCT TCT TGG GGC TTC TTG TT; reverse. GCG GAT CCT CGG ACA TAT ATG A (*see* **Note 1**).
5. Conventional PCR thermal cycler.
6. Eppendorf 5430 centrifuge with the swing-bucket rotor (A-2 MTP).
7. Two dead air box PCR workstation placed in different rooms (*see* **Note 2**).
8. Two sets of pipettes (P10, P20, P200, P1000) and filtered tips.
9. 96 well unskirted PCR plate, low profile (Bio-Rad #MLL-9601).
10. Adhesive plate seals, optically clear (Bio-Rad #MSB-1001).
11. Hand applicator squeegee.
12. iTaq™ Universal SYBR® Green One-Step Kit (Bio-Rad #1725151).
13. Nuclease-free water.

14. 1.5 mL microcentrifuge tubes.
15. Ice bucket with ice.
16. Microcentrifuge.
17. 96 well PCR plate cool rack.
18. Vortex mixer.

3 Methods

3.1 RT-qPCR Preparation

1. Resuspend the lyophilized primers with nuclease-free water to make 100μM primer stock solution under a dedicated (not used for PCR reaction setup with RNA template) dead air box PCR workstation for preparing PCR mater mixes and primers (*see* **Note 2**).
2. Create a working primer solution of 10μM by performing a 1 to 10 dilution in a 1.5μL microcentrifuge tube (*see* **Note 3**).
3. Thaw the iTaq™ Universal SYBR® Green reaction mix (2×) on ice, and protect from light.
4. Centrifuge all reagents briefly on a benchtop centrifuge to collect the solutions at the bottom of the tube. Prepare the appropriate volume of primer and master mix for the experiment in separate tubes.
   (a) 7.4μL of nuclease-free water and 0.6μL of forward and reverse primer (300 nM final concentration for each primer) per reaction.
   (b) 10μL of iTaq™ Universal SYBR® Green reaction mix (2×) and 0.25μL reverse transcriptase enzyme per reaction.
5. Move the prepared reagents to a second dead air box PCR workstation with dedicated (used for PCR reactions setup with RNA template) pipettes and filtered tips.
6. Place an unskirted PCR plate into a cool rack.
7. Vortex the primer mix thoroughly and dispense 8.6μL into each well.
8. Add 1μL of RNA (25–100 ng) to the individual PCR wells.
9. Use the hand applicator squeegee to gently seal the plate with an adhesive plate seal.
10. Place the PCR plate into the swing-bucket centrifuge, and centrifuge for 30 s at 300 × *g*.
11. Place the plate with the primers and RNA template into a conventional PCR thermal cycler, and incubate at 80 °C for 5 min (*see* **Note 4**).

12. After the completing the program, immediately place the plate on ice for at least 5 min.
13. Place the PCR plate into the swing-bucket centrifuge, and centrifuge for 30 s at 300 × *g*.
14. Gently remove the plate seal (*see* **Note 5**).
15. Mix the solution thoroughly with a vortex, and dispense 10.25μL of the SYBR Green and reverse transcriptase mix into each well. The final volume of the reaction will be 20μL.
16. Use the hand applicator squeegee to completely seal the plate with an adhesive plate seal.
17. Place the PCR plate into the swing-bucket centrifuge, and centrifuge for 30 s at 300 × *g*.
18. Place the PCR plate into the Bio-Rad CFX96™ Real-Time PCR machine, and run the program according to the corresponding citrus viroids of interests. Set the fluorophore to scan for SYBR® Green.
    (a) The conditions for "apsca" citrus viroids: reverse transcription at 50 °C for 30 min, polymerase activation at 95 °C for 5 min, denaturation at 95 °C for 10 s, annealing at 62 °C for 30 s, repeat denaturation and annealing for an additional 34 cycles, incubation at 95 °C for 1 min, incubation at 55 °C for 1 min, and melting curve analysis from 55 to 95 °C with 0.5 °C increments for 10 s plate read.
    (b) The conditions for "non-apsca" citrus viroids: reverse transcription at 50 °C for 30 min, polymerase activation at 95 °C for 5 min, denaturation at 95 °C for 10 s, annealing at 61 °C for 30 s, repeat denaturation and annealing for an additional 34 cycles, incubation at 95 °C for 1 min, incubation at 55 °C for 1 min, and melting curve analysis from 55 to 95 °C with 0.5 °C increments for 10 s plate read.
    (c) The NAD internal control primers can be run with either condition.
19. The RT-qPCR will typically take 2 to 2.5 h to complete.

3.2 Interpreting RT-qPCR Results

1. Ensure that the internal plant control gene nad5 produced C_q values and a melting temperature at 77.5 °C.
2. Ensure the positive controls produced C_q values and melting temperatures. The melting temperature will vary between depending on the citrus viroid species (*see* **Note 6**).
    (a) Primers "apsca" generated amplicons have melting temperatures with the approximate range of 84 to 87.5 °C ± 0.5 °C.

   (b) Primers "non-apsca" generated amplicons have melting temperatures with the approximate range of 81.5 to 83.5 °C ± 0.5 °C.

3. Compare the results from the tested samples to the viroid controls.
   (a) If a C_q value and melting temperature match with the positive controls, then the sample is positive for citrus viroid.
   (b) If a C_q value is present, but no melting temperature is generated, this is a questionable sample. Consider retesting the sample.
   (c) If there is no C_q value and melting temperature, the sample is negative for citrus viroids.

4 Notes

1. The RNA concentration and purity of all tested samples should be accessed by spectrophotometer (OD A_{260} and A_{280}) or any equivalent method [12, 13]. However, since in most cases it is not practical or cost-effective to access the RNA integrity of every sample tested, the user should run a subset of the tested samples (in the same or different plate) with the plant internal control NADH dehydrogenase (nad5) [11].
2. A dedicated clean room with a −20 °C freezer, dead air box PCR workstation, and pipettes should be established to be used for preparing primers and PCR reagent master mixes. This will help prevent RNA template contamination of stock PCR reagents and primers. A second room with dead air box PCR workstation should be established that is used strictly used to dispense the mixes, add the RNA template, and finalize PCR reaction setup. Use filter tips when preparing the RT-qPCR. If possible the PCR thermocycler instruments should be located in an isolated dedicated room (i.e., not in the main lab). Sanitize thoroughly all working stations in all rooms using 10% household bleach solution (*see* Chapter 22 Subheading 3.1).
3. It is good practice to prepare many aliquots of the 10μM working primer solution to prevent repeated freeze and thaw of the main stock.
4. Because of the complex secondary viroid RNA structures and high G + C content an incubation set at 80 °C is required. This will help denature the viroid secondary structures to allow the primers to anneal more efficiently and achieve proper cDNA synthesis.
5. Remove the seal in one continuous, slow, steady motion to prevent droplets from cross contamination neighboring wells.

6. Include appropriate positive and negative controls for each primer set. The user should include one of each viroid positive controls to the respective primer set to ensure the reactions are working properly and have a reference for the melting curve analysis.

Acknowledgments: *This work was supported by the California Citrus Nursery Board (CCNB) project "Implementation and streamlining of the newly developed high throughput diagnostic system for citrus nurseries registration" and by the Citrus Research Board (CRB) project "Citrus Clonal Protection Program" (6100) awarded to Vidalakis G. Additional support was provided by the United States Department of Agriculture (USDA) National Institute of Food and Agriculture (NIFA), Hatch (project 1020106), and the National Clean Plant Network (NCPN) which operates under the auspices of USDA Animal and Plant Health Inspection Service (APHIS) (12-8100-1544-CA; 14-, 15-, 16- 8130-0419-CA; AP17PPQS&T00C118; AP18PPQS&T00C107). The findings and conclusions in this publication are those of the author(s) and should not be construed to represent any official USDA or US Government determination or policy.*

URLs to sponsors' websites are listed below:

1. *CRB: https://www.citrusresearch.org*
2. *CCNB: http://www.ccnb.info/*
3. *USDA NIFA: https://nifa.usda.gov*
4. *NCPN: http://nationalcleanplantnetwork.org.*

References

1. Roistacher CN (1991) Graft-transmissible diseases of citrus: Handbook for detection and diagnosis. Food and Agriculture Organization of the United Nations, Rome
2. Romero-Durbán J, Cambra M, Duran-Vila N (1995) A simple imprint-hybridization method for detection of viroids. J Virol Methods 55:37–47
3. Palacio-Bielsa A, Foissac X, Duran-Vila N (1999) Indexing of citrus Viroids by imprint hybridisation. Eur J Plant Pathol 105:897–903
4. Rivera-Bustamante RF, Gin R, Semancik JS (1986) Enhanced resolution of circular and linear molecular forms of viroid and viroid-like RNA by electrophoresis in a discontinuous-pH system. Anal Biochem 156:91–95
5. Vidalakis G, Wang J. Molecular method for universal detection of citrus viroids, United States patent US 2013/0115591 A1, May 9, 2013
6. Chambers GA, Donovan NJ, Bodaghi S, Jelinek SM, Vidalakis G (2018) A novel citrus viroid found in Australia, tentatively named citrus viroid VII. Arch Virol 163:215–218
7. Osman F, Dang T, Bodaghi S, Vidalakis G (2017) One-step multiplex RT-qPCR detects three citrus viroids from different genera in a wide range of hosts. J Virol Methods 245:40–52
8. Bostock RM, Thomas CS, Hoenisch RW, Golino DA, Vidalakis G (2014) Plant health: how diagnostic networks and interagency partnerships protect plant systems from pests and pathogens. Calif Agric 68:117–124
9. Osman F, Hodzic E, Kwon S-J, Wang J, Vidalakis G (2015) Development and validation of a multiplex reverse transcription quantitative PCR (RT-qPCR) assay for the rapid detection of citrus tristeza virus, citrus psorosis virus, and citrus leaf blotch virus. J Virol Methods 220:64–75

10. Vidalakis G. California's citrus nursery stock pest cleanliness program: a success story of the collaborative power of industry, university, and regulatory agencies. In: Abstracts of Presentations at the 20th Conference of the International Organization of Citrus Virologists (IOCV). J. Citrus Pathol. 3 (1), 15 (iocv journalcitruspathology 32500), 2016. Retrieved from: https://escholarship.org/uc/item/7986f88j
11. Saponari M, Manjunath K, Yokomi RK (2008) Quantitative detection citrus tristeza virus in citrus and aphids by real-time reverse transcription-PCR (Taqman®). J Virol Methods 147:43–53
12. Manchester KL (1996) Use of UV methods for measurement of protein and nucleic acid concentrations. BioTechniques 20:968–970
13. Glasel JA (1995) Validity of nucleic acid purities monitored by 260nm/280nm absorbance ratios. BioTechniques 18:62–63

Chapter 19

Detection of *Avocado Sunblotch* and Other Viroids Using RNA Filter Paper Capture and RT-PCR

Deborah M. Mathews, Sohrab Bodaghi, James A. Heick, and J. Allan Dodds

Abstract

This method originated due to the need to quickly and sensitively detect *Avocado sunblotch viroid* (ASBVd) in nursery and field trees in California. Optimum sampling protocols were developed for leaf collection from different sized trees based on size and branching as well as for fruit. An ethanol containing buffered extract from 1 g of ground leaf tissue was used as the source of RNA. The extract was absorbed onto small pieces (disks) of Whatman No. 1 filter paper which were then washed and dried. RNA was eluted from the filter paper using sterile water and used as a template in a standard single-tube RT-PCR reaction. The RNA adsorbed on the filter paper disks was quite stable, and the disks could be stored for over 1 year and shipped worldwide at ambient temperature with no noticeable decline in the quality or quantity of the resulting RT-PCR products. The filter paper capture method was expanded to the detection of other viroids including *Potato spindle tuber viroid*, *Peach latent mosaic viroid*, and *Chrysanthemum stunt viroid* and was tested with some viruses as well with minor modifications of the standard protocol.

Key words Viroid detection, RNA filter paper capture, RT-PCR, *Avocado sunblotch viroid*, Avocado sampling

1 Introduction

Avocado sunblotch viroid (ASBVd) infections may either be symptomless or have foliar symptoms with bright yellow areas, mosaic, and or distortion (Fig. 1a). Fruit symptoms are expressed as white-, yellow-, or red-colored sunken lesions making them unmarketable in the USA (Fig. 1b) [1, 2]. In almost 20 years of field testing using this method, less than 10% of the ASBVd-positive trees showed any visible symptoms; therefore a laboratory detection method is required for accurate detection of infected trees. ASBVd is transmitted by grafting (both in the nursery and in nature via the roots), through pollen and seed but has no known vector other than bees transmitting infected pollen to flowers [3]. Historically, biological indexing was used for the detection of ASBVd in the breeding, field

Ayala L. N. Rao et al. (eds.), *Viroids: Methods and Protocols*, Methods in Molecular Biology, vol. 2316, https://doi.org/10.1007/978-1-0716-1464-8_19,

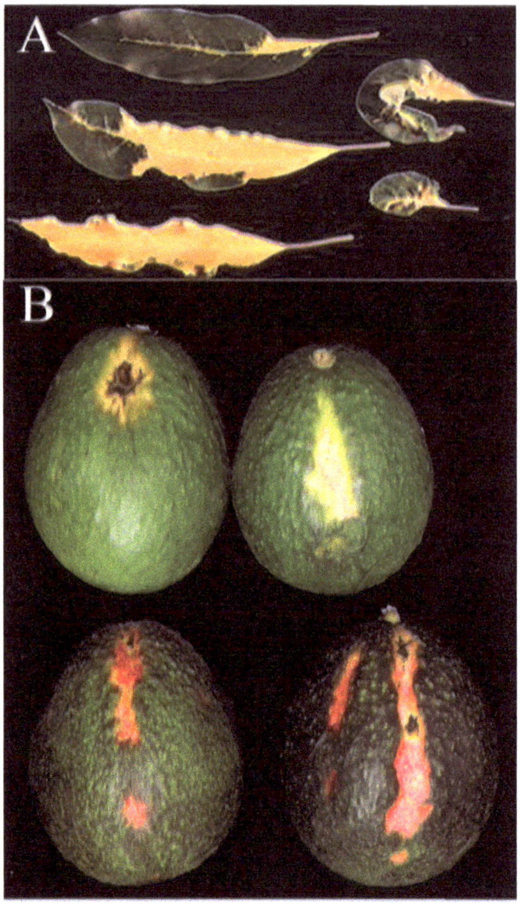

Fig. 1 Symptoms of *avocado sunblotch viroid* on leaves (**a**) and fruit (**b**). (photo in panel A courtesy of MaryLu Arpaia)

production, and nursery industries in California. This was a time-consuming, space-intensive procedure that took up to 3 years to complete, and as technicians retired and greenhouse space became more valuable, the ability to test for this pathogen was all but lost. In the late 1990s, a major nursery contacted our laboratory at the University of California, Riverside (UC Riverside), and asked if we could develop a quicker, cheaper, and more sensitive testing protocol using modern laboratory techniques. PCR had come into its own by then, so we set out to design a quick, inexpensive, and efficient extraction protocol that would enable multiple samples to be processed in less than a day with the RT-PCR result the next day. Our lab specialized in the extraction and analysis of double-stranded RNAs using CF-11 cellulose chromatography with a Tris/NaCl/EDTA buffer containing 16.5% ethanol [4, 5]. Knowing that different classes of nucleic acids will selectively bind to cellulose at different ethanol concentrations [6–8], we set out to use this phenomenon to bind viroid RNA, which spends most of its life in a double-stranded form, onto a solid filter substrate. Multiple membranes and filters were tried, but standard Whatman No. 1 filter paper cut into small squares turned out to bind and release the RNA optimally. The method produces good-quality RNA that is stable for up to 2 years at room temperature, can be shipped long distances through the mail in a standard envelope with no loss of suitability as a PCR template, and enables testing for multiple pathogens from one sample if desired. These attributes would allow preparation of RNA containing disks in one location, perhaps in a country with minimal molecular biology capabilities, and shipping them to a laboratory for analysis without requiring a tissue transfer permit. This was the first PCR method approved for pathogen detection in a regulated commercial crop by the California Department of Food and Agriculture (CDFA). We also consulted with researchers from New Zealand, Chile, and Costa Rica helping them to establish sampling and testing protocols of their own.

We present here the method that was used for 17 years in the laboratories of Mathews and Dodds at UC Riverside. Over 5000 samples were processed during that time from plants in the UC Riverside avocado breeding program and from over 40 commercial growers and nurseries. Approximately 4% of the samples tested positive for ASBVd, over 90% of which were asymptomatic, and those trees were generally removed to avoid root grafting and transmission of the viroid to adjacent trees and to prevent infected pollen from being dispersed.

2 Materials

Prepare all buffers with distilled or column purified water and analytical grade reagents. Store at room temperature unless noted.

1. Plant material: leaf tissue is primarily used for testing. For avocados, bark, fruit, and seeds can also be used in some cases. Details for collecting and processing plant tissue are in Subheading 3.
2. Plastic bags for collecting and storing cut tissue.
3. Weigh boats, inner base diameter 47 mm or similar.
4. Razor blades for cutting tissue.
5. Mortars (approximate size and volume: 80 mm outside diameter × 53 mm high, 65 mL) and pestles (approximate length: 114 mm) for grinding tissue.
6. Liquid nitrogen.
7. Concentrated stock buffers:

1 M Tris–HCl, pH 8.0: Add 157.6 g of Tris–HCl to 800 mL of water in a glass beaker. Fully dissolve and adjust pH to 8.0 with HCl. Adjust final volume to 1 L and autoclave.

5 M NaCl: Add 146.1 g NaCl to 400 mL of water, dissolve fully, adjust final volume to 500 mL, and autoclave.

10× STE buffer: 0.5 M Tris, 1 M NaCl, 10 mM EDTA, pH 6.8. Add 61 g Tris, 58.4 g NaCl, and 3.7 g Na_2 EDTA to 800 mL of distilled water, and stir until dissolved. Adjust pH to 6.8 with concentrated HCl. Add water to a final volume of 1 L.

8. Tissue extraction buffer (TXB) (*see* **Note 1**): 100 mM Tris–HCl, pH 8.0, 1.5 M NaCl, 5% polyvinylpyrrolidone 40 (PVP-40), and 0.5% 2-mercaptoethanol (*see* **Note 2**). To make 500 mL, measure 200 mL of distilled water into a glass beaker, add 50 mL 1 M Tris–HCl, pH 8.0, 150 mL 5 M NaCl, and 25 g of PVP-40 (*see* **Note 3**), and stir until PVP-40 is completely dissolved. Add 2.5 mL 2-mercaptoethanol (*see* **Note 2**), and adjust volume to 500 mL with distilled water. Sterilize the solution using a filtration unit containing a 0.2μm filter (for RNAse exclusion), and store at 4 °C.
9. Wash buffer: 1× STE buffer containing 28% ethanol. Mix 100 mL of 10× STE buffer with 620 mL distilled water, stir until mixed, then add 280 mL of 100% (200 proof) ethanol (*see* **Note 4**) for a final volume of 1 L, and stir. Transfer to a tightly capped glass bottle.
10. Filter disks: Whatman #1 filter paper circles, 90 mm diam (*see* **Note 5**). Using gloves and sharp scissors that have had the blades wiped with ethanol, cut the filter circles into small pieces approximately 4–5 mm square. The exact size and shape are not overly important but should be fairly consistent (Fig. 2). Handle in a sterile manner at all times. Store cut disks in a screw capped sterile centrifuge tube, and remove as needed.
11. Disposable glass tubes: 12 mm × 75 mm or 13 mm × 100 mm.
12. Disposable polyethylene transfer pipets.
13. Vortex mixer.
14. Aluminum foil.
15. PCR primers: *Avocado sunblotch viroid* – forward primer 5′-A AGTCGAAACTCAGAGTCGG-3′, reverse primer 5′-GTGA GAGAAGGAGGAGT-3′, (247 bp amplicon) [9]. A secondary back up set of primers was used to confirm sample infection status when background bands near 247 bp were present or amplicon levels were very weak: forward primer AR2 5-TCT TTCCCTGAAGAGACGA-3′, reverse primer AR2 5′-TGGG AAGAACACTGATGAG-3′ (247 bp amplicon) [10].

    Pospiviroid group primers for the detection of *Potato spindle tuber viroid* (PSTVd, 199 bp amplicon) and

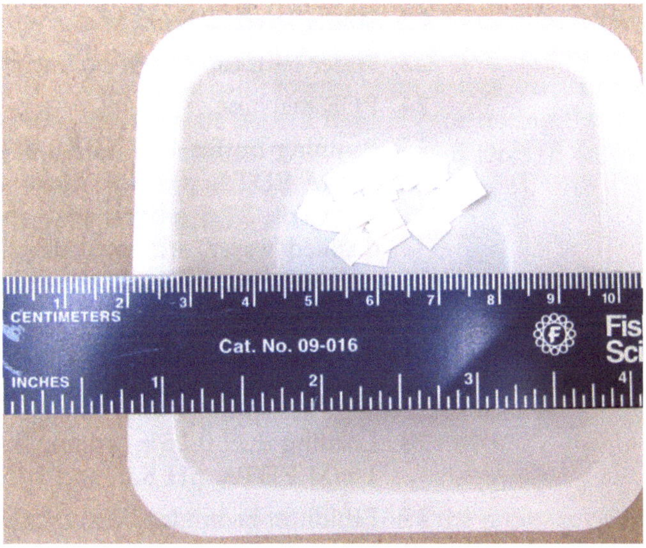

Fig. 2 Filter paper disks for capturing viroid RNA. Disks approximately 4–5 mm square are cut from Whatman #1 filter paper circles and stored in a sterile tube until needed

Chrysanthemum stunt viroid (CSVd, 199 bp amplicon) among others; forward primer 5′-ATTAATCCCCGGGGAAACCTG GAG -3′, reverse primer 5′- AGCTTCAGTTGTTTC CACCGGGT-3, [11]. PSTVd has been reported in avocado in Peru, [12], and some countries require testing for this viroid before plants can be exported from California. We performed dual ASBVd and PSTVd tests on several nursery trees over the years by producing two disks from a single sample and using each for a separate RT-PCR reaction. PSTVd was never found in any avocado sample in our laboratory. However, PSTVd was found in a sample of the ornamental plant *Cestrum* spp. obtained from a grower.

Peach latent viroid: Forward primer 5′- CCCGATA GAAAGGCTAAGCACCTCG-3′, reverse primer 5′-AACTG CAGTGCTCCGT-3′, (336 bp amplicon) [13].

Satellite tobacco mosaic virus (*see***Note 1**): forward primer 5′- CAAACCGTAAATCGACGGGTGACAATTCGA -3′, reverse primer 5′- TTACTTCTTACTTCGCATA CATTGTCGGCT-3′ (Mathews unpublished data).

16. 0.6 mL microcentrifuge tubes, sterile.
17. Taq DNA polymerase and buffer (Promega Go Taq system).
18. AMV-Reverse Transcriptase.
19. RNasin ribonuclease inhibitor.
20. Dithiothreitol, 10 mM.
21. $MgCl_2$, 25 mM.

DTT, 0.25μL RNasin (10 units, Promega), 1.5μL AMV-RT (15 units, Promega), and 0.5μL Taq DNA polymerase (2.5 units, Promega); store on ice (*see* **Note 15**).

2. Open the microfuge tubes containing a dried filter disk, and pipet 42μL of sterile water onto the disk, and then use the tip of the pipet to gently swirl the water and disk together to ensure complete hydration of the disk and elution of the RNA; place on ice. When all samples have been hydrated, place in a boiling water bath for 3 min to denature the RNA, and immediately place on ice. After a few minutes of cooling, briefly centrifuge the samples to remove any condensate from the walls of the tubes.
3. To each sample tube, add 58μL of PCR master mix cocktail leaving the disk in place and mixing well with the pipet tip. Final volume is now 100μL. As an alternative, 20μL of the eluted and denatured extract may be removed and added to 30μL of PCR master mix (*see* **Note 16**).
4. Perform single-tube RT-PCR using the following program: 1 first-strand cycle of 45 min at 42 °C, 1 denaturation cycle of 2 min at 95 °C, and 40 amplification cycles of 1 min at 94 °C, 2 min at 55 °C, and 3 min at 72 °C, with a final extension cycle of 5 min at 72 °C [9]. PSTVd and PLMVd have optimal annealing temperatures of 62 °C; adjust for other pathogens as necessary.

3.5 Gel Electrophoresis

1. Cast a 2% agarose gel in running buffer, 1× TBE, size and number of wells is dependent on number of samples. For an 8 cm × 15 cm gel, add 2 g of agarose to 100 mL of 1× TBE buffer in a glass flask. Boil in a microwave oven until melted. Cool to approximately 65 °C, and pour into the casting tray with a comb containing the desired number of wells. Cool until solidified, place in electrophoresis apparatus, and add 1× TBE buffer until gel is barely covered.
2. Remove a 10μL aliquot of the RT-PCR sample and mix with 1μL of loading dye. Load the entire aliquot into a well. Electrophorese at 175 mA for 90 min or until the bromophenol blue dye is 2 cm from the bottom of the gel (*see* **Note 17**).
3. Remove the gel from the apparatus and place in a plastic container with 100 mL water. Add 10μL of ethidium bromide and gently shake for 20 min to stain the DNA. Drain the ethidium mixture and rinse once quickly with water, and then destain the gel fully using two 10 min washes with water.
4. Place the destained gel on a UV transilluminator, and photograph any bands that are present (Fig. 4).

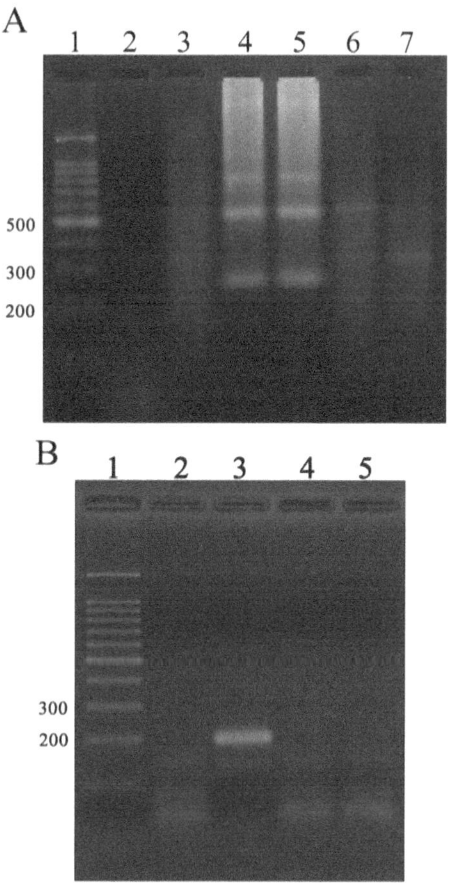

Fig. 4 Agarose gel electrophoresis of ASBVd (**a**) and PSTVd (**b**) RT-PCR products. (Panel **a**) lanes: (1) 100 base pair ladder; (2) water control; (3) healthy avocado control; (4) ASBVd-positive control; (5) ASBVd-positive field sample; (6) ASBVd-negative field sample; (7) ASBVd-negative field sample. Note extraneous background bands in lanes 6 and 7 that are not of expected 247 bp size for ASBVd (*see* **Note 17**). Also note dimers (500 bp) and trimers (750 bp) of monomer RT-PCR product in lanes 4 and 5 (*see* **Note 17**). (Panel **b**) lanes: (1) 100 base pair ladder; (2) healthy control; (3) PSTVd-positive control from *Cestrum* spp.; (4 and 5) avocado field samples, PSTVd-negative

4 Notes

1. When using this method for viruses, we found that adding sodium dodecyl sulfate (SDS) to a final concentration of 0.6% increased the amount of resulting PCR amplicons tenfold. The negatively charged detergent presumably helps disrupt the coat protein thus releasing capsid bound ssRNA which can then be bound to the filter paper disk. This method worked well for the small (17 nm diameter) spherical virus *Satellite tobacco mosaic virus* [14, 15]. When analyzing citrus tissues infected with

Citrus tristeza virus [16, 17], a long, flexuous rod-shaped virus (2000 nm), we had variable, unrepeatable results even after heating the extract to 65 °C prior to binding onto the filter paper disk and ultimately did not use this method for routine screening of CTV. Many factors could have been responsible for this unsatisfactory result. If researchers wish to use this method with other viruses, they should be prepared to modify the buffer and perform preliminary controlled testing with known positives and healthy tissues before using it routinely for testing. In contrast, all viroids tested using the standard method presented here were detected readily regardless of host tissue type or size of viroid.

2. 2-Mercaptoethanol acts as a reducing agent in the extraction and has a relatively short half-life once diluted into a buffer. Unless using a large amount of TXB extraction buffer within 48 hrs, it is best to prepare the rest of the working buffer as listed without 2-mercaptoethanol (Tris–HCl, NaCl, PVP-40), filter sterilize and store at 4 °C for up to 6 months, remove aliquots as needed, and add the 2-mercaptoethanol within an hour of use. For example, to make 60 mL of TXB extraction buffer to process 14 samples, add 300μL of 2-mercaptoethanol to 59.7 mL of the Tris–HCl, NaCl, PVP-40 working buffer. This can be scaled up or down depending on the number of samples being processed that day.
3. To aid in the dissolving of PVP-40, grind it first to a fine powder using a mortar and pestle.
4. Be sure to add the water and Tris together first before adding the ethanol. If the ethanol is added directly to the 10× Tris first, it becomes cloudy and may not resolve after adding the water to volume. 95% ethanol may be used if 100% is not available: for 1 L add 295 mL of 95% ethanol, and reduce the volume of water to 605 mL; the 100 mL of 10× STE remains the same.
5. Any size of Whatman No. 1 filter paper can be used, but 90 mm circles were the easiest to handle, and cut with scissors while optimizing the number of disks obtained per sheet.
6. Avocado trees infected with ASBVd which do not express symptoms are called "symptomless carriers." These strains of ASBVd are usually 100% transmitted by seed as opposed to 5% seed transmission with symptomatic strains. It depends on the strain of the viroid and also the scion/rootstock combination of the avocado tree. Some graft combinations more readily develop symptoms than others. Environmental stress also plays a factor on symptom development. Trees showing no symptoms for many years suddenly began to show distinct foliar symptoms when new tissue developed after a wildfire.

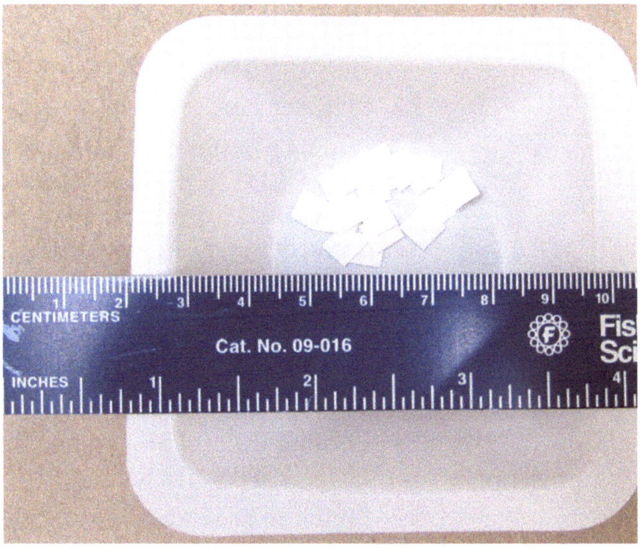

Fig. 2 Filter paper disks for capturing viroid RNA. Disks approximately 4–5 mm square are cut from Whatman #1 filter paper circles and stored in a sterile tube until needed

Chrysanthemum stunt viroid (CSVd, 199 bp amplicon) among others; forward primer 5′-ATTAATCCCCGGGGAAACCTG GAG -3′, reverse primer 5′- AGCTTCAGTTGTTTC CACCGGGT-3, [11]. PSTVd has been reported in avocado in Peru, [12], and some countries require testing for this viroid before plants can be exported from California. We performed dual ASBVd and PSTVd tests on several nursery trees over the years by producing two disks from a single sample and using each for a separate RT-PCR reaction. PSTVd was never found in any avocado sample in our laboratory. However, PSTVd was found in a sample of the ornamental plant *Cestrum* spp. obtained from a grower.

Peach latent viroid: Forward primer 5′- CCCGATA GAAAGGCTAAGCACCTCG-3′, reverse primer 5′-AACTG CAGTGCTCCGT-3′, (336 bp amplicon) [13].

Satellite tobacco mosaic virus (*see***Note 1**): forward primer 5′- CAAACCGTAAATCGACGGGTGACAATTCGA -3′, reverse primer 5′- TTACTTCTTACTTCGCATA CATTGTCGGCT-3′ (Mathews unpublished data).

16. 0.6 mL microcentrifuge tubes, sterile.
17. Taq DNA polymerase and buffer (Promega Go Taq system).
18. AMV-Reverse Transcriptase.
19. RNasin ribonuclease inhibitor.
20. Dithiothreitol, 10 mM.
21. $MgCl_2$, 25 mM.

22. Sterile water, RNase-free.
23. Water bath capable of boiling temperatures.
24. PCR machine.
25. Running buffer (1× TBE): 89 mM Tris, 89 mM boric acid, 2 mM EDTA, pH 8.3. Make a 5× stock solution by adding 54 g Tris, 27.5 g boric acid, and 3.7 g Na_2EDTA to 800 mL distilled water, stir until dissolved, check pH, and adjust if necessary, but it should be very close to 8.3, adjust final volume to 1 L.
26. Agarose.
27. Microwave oven.
28. Loading dye: 0.25% bromophenol blue, 0.25% xylene cyanol, 1 mM EDTA, pH 8.0, and 50% glycerol in water.
29. Ethidium bromide, 10 mg/mL in water.
30. UV transilluminator.
31. Camera or digital imager.

3 Methods

3.1 Tissue Collection

Viroids and viruses are not always distributed uniformly within a large tree, especially early in the infection cycle, and viroids can cause non-symptomatic diseases (*see* **Note 6**). It is critical to collect samples in a manner that will provide the best chance of finding infected tissue. For avocados, we have found that the collection strategy depends on how large the plant is. Repeated analysis showed that ASBVd can easily be detected with this protocol at a dilution level of one infected leaf within a total of 40 leaves (i.e., 1 infected leaf and 39 uninfected leaves). We present methods for small, primarily unbranched nursery trees, 3- to 5-year-old field trees under 5 feet tall with minimal branching, and mature field trees. Ideal leaves should be turgid, fully expanded, and hardened off, but not wind or insect damaged or ready to senesce.

3.1.1 Nursery Trees

Remove two leaves from each tree. If seedlings or young grafted trees are quite small, two half leaves may be cut and removed so that excessive tissue loss does not occur. If screening large numbers of seedlings, pooling of multiple trees is possible. In order to ensure detection of a viroid with this method, limit the total number of leaves in a pool to 40 (20 seedlings per pool, 2 leaves per seedling). If a pool is scored as positive, then each individual tree within that pool can be tested, or it may be cheaper and more time-efficient to discard the entire pool of trees. If a tested pool is negative, no further testing is required.

3.1.2 Young Field Trees

Trees that are 3 to 5 feet tall with a few main branches fall into this category. One leaf from each branch that traces back to the trunk should be collected, usually five to eight leaves. As with nursery trees, several trees can be pooled together for testing. In this case, 5–8 trees can be combined (8 leaves from 5 trees, 5 leaves from 8 trees, and so on depending on branch structure), with a total of 40 leaves or less comprising the pool.

3.1.3 Mature Field Trees

Ten leaves from each of the 4 compass points should be collected for a total of 40 leaves (10 from each of the north, south, east, and west quadrants of the tree). Each leaf should be removed from a different branch, being sure to take some from upper and lower areas of the canopy. No pooling of full-sized field trees is recommended since 40 leaves comprise 1 sample.

3.1.4 Fruit

Symptomatic fruit can be tested to confirm the presence of ASBVd. It is important however to determine if the fruit became infected through pollen transmission or if the tree itself is infected leading to symptomatic fruit. In the case of pollen transmission, it has been shown that a previously uninfected recipient tree does not become infected and ASBVd is isolated to the pollinized fruit [3]. It can be important to distinguish between these two scenarios to avoid removing mature field trees that are ASBVd-negative, when only some of the fruits are infected. Harvest a branch with one or more symptomatic fruit and several leaves immediately behind the fruit. In the laboratory, analyze both the fruit and the leaves separately. If both samples are positive for ASBVd, the tree itself is infected and should be removed. If the fruit is positive and the leaves are negative, then it is a pollen transmission event, and the tree can be spared from removal.

3.2 Tissue Preparation

Leaves should be stacked, lining up the leaf bases. It is convenient to stack 7–12 leaves at once depending partly on size. Using a new, sterile razor blade, cut and remove a slice of tissue approximately 2–3 cm up from the leaf bases. Starting at the edges of the leaves, cut inward just past the midrib, then make a parallel cut approximately 6–7 mm away, and remove the slices of tissue by tearing them away from the remainder of the leaves (Fig. 3a). Each slice should contain both laminar and midrib tissues. Place cut tissue in a weigh boat, plastic bag, or other clean vessel until ready to grind (*see* **Note 7**). Continue to stack and cut each leaf until all leaves in a given sample are done. Final weight of the combined leaf slices from a 40-leaf sample should be between 1 and 2 g, less if fewer leaves were collected. In the case of testing other plants such as ornamentals or vegetables, tissue weights will normally be less than 1 gram due to the nature of the tissues. However, when collecting such samples, a similar strategy of harvesting a representative sample from around the entire plant should be followed.

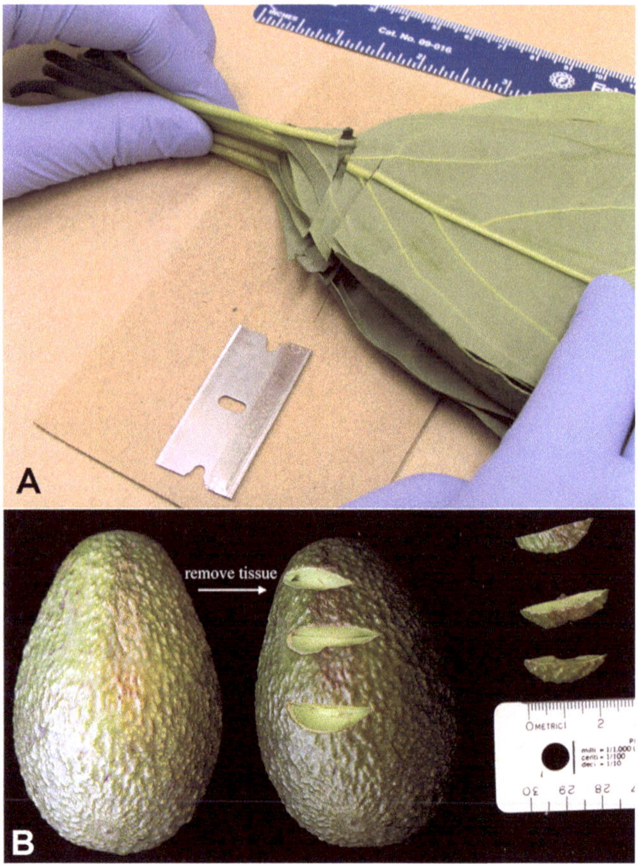

Fig. 3 Avocado sample preparation. (Panel **a**) For analyzing leaves, remove a slice of tissue approximately 6 mm wide containing both laminar and midrib sections which is cut with a razor blade and removed for analysis; (Panel **b**). For sampling fruit, remove three wedges of tissue composed of skin and flesh, approximately 2 cm long × 6 mm wide × 5 mm deep

Fruit sampling requires cutting small wedge-shaped pieces of skin and flesh approximately 2 cm long × 6 mm wide × 5 mm deep (Fig. 3b). For a medium-sized fruit, three such wedges are removed. We also found that from symptomatic fruit, ASBVd could be readily detected in similar wedges from the outer seed coat and in dissected embryos within the seed. This could be useful when testing nurse seeds for possible ASBVd infection. When testing other plant species, multiple young to medium age leaves from around the canopy should be collected, including any with visible symptoms. If leaves are small, they do not need to be cut and subsampled, entire leaves may be ground. Other plants with larger leaves may be subsampled similar to the avocado method above, keeping total tissue weight to between 1 and 2 g.

3.3 RNA Extraction

We have found it convenient for one person to process 14 samples or less at a time so that the mortars will stay cold during the procedure until the sample is removed to the glass tube with the filter disk.

1. Place a set of leaf slices in a mortar and using a pestle, and crush the midribs while at room temperature. Add liquid nitrogen and grind the sample to a fine powder. More than one addition of liquid nitrogen may be required to achieve this; be careful when adding to avoid splashing of tissue and possible cross contamination of adjacent samples.
2. Add 4 mL of tissue extraction buffer (TXB) to each mortar, and mix into the ground tissue using the pestle. The mixture may temporarily freeze; continue to periodically mix using the pestle until thawed but still cold (*see* **Note 8**).
3. Remove approximately 150–200μL of sample with a disposable transfer pipet (*see* **Note 9**), and place in a small disposable glass tube (*see* **Note 10**) containing one or more filter disks. It is convenient to prepare at least two filter disks at a time in this manner so that a replicate sample is available. When making controls, up to six disks may be placed in the tube and then used in multiple experiments.
4. Add 2 mL of room temperature wash buffer to the tube (*see* **Note 11**), vortex to mix, then incubate at room temperature for 5 min, vortex again, and incubate another 5 min.
5. Remove the liquid and as much physical plant debris from the tube as possible using a transfer pipet leaving the filter disk inside. Wash the disk(s) by adding successive aliquots of wash buffer (2–3 mL each) until all plant residue has been removed and the disks retain little or no color (*see* **Note 12**).
6. Remove the disks from the tube with a clean wooden applicator stick (*see* **Note 13**), and place on a sheet of aluminum foil to dry at room temperature (*see* **Note 14**).
7. Once dry, place filters in a labeled, sterile 0.6 mL microcentrifuge tube (one disk per tube). These may be stored at room temperature or at 4 °C (preferred) until needed for RT-PCR. We have stored disks at 4 °C and at room temperature for up to 2 years before performing the RT-PCR reactions with little decrease in amplification. However, we normally used positive control disks within 6 months of RNA capture to ensure a reliable reaction.

3.4 RT-PCR

1. Prepare a PCR master mix cocktail sufficient for the number of samples being analyzed. For each sample mix: 20μL 2 mM dNTPs, 1μL forward primer, 1μL reverse primer, 20μL 5× Go Taq buffer (Promega), 4μL 25 mM $MgCl_2$, 10μL 10 mM

DTT, 0.25μL RNasin (10 units, Promega), 1.5μL AMV-RT (15 units, Promega), and 0.5μL Taq DNA polymerase (2.5 units, Promega); store on ice (*see* **Note 15**).

2. Open the microfuge tubes containing a dried filter disk, and pipet 42μL of sterile water onto the disk, and then use the tip of the pipet to gently swirl the water and disk together to ensure complete hydration of the disk and elution of the RNA; place on ice. When all samples have been hydrated, place in a boiling water bath for 3 min to denature the RNA, and immediately place on ice. After a few minutes of cooling, briefly centrifuge the samples to remove any condensate from the walls of the tubes.
3. To each sample tube, add 58μL of PCR master mix cocktail leaving the disk in place and mixing well with the pipet tip. Final volume is now 100μL. As an alternative, 20μL of the eluted and denatured extract may be removed and added to 30μL of PCR master mix (*see* **Note 16**).
4. Perform single-tube RT-PCR using the following program: 1 first-strand cycle of 45 min at 42 °C, 1 denaturation cycle of 2 min at 95 °C, and 40 amplification cycles of 1 min at 94 °C, 2 min at 55 °C, and 3 min at 72 °C, with a final extension cycle of 5 min at 72 °C [9]. PSTVd and PLMVd have optimal annealing temperatures of 62 °C; adjust for other pathogens as necessary.

3.5 Gel Electrophoresis

1. Cast a 2% agarose gel in running buffer, 1× TBE, size and number of wells is dependent on number of samples. For an 8 cm × 15 cm gel, add 2 g of agarose to 100 mL of 1× TBE buffer in a glass flask. Boil in a microwave oven until melted. Cool to approximately 65 °C, and pour into the casting tray with a comb containing the desired number of wells. Cool until solidified, place in electrophoresis apparatus, and add 1× TBE buffer until gel is barely covered.
2. Remove a 10μL aliquot of the RT-PCR sample and mix with 1μL of loading dye. Load the entire aliquot into a well. Electrophorese at 175 mA for 90 min or until the bromophenol blue dye is 2 cm from the bottom of the gel (*see* **Note 17**).
3. Remove the gel from the apparatus and place in a plastic container with 100 mL water. Add 10μL of ethidium bromide and gently shake for 20 min to stain the DNA. Drain the ethidium mixture and rinse once quickly with water, and then destain the gel fully using two 10 min washes with water.
4. Place the destained gel on a UV transilluminator, and photograph any bands that are present (Fig. 4).

Symptoms have also been noted to develop after the original scion was removed from a rootstock and a new one was grafted by topworking.

7. Ideally once the tissue is cut, it should be ground and processed the same day. If this is not possible, place the weigh boats containing the cut tissue in sealed plastic bags and store at 4 °C overnight. We do not recommend storing cut tissue more than 24 h.
8. After the ground buffer/tissue has thawed, some varieties of avocado will have a very thick, mucous-like texture which makes removing an aliquot of liquid difficult. In these cases, you may add up to 1 mL extra TXB buffer, in aliquots of 0.5 mL at a time with grinding to try and loosen the mixture. If this still does not allow for easy liquid removal, let it sit for 3–5 min, and use a transfer pipet at the edge of the grindate to remove an acceptable amount of sample. In the case of more herbaceous plant samples with tissue weights less than 1 g, the amount of TXB buffer may be reduced to prevent excessive dilution of the sample.
9. Transfer pipets with a built in bulb made of polyethylene (Fisher Scientific catalog # 13-711-7M) were ideal for transferring samples and washing the disks. Use a clean pipet for each manipulation to avoid cross contamination of samples and buffers. Discard after a single use.
10. Disposable glass tubes were selected because it was easy to see the disks while washing and assess their cleanliness. Tubes measured 12 mm × 75 mm (6 mL volume) or 13 mm × 100 mm (10 mL volume). Six milliliters is the minimum acceptable volume due to the need to vortex a little over 2 mL of solution when binding the RNA to the filters.
11. It is critical that the binding and washing steps be performed at room temperature with room temperature wash buffer. Binding of RNA to cellulose in the presence of ethanol is greatly reduced at cold temperatures.
12. Usually three washes with 2–3 mL of wash buffer each is sufficient to remove all physical plant debris and most of the pigment from the disks. Some samples have pigments that never completely disappear, so four washes is the maximum needed.
13. A new wooden applicator stick, 5 3/4 in × 1/12 in (Fisher Scientific catalog # 22-029-491) should be used for each sample; they are long enough to go into the tube and "grab" the disk while not damaging it. Alternatively, a pipet or other sterile object can be used.

14. The sheet of aluminum foil should be pre-labeled with a lab marker so that each disk can be placed in a specific position. Disks may either be dried on the bench top or placed in a low rate of flow fume hood to accelerate drying. Drying usually takes approximately 45 min at room temperature. Disks should be fully dried before subsequent steps of elution and RT-PCR.
15. This PCR cocktail master mix uses the Promega Go-Taq buffers and enzymes. AMV-RT will work with this PCR buffer in a single-tube system, so there is no need to do a first-strand reaction separately which increases the time and handling required and introduces one more place for cross contamination between samples to occur. If you wish to use another manufacturer's enzymes, be sure to test them for compatibility in a single-tube RT-PCR system.
16. We have found that some PCR machines with very fast ramp times do not produce reliable results when the filter disk is left in the tube during the RT-PCR reaction. Older, water-cooled PCR machines were preferable to modern, Peltier-cooled machines.

    Our assumption is that primers and extension products may get "tangled up" with the disk and are not readily available for denaturation, annealing, and extension if heating and cooling occur too rapidly. You should do a controlled test first to assess your machine's ability to perform using this system. As an alternative, after eluting the RNA from the disk with water and boiling to denature, an aliquot, usually 20μL, can be removed and added to 30μL of RT-PCR master mix and then cycled as described.
17. Since the amplicon of this ASBVd RT-PCR reaction has an average length of 247 bp, depending on the strain, it is important to use 2% agarose and not run the sample off the end of the gel. Bromophenol blue dye runs at approximately 300 bp, so keep it well above the bottom of the gel matrix by the end of electrophoresis. Samples with very high titers of the viroid will also have dimer and trimer amplicon bands of approximately 500 and 750 bp. It is very common to have a variety of weak background bands present in the gel whether or not there are any ASBVd specific bands. This seems to be variety dependent, and since we tested the University of California's avocado variety collection, there were dozens of varieties and accessions. We did extensive testing to ensure that these bands were not related to the viroid, including using an alternate primer pair and hybridization with a labeled, internal ASBVd specific probe. There was no evidence that bands in the gel that were not 247 bp were at all related to ASBVd infection. A 100 bp ladder with bands representing 100 to 1000 bp is the optimum size marker for this PCR product.

References

1. Wallace JM, Drake RJ (1962) High rate of seed transmission of avocado sun-blotch virus from symptomless trees and origin of such trees. Phytopathology 52:237–241
2. Semancik JS, Szychowski JA (1994) Avocado sunblotch disease: a persistent viroid infection in which variants are associated with differential symptoms. J Gen Virol 75:1543–1549
3. Desjardins PR (1987) Avocado sunblotch. In: Deiner TO (ed) The viroids. Plenum, New York
4. Dodds JA, Bar-Joseph M (1982) Double-stranded RNA from plants infected with Closteroviruses. Phytopathology 73(3):419–423
5. Dodds JA, Morris TJ, Jordan RL (2003) Plant viral double-stranded RNA. Annu Rev Phytopathol 22(1):151–168
6. Franklin RM (1966) Purification and properties of replicative intermediate of the RNA bacteriophage R17. Proc Natl Acad Sci U S A 55:1504–1511
7. Morris TJ, Dodds JA (1979) Isolation and analysis of double-stranded RNA from virus -infected plant and fungal tissue. Phytopathology 69:854–858
8. Semancik JS (1986) Separation of viroid RNAs by cellulose chromatography indicating conformational distinctions. Virology 155:39–45
9. Schnell RJ, Kuhn DN, Ronning CM, Harkins D (1997) Application of RT-PCR for indexing avocado sunblotch viroid. Plant Dis 81:1023–1026
10. Rakowski AG, Symons RH (1989) Comparative sequence studies of variants of avocado sunblotch viroid. Virology 173:352–356
11. Bostan H, Nie X, Singh RP (2004) An RT-PCR primer pair for the detection of Pospiviroid and its application in surveying ornamental plants for viroids. J Virol Methods 116:189–193
12. Querci M, Owens RA, Vargas C, Salazar LF (1995) Detection of potato spindle tuber viroid in avocado growing in Peru. Plant Dis 79:196–202
13. Hadidi A, Giunchedi L, Shamloul AM, Poggi-Pollini C, Amer MA (1997) Occurrence of peach latent mosaic viroid in stone fruits and its transmission with contaminated blades. Plant Dis 81:154–158
14. Valverde RA, Dodds JA (1987) Some properties of isometric virus particles which contain the satellite RNA of tobacco mosaic virus. J Gen Virol 68:965–972
15. Mirkov TE, Mathews DM, Du Plessis DH, Dodds JA (1989) Nucleotide sequence and translation of satellite tobacco mosaic virus RNA. Virology 170:139–146
16. Karasev AV, Boyko VP, Gowda S, Nikolaeva OV, Hilf ME, Koonin EV, Niblett CL, Cline K, Gumpf DJ, Lee RF, Garnsey SM, Lewandowski DJ, Dawson WO (1995) Complete sequence of the citrus tristeza virus RNA genome. Virology 208:511–520
17. Mathews DM, Riley K, Dodds JA (1997) Comparison of detection methods for citrus tristeza virus in field trees during months of nonoptimal titer. Plant Dis 81:525–529

Part IV

Cloning and Sequencing Techniques

Chapter 20

Cloning and Sequencing of Viroids

Rosemarie W. Hammond

Abstract

Determining the sequence identity of viroid RNAs present in symptomatic or asymptomatic plant tissues is critical to obtain knowledge of their distribution. It enables the development of tools for diagnostics and for studying the basic biology of viroids. With the advent of cDNA-based methods for cloning RNAs and cloning strategies that do not require prior knowledge of the viroid sequence, characterization of several newly discovered viroids has rapidly expanded our knowledge of these unusual pathogenic RNAs. This chapter describes two methods, using random primers or viroid-specific primers, to generate complementary DNA (cDNA) copies of viroid RNAs for subsequent cloning and sequence analysis.

Key words Viroids, Complementary DNA (cDNA), Reverse transcription, RT-PCR (reverse transcription-polymerase chain reaction)

1 Introduction

Cloning copies of viroid RNAs was initially performed to produce complementary DNAs (cDNAs) for use as probes to study the replication cycle of viroids [1, 2]; it soon became apparent that the cDNAs could be used for practical screening tests and diagnostics [3], for characterization, and for reverse genetics using cloned, infectious copies of the viroid molecule [4]. Cloning of viroids can be performed by synthesizing cDNA using random primers for viroids whose sequence is unknown [5] and viroid-specific oligonucleotide primers for known viroids and/or degenerate primers for a group of viroids (e.g., pospiviroids, [6] followed by cloning of the DNA product into a plasmid vector for subsequent sequence analysis. Assuming the viroid RNA has been gel-purified by return-gel electrophoresis using standard protocols and the relationship to other viroids is unknown; the random primer approach is useful for cDNA cloning. The ds cDNAs, whether generated using the traditional method or using RT-PCR, the PCR products can be cloned

Ayala L. N. Rao et al. (eds.), *Viroids: Methods and Protocols*, Methods in Molecular Biology, vol. 2316, https://doi.org/10.1007/978-1-0716-1464-8_20, © Springer Science+Business Media, LLC, part of Springer Nature 2022

into a bacterial plasmid vector; alternatively, the RT-PCR products can be sequenced directly. Infectious cDNA clones can be generated from the cloned viroid copies for further experimentation [4, 7].

2 Materials

Prepare all solutions using deionized for distilled water and analytical grade reagents.

2.1 cDNA Cloning of Viroid RNAs Using Random Primers

1. RNA template (0.5–1.0 μg).
2. pd(N)$_6$ primers (5 μg/ μL).
3. 10× cDNA buffer: 100 mM Tris–HCl, pH 8.3, 500 mM KCl, 15 mM $MgCl_2$.
4. 10× second-strand buffer: 300 mM Tris–HCl, pH 7.5, 100 mM $MgCl_2$, 700 mM KCl, 100 mM DTT, 4 mM dNTPs.
5. 10 mM dNTPs (dG, dA, dT, dC).
6. 15 mM $MgCl_2$.
7. 100 mM dithiothreitol (DTT).
8. M-MuMLV reverse transcriptase, 200 μ/ μL, Invitrogen).
9. Large fragment of DNA polymerase I (Klenow fragment).
10. Buffer-saturated phenol (Thermo Fisher Scientific).
11. Chloroform.
12. 3 M Na acetate, pH 5.2.
13. 100% ethanol.
14. S1 nuclease to create blunt ends.
15. Plasmid vector digested with *Sma*I or *Eco*RV.
16. T4 DNA ligase.
17. Microfuge and microfuge tubes.

2.2 cDNA Cloning of Viroid RNAs Using Specific Primers and PCR Amplification

1. RNA template (50–100 ng purified RNA).
2. Viroid-specific oligonucleotide primers, 18–25 nucleotides in length, 20 pmol/ μL. As an example, amplification of most members of the pospiviroids can be performed using the primer pair Pospi1-RE (5′-AGC TTC AGT TGT (T/A)TC CAC CGG GT-3′; complementary to nt 261–283) and Pospi1- FW (5′-GGG ATC CCC GGG GAA AC-3′; identical to nt 86–102) [6].
3. Reverse transcriptase (e.g., AffinityScript MltTempDNA Synthesis Kit, Agilent Technologies, Wilmington DE).

4. RNase Out Recombinant Ribonuclease Inhibitor (Invitrogen, Carlsbad, CA).
5. Titan One Tube RT-PCR System with AMV and Expand High Fidelity (Roche Molecular Biochemicals, Chicago, IL).
6. AmpliTaq DNA polymerase (Applied Biosystems, Foster City, CA).
7. Thermal cycler.
8. Agarose gel (1.5–2.0% Bio Rad agarose, 1×TBE) and SYBR Safe gel stain (Invitrogen).
9. Certified Molecular Biology Agarose (Bio Rad Laboratories, Hercules, CA).
10. 10× TBE: 121.1 g Tris base, 61.8 g boric acid, 7.4 g EDTA (sodium salt) in 1 L (final concentration 1 M Tris, 1 M boric acid, 0.02 M EDTA). Prepare with RNase-free water prepare or obtain commercially.
11. SYBR® Safe DNA gel stain, Invitrogen, USA.
12. x-tracta Gel Extractor Tool (Promega Corporation, Madison, WI).
13. Gel purification (QIAquick Gel Extraction Kit, Qiagen GmbH, Hilden, Germany).
14. Bacterial plasmid vector for cloning amplicons containing A overhangs that result from Taq polymerase (e.g. pCR4®-TOPO, Invitrogen).
15. Competent bacteria for transformation with the plasmid vector: prepare chemically competent *Escherichia coli* bacterial cells or purchase.
16. Luria Broth and Luria agar (5 g bacto tryptone, 2.5 g yeast extract, 2.5 g NaCl in 500 mL water, add 7.5 g bacto agar for Luria agar) Autoclave. Add appropriate antibiotics to the Luria agar when cooled before pouring plates.
17. Plasmid DNA purification kit -QIAprep Spin Miniprep Kit (Qiagen, Hilden, Germany).

3 Methods

For cDNA cloning, viroid RNAs requires the introduction of reverse transcriptase (RT) step to synthesize complementary DNA (cDNA) prior to the synthesis of the second-strand DNA. This can be achieved by using either random primer if the viroid sequence is unknown or specific primers if the sequence is known or the viroid is suspected to be a member of a viroid family for which primers have been designed for conserved regions (e.g., pospiviroids, [6]). If the viroid sequence is known or known to be related to characterized viroids, a two-step RT-PCR reaction can be performed to obtain an amplicon for sequence analysis, or a one-step RT-PCR reaction can be performed using a thermostable RT.

3.1 cDNA Cloning of Viroid RNAs Using Random Primers

1. For the synthesis of first-strand cDNA using random primers, the primers are first annealed to the RNA template by mixing 5 μL of RNA template, 6 μL sterile water, 2.5 μL 10 × cDNA buffer, and 2 μL pd(N)$_6$ primers (5 μg/ μL). The reaction is heated at 92 °C for 3 min and then placed on ice.
2. Synthesize the first-strand cDNA by adding to the annealed mixture above: 2.5 μL 10 mM dNTP mix, 2.5 μL 15 mM $MgCl_2$, 2.5 μL 100 mM DTT, and 2.0 μL M-MuMLV reverse transcriptase. Incubate the reaction mixture at 37–42 °C for 1–2 h. Chill on ice.
3. Second-strand synthesis: The RNA/DNA template must be separated prior to second-strand synthesis. This can be done in three ways—using RNase H, using NaOH, or by boiling. The boiling method is described.
4. Boil the first-strand synthesis reaction for 3 min. Chill on ice. Add 5 μL of 10× second-strand buffer and 5–0 units of large fragment of DNA polymerase I (Klenow fragment) and water to 50 μL. Incubate at 15 °C for 3 h for second-strand synthesis. Add 25 μL buffered phenol and 25 μL chloroform. Mix well, and centrifuge for 5 min on high speed at 4 °C in the microfuge to separate the phases. Remove the upper phase into a new tube, and add 1/10th volume 3 M NaAc pH 5.2 and 2 volumes of 100% EtOH. Precipitate the DNA at −20 °C. Centrifuge on high speed at 4 °C for 10–15 min to pellet the nucleic acid. Air dry. Add sterile water (20 μL) (*see* **Note 1**).
5. Non-base paired regions are removed by incubation with S1 nuclease at 200 units/mL in a reaction buffer containing 30 mM sodium acetate (pH 4.6), 280 mM NaCl, and 1 mM $ZnSO_4$ [8]. The resulting cDNA is incubated with 1 unit of Klenow DNA polymerase at 37 °C for 15 min to create blunt ends. To stop the reaction, add EDTA to a final concentration of 10 mM, and incubate at 70 °C for 20 min to inactivate the enzyme. The blunt-end cDNA can be cloned into the *Sma*I or *Eco*RV restriction site of an appropriate plasmid vector.

3.2 cDNA Cloning of Viroid RNAs Using Specific Primers and PCR Amplification

For the synthesis of double-stranded cDNA using viroid-specific primers, two methods are used. In the first method, first-strand cDNA is synthesized from total nucleic acid using a viroid-specific primer and reverse transcriptase, followed by PCR using two primers (two-step). In the second method, a primer set is incubated in an RT-PCR reaction (one-step) (*see* **Notes 2** and **3**).

Two-step RT-PCR using viroid-specific primers. First-strand cDNA can be prepared from 200–500 ng of purified RNA from viroid-infected plants using a viroid-specific primer that is complementary to the viroid RNA. The Advantage RT for PCR cDNA synthesis kit or the AffinityScript MltiTemp DNA Synthesis kit can be used to prepared first-strand cDNA following manufacturers' instructions. Double-stranded cDNA can then be synthesized using the complementary primer and a second primer to amplify the viroid-specific amplicon using 1 μL of the first-strand cDNA in a 50 μL reaction containing 5 μL of 10× buffer (containing 15 mM $MgCl_2$), 1 μL 2.5 mM dNTPs, 20 pmol each of the forward and reverse primers, and 0.6 units of a thermostable DNA polymerase, such as AmpliTaq DNA polymerase. Suggested cycling conditions are 95 °C for 5 min; 35 cycles of 95 °C, 45 s, 50 °C, 1 min, and 72 °C, 2 min, followed by 72 °C for 7 min. The annealing temperature depends upon the melting temperature of the primers. The RT-PCR products are analyzed by electrophoresis through a 1.5% TBE agarose gel, stained with SYBR safe DNA gel stain and visualized on a UV transilluminator.

One-step RT-PCR using viroid-specific primers. To specifically amplify the target region using viroid-specific primers, 1 μL (200–500 ng) of purified RNA from viroid-infected plants is combined in a final volume of 50 μL with 1× reaction buffer, 1.5 mM $MgCl_2$, 0.2 mM if each of the four dNTPs, 1 μL of each primer at 20 pmol/ μL, 0.5 μL of enzyme mix, 5 mM DTT, and 20 units of RNase Out Recombinant Ribonuclease Inhibitor using the Titan One tube RT-PCR system. The cycling parameters are 50 °C for 30 min; 95 °C for 5 min; and 35 cycles of 95 °C, 45 s, 50 °C, 1 min, and 68 °C, 2 min, followed by 68 °C for 7 min. The RT-PCR products are analyzed by electrophoresis through a 1.5% TBE agarose gel, stained with SYBR safe DNA gel stain and visualized on a UV transilluminator.

3.3 Cloning of cDNA Products Produced in Subheadings 3.1 and 3.2 for Sequence Analysis

Once amplified, the amplicons can be gel-purified from the agarose gels using the Qiagen Quick Gel Extraction kit. Briefly, fragments identified using gel staining are excised from the gel using a razor blade or the x-tracta Gel Extractor tool and placed into a 1.5 mL microfuge tube. The DNA is extracted from the gel using the Qiagen gel extraction kit per manufacturer's instructions and eluted in 10–15 μL of sterile water. To clone into the pCR4 TOPO vector, 4 μL of the purified DNA is combined with 1 μL of salt and 1 μL of the TOPO plasmid. After a 5-min incubation at room temperature, the ligation mixture is added to competent *E. coli* cells, and the mixture is incubated on ice for 25 min, followed by a heat shock at 42 °C for 30 s. 400 μL of LB broth is added, and the mixture is incubated with shaking for 1 h at 37 °C. 100 μL of cells are plated on an LB agarose containing the appropriate antibiotic an incubated at 37 °C overnight. The next day, 2–5 mL cultures of LB broth are inoculated with single colonies from the resulting transformation and grown overnight with shaking at 37 °C. Plasmid purification can be performed using standard protocols [8] or by using a commercial kit such as the Qiagen QIAprep Spin Miniprep Kit per manufacturer's instructions. The concentration of DNA is determined using a spectrophotometer and absorbance at 280 nm, and the resulting plasmid DNA can be sequenced in-house or sent to a commercial provider.

4 Notes

1. Caution: Phenol and chloroform are toxic. Wear gloves and safety goggles when handling in the chemical fume hood, and adhere to safety guidelines for disposal of hazardous waste. Oxidized phenol (pink or red) can damage nucleic acids.
2. Avoid RNase; glassware and plastics can be treated with RNaseZap (Ambion Inc., Austin TX) to remove RNase contamination.
3. Synthesis of first-strand cDNA from an unknown viroid template using a random primer can also be performed as described by [9]. In this case, a primer containing six randomized positions at the 3′ end is used in the presence of AMV-RT for the synthesis of first-strand cDNA. The same primer is used for the synthesis of second-strand cDNA using the Klenow fragment, and the resulting ds cDNAs were amplified using the identical primer without the six randomized nucleotides.

References

1. Owens RA (1978) In vitro synthesis and characterization of DNA complementary to potato spindle tuber viroid. Virology 89:380–387
2. Palukaitis P, Symons RH (1978) Synthesis and characterization of a complementary DNA probe for chrysanthemum stunt viroid. FEBS Lett 92:262–272
3. Owens RA, Diener TO (1981) Sensitive and rapid diagnosis of potato spindle tuber viroid by nucleic acid hybridization. Science 213:670–672
4. Owens RA, Cress DE (1980) Molecular cloning and characterization of potato spindle tuber viroid cDNA sequences. Proc Natl Acad Sci U S A 77:5302–5306
5. Navarro B, Daros JA, Flores R (1996) A general strategy for cloning viroids and other small circular RNAs that uses minimal amounts of template and does not require prior knowledge of its sequence. J Virol Methods 56:59–66
6. Verhoeven JTJ, Jansen CCC, Willemen TM, Kox LFF, Owens RA, Roenhorst JW (2004) Natural infections of tomato by *Citrus exocortis viroid*, *Columnea latent viroid*, *Potato spindle tuber viroid* and *Tomato chlorotic dwarf viroid*. Eur J Plant Pathol 110:823–831
7. Yoon J-H, Cho I-S, Choi G-S, Choi S-K (2014) Construction of an infectious cDNA clone of a *Chrysanthemum stunt viroid* Korean isolate. Plant Pathol J 30:68–74
8. Sambrook J, Fritsch EF, Maniatis T (1989) Molecular cloning: a laboratory manual. Cold Spring Harbor Laboratory Press, Cold Spring Harbor, New York. 574.87324 1/1989
9. Verhoeven JTJ, Meekes ETM, Roenhorst JW, Flores R, Serra P (2013) Dahlia latent viroid: a recombinant new species of the family *Pospiviroidae* posing intriguing questions about its origin and classification. J Gen Virol 94:711–719

Chapter 21

QuantiGene Plex Assay: A Method for High-Throughput Multiplex Citrus Viroid Detection and Identification

Tyler Dang, Jinbo Wang, Tavia Rucker, Sohrab Bodaghi, Irene Lavagi-Craddock, and Georgios Vidalakis

Abstract

The QuantiGene Plex assay is a molecular non-polymerase chain reaction (PCR)-based multiplex method adapted for citrus viroid detection and identification. Here, we describe the procedures to utilize the QuantiGene Plex assay as a high-throughput screening tool for viroids in purified or crude RNA extracts from citrus tissues.

Key words Multiplex, High-throughput, Branched amplification assay, Branched DNA, Luminex MAGPIX®

1 Introduction

The QuantiGene Plex assay is a non-PCR-based multiplex method that was previously developed for various applications such as quantification of mRNA from a human lung epithelial cell line, the quantification of dsRNA from plants [1, 2], gene expression analysis [3], and pathogen detection [4]. The technology has been adapted for detection and identification of up to seven citrus viroids as well as a plant housekeeping gene (*nad5*) in a single reaction.

Two universal probes have been designed: the "Apsca" universal probe, that can detect citrus leaf bent viroid, citrus dwarfing viroid, citrus viroid V, and citrus viroid VI, and the "NonApsca" universal probe that can detect citrus exocortis viroid (CEVd), hop stunt viroid (HSVd), and citrus bark cracking viroid. Two additional probes were designed specifically to detect and identify CEVd and HSVd as they are the causal agents of the important exocortis and cachexia disease, respectively.

QuantiGene Plex uses branch DNA (bDNA) amplification technology, which amplifies the signal that is hybridized to the targeted nucleic acids rather than amplifying the target itself as in

Ayala L. N. Rao et al. (eds.), *Viroids: Methods and Protocols*, Methods in Molecular Biology, vol. 2316, https://doi.org/10.1007/978-1-0716-1464-8_21,

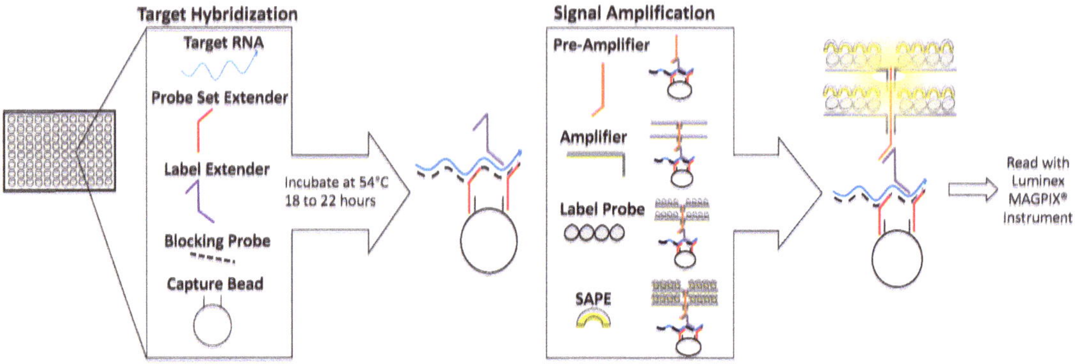

Fig. 1 Diagram of the QuantiGene Plex assay protocol. The protocol is conducted in a 96-well high-throughput manner. Purified RNA or crude extracts from citrus tissues are combined with the Probe Extender, Label Extender, Blocking Probe, and Capture Bead in each well. The samples are then incubated at 54 °C for 18–22 h to allow the components to hybridize with the target of interest. Additional hybridization steps occur to amplify the signal with the Pre-Amplifier, Amplifier, Label Probe, and SAPE. The final complex is then read by the Luminex MAGPIX® machine

PCR (Fig. 1). Other methods such as SYBR Green [5] or probe [6]-based multiplex RT-qPCR have been developed for the identification and detection of citrus viroids but are limited in their multiplex capacity.

The QuantiGene Plex assay is performed in a high-throughput 96-well format and requires 2 days to complete. The first day is dedicated to target hybridization, which requires an overnight incubation step, while the signal amplification procedure, which involves a series of incubation and wash steps using a handheld magnetic plate washer, is performed on the second day. Samples are analyzed using the Luminex MAGPIX® instrument. The QuantiGene Plex assay can utilize purified RNA or crude extracts directly from citrus tissue as a template for citrus viroid detection.

2 Materials

1. Purified or crude RNA extracts.
2. Micropipettes (P10, P20, P200, and P1000) and tips.
3. QuantiGene 2.0 Plex Set (*see* **Note 1**): nuclease-free water, lysis mixture, 2.0 Probe Set, blocking reagent, Proteinase K, and Capture Bead.
4. QuantiGene Plex 2.0 Assay Kit (catalog number: QP1013): Pre-amplifier Solution, Amplifier Solution, Amplifier Diluent, Label Probe Solution, Label Diluent, Streptavidin, R-Phycoerythrin Conjugate (SAPE), SAPE Diluent, Wash Buffer I, and Wash Buffer II.
5. Diethyl pyrocarbonate (DEPC)-treated water.

6. Adhesive pressure seals.
7. Adhesive aluminum foil seals.
8. Aluminum foil.
9. Round bottom 96-well polypropylene microtiter plate.
10. Microtiter plate shaking incubator.
11. Flat bottom 96-well polystyrene microtiter plate.
12. Handheld magnetic plate washer.
13. Microtiter plate shaker.
14. Luminex MAGPIX® instrument.
15. Microplate centrifuge.
16. Incubator or water bath set to 37 °C.
17. MAGPIX® Drive Fluid.
18. Paper towels.

3 Methods

Day 1: Target Hybridization

All reagents should be prepared fresh for every experiment.

1. Pre-warm the lysis mixture to 37 °C for 30 min.
2. Prepare the appropriate volume of the Working Bead Mix by combining the reagents based on the purification method (*see* **Note 2**).
   (a) For purified RNA samples, mix 38.7 μL of nuclease-free water, 33.3 μL of lysis mixture, 2 μL of blocking reagent, 1 μL of Capture Beads, and 5 μL of 2.0 Probe Set per reaction.
   (b) For crude extracts, mix 18.5 μL of nuclease-free water, 33.3 μL of lysis mixture, 2 μL of blocking reagent, 1 μL of Capture Beads, 0.2 μL of Proteinase K, and 5 μL of 2.0 Probe Set per reaction.
3. Vortex the bead mix for 30 s. Dispense 80 μL for purified RNA or 60 μL for crude extracts of the bead mix into the round bottom 96-well microtiter plates (*see* **Note 3**).
4. Dispense 20 μL of purified RNA or 40 μL of crude extracts into each well.
5. Seal the plate with an adhesive pressure seal, and ensure the seal is secured tightly to prevent evaporation during the incubation process.
6. Place the plate into the microtiter plate shaking incubator at 54 °C ± 1 °C at 600 rpm for 18–22 h.

Day 2: Signal Amplification

All reagents should be prepared fresh for every experiment.

7. Prepare the wash buffer mixture fresh by first adding 150 mL of DEPC-treated water, 0.6 mL of Wash Buffer I and 10 mL of Wash Buffer II, and bring the final volume to 200 mL with DEPC-treated water (*see* **Note 4**).
8. Remove the round bottom microtiter plate from the shaking incubator and centrifuge the plate for 60 sec at 240 RCF in the microplate centrifuge.
9. Transfer the microtiter plate contents into a flat bottom 96-well microtiter plate. Place the plate onto the handheld magnetic plate washer, and allow it to sit for at least 1 min (*see* **Note 5**).
10. Fasten the plate onto the magnetic plate washer. Decant the mixture, then dry the plate by turning the plate with the magnetic plate washer upside down and gently press against clean paper towels to remove excess fluid.
11. Add 100 μL of wash buffer mixture to each well. Allow the buffer to sit in the well for at least 1 min (*see* **Note 5**).
12. Decant the buffer, dry the plate by turning the plate with the magnetic plate washer upside down, and gently press against clean paper towels to remove excess buffer. Repeat the wash step two more times.
13. Use the incubator or water bath to pre-warm the Amplifier Diluent to 37 °C for 30 min.
14. Prepare the Pre-Amplifier mixture by adding 36 μL of Pre-Amplifier to 12 mL Amplifier Diluent and invert three to four times to mix (*see* **Note 4**).
15. Pipette 100 μL of the Pre-Amplifier mixture to each well, and cover the plate with an aluminum foil seal. Ensure that the seal is tightly secured.
16. Place the plate into the microtiter plate shaking incubator at 50 °C ± 1 °C for 1 h at 600 rpm.
17. Remove the microtiter plate from the incubator. Fasten the plate onto the magnetic stand and allow the plate to sit for 1 min.
18. Decant the buffer, dry the plate by turning the plate with the magnetic plate washer upside down, and gently press against clean paper towels twice.
19. Add 100 μL of the freshly prepared wash buffer mixture to each well. Allow the buffer to sit in the well for at least 1 min and no more than 2 min.

20. Decant the buffer, dry, and repeat the wash steps (**steps 12** and **13**) two more times.
21. Prepare the amplifier mixture by adding 36 μL of Amplifier Solution to 12 mL Amplifier Diluent, and invert three to four times to mix (*see* **Note 4**).
22. Pipette 100 μL of the amplifier mixture to each well, and cover the plate with a new adhesive aluminum foil seal. Ensure that the seal is tightly secured.
23. Place the plate into the microplate shaking incubator at 50 °C ± 1 °C for 1 h at 600 rpm.
24. Remove the microtiter plate from the incubator, and fasten plate onto magnetic stand. Allow the plate to sit for 1 min.
25. Decant the buffer, dry the plate by turning the plate with the magnetic plate washer upside down, and gently press against clean paper towels to remove excess buffer.
26. Add 100 μL of wash buffer mixture to each well. Allow the buffer to sit in the well for at least 1 min and no more than 2 min.
27. Decant the buffer, dry, and repeat the wash steps (**steps 19** and **20**) two more times.
28. Prepare the Label Probe mixture by adding 36 μL of Label Probe to 12 mL Label Diluent, and invert three to four times to mix (*see* **Note 4**).
29. Pipette 100 μL of the Label Probe mixture to each well, and cover the plate with a new adhesive aluminum foil seal. Ensure that the seal is secured on tightly.
30. Place the plate into the microtiter plate shaking incubator at 50 °C ± 1 °C for 1 h at 600 rpm.
31. Remove the microtiter plate from the incubator, and fasten the plate onto magnetic stand. Allow the plate to sit for at least 1 min and no more than 2 min.
32. Decant the buffer, dry the plate by turning the plate with the magnetic plate washer upside down, and gently press against clean paper towels twice.
33. Add 100 μL of wash buffer mixture to each well. Allow the buffer to sit in the well for at least 1 min and no more than 2 min.
34. Decant the buffer, dry, and repeat the wash steps (**steps 26** and **27**) two more times.
35. Prepare the SAPE mixture by adding 36 μL of SAPE to 12 mL SAPE Diluent and invert three to four times to mix (*see* **Note 4**).

36. Pipette 100 μL of the SAPE Diluent mixture to each well. Cover the plate with new adhesive aluminum foil seal, and then wrap the entire plate with aluminum foil (*see* **Note 6**).
37. Place the plate into the microtiter plate shaker for 30 min at 600 rpm at room temperature.
38. Remove magnet separation plate from the shaker, and fasten the plate onto the magnetic stand. Allow the plate to sit for 1 min.
39. Decant the buffer, dry the plate by turning the plate with the magnetic plate washer upside down, and gently press against clean paper towels to remove excess buffer.
40. Add 100 μL of SAPE Wash Buffer to each well. Allow the buffer to sit in the well for at least 1 min and no more than 2 min.
41. Decant the buffer, dry, and repeat the SAPE Wash Buffer wash step (**steps 33** and **34**) two more times.
42. Resuspend the wells with 65 μL of SAPE Wash Buffer two times. The final volume will be 130 μL for each well.
43. Wrap the entire plate with aluminum foil to minimize exposure to light.
44. Place the plate on the microplate shaker for 3 min at 600 rpm.
45. The plate is ready to be read via Luminex MAGPIX® instrument (*see* **Note 7**).

3.1 Target Detection with Luminex MAGPIX® Instrument

1. Warm up the instrument and ensure the fluid reservoir is filled with the Drive Fluid.
2. Open the xPONENT® software and follow the startup routine.
3. Ensure the probe height is set to the proper height (*see* **Note 8**).
4. Follow steps to run the Calibration and Performance Verification protocols before starting the plate read (*see* **Note 9**).
5. Under the "Protocol" tab select "Create New Protocol." Input all relevant information such as Name, Description, Version, and Manufacturer.
6. Select the "Analytes" tab. Input the bead ID provided by the kit manufacturer in the analyte grid. Name the target to the corresponding bread ID in the Name column and set the Count to 100.
7. Select the "Layout" tab. Highlight the wells that correspond to the sample plate, and label as unknown, standards, control, background, or wash (*see* **Note 10**).

8. Under "Batches" tab select "Create a new Batch from an Existing Protocol." Select the protocol that was just created. Click "Run Batch" to start.
9. Once the run has been completed, select the "Results" tab and then "Saved batched" and "Export CSV" (*see* **Note 11**).
10. On the "Home" page, perform the "Shutdown" routine.

4 Notes

1. The assay is a custom order obtained through Thermo Fisher Scientific (Waltham, MA); make sure to include the Capture Probe ID or symbol ("NonApsca," "Apsca," "CEVd," "CVdII," and "Nad5") to the company in order to receive the proper probes for this protocol.
2. The values provided are based on one reaction. Scale the volumes according to the number of reactions in the experiment. Vortex the Capture Beads regularly to keep the beads in suspension before pipetting.
3. Pipette up and down periodically to keep the beads to remain suspended allowing a more even distribution on the plate. It is also important to dedicate at least two wells (position A1 and A2 preferably) as a background control. To prepare the control, add 80 μL of bead mix for purified RNA or 60 μL for crude extracts, and adjust the final volume to 100 μL with DEPC water.
4. The volumes for the wash buffer, Pre-Amplifier, Amplifier, Label Probe, and SAPE mixtures are scaled for a full 96-well plate. Adjust the volume proportionally if performing the assay with less than 96 samples.
5. It is important to allow time for the magnetic beads to settle and bind to the bottom of the plate with the magnetic stand. Omitting this crucial step may result in lower signals when the plates are read by the Luminex MAGPIX® instrument.
6. Keep the aluminum foil on the plate at all times. SAPE is light sensitive.
7. The plate can be stored at room temperature for up to 2 h or at 4 °C for 24 h. However, it is highly recommended that plate be read and analyzed immediately with the Luminex MAGPIX® instrument upon completion. If the plate has been sitting for an extended period of time, place the plate on a microplate shaker at room temperature for 3 min at 800 rpm before reading.
8. If the height is not set correctly, it will affect how the beads are analyzed. Refer to the user manual for detailed instructions.

9. Perform the Calibration and Performance Verification before every run.
10. It is important to designate the wells for the background controls in order to allow for the xPONENT® software to adjust for any background noise.
11. Results are based on net median fluorescent intensity (MFI). Samples are considered positive when the net MFI is at least three times greater than the background controls.

Acknowledgments

This work was supported by the Citrus Research Board (CRB) project "High Throughput QuantiGene Plex Based Assay for Rapid and Accurate Detection of Citrus Pathogens" (5100-147) awarded to G. Vidalakis. Additional support was provided by the United States Department of Agriculture (USDA) National Institute of Food and Agriculture (NIFA), Hatch (project 1020106), and the National Clean Plant Network (NCPN) which operates under the auspices of USDA Animal and Plant Health Inspection Service (APHIS) (12-8100-1544-CA; 14-, 15-, 16- 8130-0419-CA; AP17PPQS&T00C118; AP18PPQS&T00C107). The findings and conclusions in this publication are those of the author (s) and should not be construed to represent any official USDA or US Government determination or policy. URLs to sponsors' websites are listed below:

1. CRB: https://www.citrusresearch.org
2. USDA NIFA: https://nifa.usda.gov
3. NCPN: http://nationalcleanplantnetwork.org.

References

1. Warrior U, Fan Y, David CA et al (2000) Application of QuantiGene nucleic acid quantification technology for high throughput screening. J Biomol Screen 5:343–352
2. Armstrong TA, Chen H, Ziegler TE et al (2013) Quantification of transgene-derived double-stranded RNA in plants using the QuantiGene nucleic acid detection platform. J Agric Food Chem 61:12557–12564
3. Zhang A, Pastor L, Nguyen Q et al (2005) Small interfering RNA and gene expression analysis using a multiplex branched DNA assay without RNA purification. J Biomol Screen 10:549–556
4. Cha W, Ma Y, Saif YM et al (2010) Development of microsphere-based multiplex branched DNA assay for detection and differentiation of avian influenza virus strains. J Clin Microbiol 48:2575–2577
5. Vidalakis G, Wang J (2013) Molecular method for universal detection of citrus viroids, US Patent 9,574,244
6. Osman F, Dang T, Bodaghi S et al (2017) One-step multiplex RT-qPCR detects three citrus viroids from different genera in a wide range of hosts. J Virol Methods 245:40–52

Chapter 22

Culture-Independent Discovery of Viroids by Deep Sequencing and Computational Algorithms

Ali Raza, Shou-Wei Ding, and Qingfa Wu

Abstract

Viroids are single-stranded circular RNA molecules that cause diseases in plants and do not encode any protein. Classical approaches for the identification of new viroids are challenging for many plant pathology laboratories as viroid cDNA synthesis and sequencing require purification and enrichment of the naked viroid RNA by two-dimensional gel electrophoresis. Conventional metagenomic approaches are not effective for viroid discovery because the total number of known viroids is small, and distinct viroids share limited nucleotide sequence similarity. In this chapter, we describe a homology-independent approach for the identification of both known and new viroids in disease samples. It is known that viroid infection of plants triggers production of overlapping viroid-derived small interfering RNAs (siRNAs) targeting the entire genome with high densities and that replication of viroids occurs via a rolling-circle mechanism to yield head-to-tail multiple-repeat replicative intermediates. Our approach involves deep sequencing of either long or small RNAs in a disease sample followed by viroid identification with a unique computational algorithm, progressive filtering of overlapping small RNAs (PFOR). Among the sequenced total small RNAs, PFOR retains viroid-derived siRNAs for viroid genome assembly by progressively eliminating nonoverlapping small RNAs and those that overlap but cannot be assembled into a direct repeat RNA, a unique feature of viroid RNA replication. In contrast, long RNAs sequenced after depletion of ribosomal RNAs are cut into 21-nucleotide virtual overlapping small RNAs with the algorithm SLS (splitting longer read into shorter fragments) before PFOR. We show that new viroids or viroids from the two known families are readily identified and their full-length sequences recovered by PFOR from long or small RNAs sequenced directly from infected plants. We propose that our approach can be used for viroid discovery in both plants and potentially animals since PFOR identifies viroids by searching for circular RNAs or a unique replication intermediate of the viroid genome in a sequence homology-independent manner.

Key words Progressive filtering of overlapping small RNAs (PFOR), Viroids, Splitting longer read into shorter fragments (SLS), Homology-independent pathogen discovery

1 Introduction

Metagenomics, especially meta-transcriptomics, a culture-independent approach for the identification of microorganisms including viruses, is transforming our understanding of the microbial diversity in disease and environment [1]. However,

Ayala L. N. Rao et al. (eds.), *Viroids: Methods and Protocols*, Methods in Molecular Biology, vol. 2316,
https://doi.org/10.1007/978-1-0716-1464-8_22,

conventional metagenomic approaches for the identification of viral and subviral pathogens are homology-dependent. For example, a new virus is identified only when its genomic DNAs or RNAs or the encoded proteins share detectable sequence similarities with a known species in the databases [1–4]. Thus, culture-independent discovery of new pathogens with undetectable sequence similarity to a known species requires the development of new strategies. In 2012, we reported the first metagenomic approach for homology-independent discovery of a class of subviral pathogens, viroids [5].

Viroids contain single-stranded circular RNA genomes ranging in size from 220 to 457 nucleotides (nt) in length and are known to cause diseases only in plants [6–10]. These subviral pathogenic RNAs lack protein-coding capabilities and thus replicate by host encoded DNA-dependent RNA polymerase. The replication of viroids and some circular satellite RNAs of viruses occurs via a rolling-circle mechanism, yielding head-to-tail multiple-repeat replicative intermediates in both polarities [5–10]. Similar to the induction of antiviral RNA interference (RNAi) in plants and animals [11], viroid infection of plants triggers Dicer processing of viroid dsRNA replicative intermediates into small interfering RNAs (siRNAs). Plants encode at least four Dicer-like enzymes (DCL1 to DCL4), among which DCL4, DCL2, and DCL3 have been shown to produce 21-, 22-, and 24-nt viroid siRNAs, respectively [5, 11].

The core component of our approach for viroid discovery is based on the computational mining of the deep sequenced small RNAs by a new algorithm [5] designated progressive filtering of overlapping small RNAs (PFOR). PFOR classifies overlapping small RNAs in a library into two groups. A terminal small RNA (TSR) overlaps at least one other small RNA in the library (default k-mer = 17) at only one end to yield a contig longer than the TSR. In contrast, an internal small RNA (ISR) overlaps at least one other small RNA in the library at both ends of the ISR to yield a longer contig. The algorithm then removes nonoverlapping small RNAs and all of those TSRs identified in the first step and reclassifies the small RNAs left in the pool into TSRs and ISRs. PFOR is repeated iteratively until no TSR is found and all of the small RNAs left in the pool are true ISRs. At the final step of PFOR, the remaining true ISRs from the small RNA library are assembled into contigs. All of the siRNAs processed by Dicer from a linear dsRNA would progressively be defined as TSRs and removed from the pool. However, overlapping siRNAs processed from an incomplete or complete head-to-tail direct repeat dsRNA would be classified as true ISRs because of the presence of the small RNAs derived from the junction and progressively enriched by PFOR. Since rolling-circle replication of viroids produces head-to-tail repeat dsRNA replicative intermediates that are diced into true ISRs [5], PFOR retains viroid-derived siRNAs for viroid genome assembly by progressively eliminating nonoverlapping endogenous small RNAs and those that overlap but cannot be assembled into a direct repeat RNA, a unique feature of viroid RNA replication.

We introduced several major improvements into PFOR written in Perl language [5] and designated the new package as PFOR2 [12]. PROR2 was written in C^{++} language and adopted the parallel programming technology of OpenMP, an Application Programming Interface (API) that supports multi-platform shared memory multiprocessing programming [13]. A vector was also used in PFOR2 to store all sequences temporarily in the filtering process to simplify OpenMP parallelization. Importantly, a new program SLS (splitting longer read into shorter fragments) was developed and incorporated into PFOR2, which cuts long RNA reads into 21-nucleotide virtual overlapping small RNAs. In addition to the small RNA libraries, therefore, PFOR2 can identify viroids from libraries of long RNA reads sequenced after ribosomal RNAs are removed. Similar to Dicer processing of head-to-tail repeat dsRNA replicative intermediates produced during viroid replication, virtual dicing of single-stranded long circular RNAs will also generate true ISRs that are retained by PFOR2 for the assembly of new, full-length circular RNAs. Thus, application of PFOR2 can identify not only viroids from host organisms that do not produce overlapping viroid siRNAs but also novel circular RNAs that do not replicate to yield head-to-tail repeat dsRNA replicative intermediates. The development of PFOR and PFOR2 for viroid discovery represents conceptual advances because, unlike NGS and several available classical approaches, the new strategy does not depend on sequence homology with a known viroid [12].

Pipeline data is available at http://staff.ustc.edu.cn/~wuqf/resources.html

2 Materials

2.1 Sample, Reagents, and Equipment for RNA Extraction and Purification

1. Grapevine (*Vitis vinifera L.*) leaves.
2. Potato (*Solanum tuberosum*) leaves.
3. DEPC-treated ddH_2O: Add 0.1% diethyl pyrocarbonate (DEPC) to double distilled H_2O (ddH_2O), mix thoroughly and incubate at 37 °C for 12 h, and then autoclave for 15 min to remove DEPC.
4. RNAlater, stabilization solution (Ambion, USA).
5. TRIzol™ reagent (Invitrogen, USA).
6. Centrifuge and rotor capable of reaching 12,000 × *g* and 4 °C.
7. Polypropylene microcentrifuge tubes.

8. Water bath or heat block at 55–60 °C.
9. Solution of Tris-buffered phenol (pH < 5.0): chloroform (1:1).
10. DNase 1 (Ambion, USA).
11. RNase-free (DEPC-treated) water.
12. RNase inhibitor.
13. Agilent Technologies 2100 bioanalyzer.
14. RiboMinus Plant Kit (Invitrogen, USA).
15. RNeasy Mini Kit (Qiagen, USA).
16. RNeasy MinElute Cleanup Kit (Qiagen, USA).
17. MinElute Gel Extraction Kit (Qiagen, USA).
18. Potassium phosphate (1 M).
19. Potassium acetate (5 M, pH 4.8).
20. Sodium acetate (0.2 M, pH 5.2).
21. Ethanol (75%, 70%, 80%, 100%).
22. 2% cetyltrimethylammonium bromide (CTAB).
23. 2× RNA Gel loading dye (Thermo Fisher Scientific, USA).
24. Random hexamer primers (Fermentas, USA).
25. RNA-seq 3′primer CTCTATGGGCAGTCGGT GATNNNNNN.
26. RNA-seq 5′primer 1 GCGTGTCTCCGACTCAGCATNNNNNN.
27. RNA-seq 5′primer 2 GCGTGTCTCCGACTCAGG TANNNNNN.
28. Ion 5′ PCR primer CCATCTCATCCCTGCGTGTCTCC GACTCAG.
29. Ion 3′ PCR primer CCGCTTTCCTCTCTATGGG CAGTCGGTGAT.
30. 5× first-strand buffer (Thermo Fisher Scientific, USA).
31. 0.1 M DTT (Thermo Fisher Scientific, USA).
32. 25 mM $MgCl_2$ (Thermo Fisher Scientific, USA).
33. 10 mM equimolar dNTPs (Thermo Fisher Scientific, USA).
34. Reverse Transcriptase (Thermo Fisher Scientific, USA) and RNase H (Thermo Fisher Scientific, USA).
35. 10× reaction buffer: 200 mM Tris–HCl (pH 7.8), 400 mM KCl, 80 mM $MgCl_2$, and 10 mM DTT.
36. 10× PCR buffer (Thermo Fisher Scientific, USA).
37. RNase R (Epicenter Biotechnologies, USA).

38. RNase R 10× reaction buffer: 0.2 M Tris–HCl (pH 8.0), 1 M KCl, and 1 mM $MgCl_2$.
39. Prehybridization buffer: 50% formamide, 10% dextran sulfate, 0.18 M NaCl, 20 mM sodium cacodylate, 0.1% SDS, 500 μg/mL yeast t-RNA, and 25 μg/mL sonicated salmon sperm DNA.
40. 2× SSC (3 M NaCl, 0.3 M sodium citrate, adjust to pH 7.0 with NaOH).
41. 4 M LiCl (Ambion, USA).
42. Nylon membrane (Hybond N, Amersham).
43. Marker dye solution: 0.02% bromophenol blue, 0.02% xylene cyanol, and 50% glycerol.
44. Qubit™ RNA HS Assay Kit.
45. Wizard SV Gel and PCR Clean-Up System.
46. Phanta Super-Fidelity DNA Polymerase.
47. Taq DNA polymerase.
48. Klenow Fragment.
49. Phusion High-Fidelity DNA Polymerase (Thermo Scientific).
50. Ion torrent PGM™ 200 Sequencing Kit.
51. Spectrophotometer (Thermo Scientific).
52. DNAMAN.
53. Primer 5.0 software.

2.2 Hardware

PFOR/PFOR2 should be installed on Linux server. To run this algorithm, ~8 GB RAM and above would be sufficient (*see* **Note 1**).

2.3 Software

Perl $\geq$ 5.8 (*see* **Note 2**).

Bowtie2 (for alignment).

SAMtools (processing aligned files).

Text editor (to open the scripts).

2.4 Data Source

The small RNA library grapevine cultivar "Pinot noir" ENTAV115 infected with grapevine yellow speckle viroid (GYSVd) can be constructed in lab or downloaded from the Gene Expression Omnibus (GEO) database of the NCBI. Additionally, construct RNA-seq libraries of PSTVd-infected potato with a modified Not Not So Random (NNSR) sequencing protocol [14] after either depletion of the abundant ribosomal RNAs [15, 16] or enrichment for circular RNAs following the degradation of linear RNAs by RNase R [17] (*see* **Note 3**).

3 Methods

The protocol for enrichment of viroid-specific small RNAs is based on [12, 18] and is described in detail below.

3.1 RNA Extraction and Purification

3.1.1 Grapevine

The centrifugation conditions for grapevine RNA extraction are 12,000 × *g* for 15 min at 4 °C for all steps described below unless if otherwise stated:

1. Quickly grind 5 g of leaves to a powder in liquid nitrogen with a pestle and mortar. Then transfer 10 mL/g of leaf extract to a room temperature mortar and homogenize. Vortex for 30 s, and allow to stand at room temperature for 5 min (liquid nitrogen is evaporated).
2. Add two volumes (w/v) of 1 M potassium phosphate (K_2HPO_4) containing 0.1% mercaptoethanol and 0.5% sodium dodecyl sulfate (SDS) and with 2 volumes of phenol: chloroform (1:1, pH $<$ 5.0) in a blender.
3. After homogenizing for 5 min, centrifuge the homogenate for 15 min at 3000 × *g*.
4. Collect the aqueous phase, and add 2.5 volumes (v/v) of absolute ethanol. Mix gently, and stand for 1 h at −30 °C.
5. Precipitate total nucleic acid by centrifugation at 12,000 × *g* for 10 min at 4 °C, air-dry for 10 min at room temperature.
6. Dissolve the pellet in 1 mL of DEPC-treated water, and mix with one volume each of 2.5 M K_2HPO_4 (pH 8.3) and 2-methoxyethanol.
7. After shaking vigorously for 1 min and centrifuging for 15 min at 3000 × *g*, mix the supernatant with ten volumes of DEPC-treated water and 1/10 volumes of 2% cetyltrimethylammonium bromide (CTAB) solution. Keep the mixture for 1 h on ice.
8. Centrifuge for 15 min at 3000 × *g*, discard the supernatant, gently wash the pellet twice with 70% ethanol containing 0.2 M sodium acetate (NaAc), dry under vacuum for 5 min, dissolve in 200 μL DEPC-treated water, and store at −80 °C.

3.1.2 Potato

RNA extraction of potato seedlings can be carried out using TRIzol™ following manufacturer's guideline (*see* **Note 3**):

1. Quickly grind 0.1 g of leaves to a powder in liquid nitrogen.
2. Resuspended leaf powder in 1 mL TRIzol™, homogenize using a homogenizer, and incubate at room temperature for 15 min to make a thick paste allowing it to thaw.

3. Add 200 μL of chloroform per 1 mL of TRIzol™ Reagent for lysis, and then tightly cap the tube. Incubate for 2–3 min at room temperature.
4. Centrifuge the sample for 15 min at 12,000 × *g* at 4 °C.
5. Transfer the aqueous phase containing the RNA to a new tube.
6. Transfer the aqueous phase containing the RNA to a new tube by angling the tube at 45 °C and pipetting the solution out.
7. Add 0.5 mL of isopropanol to the aqueous phase, per 1 mL of TRIzol™ Reagent used for lysis. Incubate for 10 min at room temperature.
8. Centrifuge for 10 min at 12,000 × *g* at 4 °C. Total RNA precipitate forms a white gel-like pellet at the bottom of the tube. Discard the supernatant with a micropipettor.
9. Resuspend the pellet in 1 mL of 75% ethanol per 1 mL of TRIzol™ Reagent, used for lysis.
10. Vortex the sample briefly, and then centrifuge for 5 min at 7500 × *g* at 4 °C. Discard the supernatant with a micropipettor.
11. Vacuum or air-dry the RNA pellet for 5–10 min.
12. Resuspend pellet at 200 μL of DEPC water.

3.2 Enrichment of Viroid and Viroid-Specific Small RNAs

1. The 200 μL of the dissolved nucleic acid pellet (*see* **step 8** in Subheading 3.1.1 for grapevine and **step 12** in Subheading 3.1.2 for potato) will be treated with LiCl.
2. Slowly add an equal volume (200 μL) of 4 M LiCl with gentle shaking, and allow to stand on ice for 4 h to precipitate high molecular weight RNAs (*see* **Note 4**).
3. Collect the supernatant (i.e., 2 M LiCl soluble nucleic acids fraction) by centrifugation at 12,000 × *g* for 5 min at room temperature, add 2.5 volumes (1 mL) of absolute ethanol, mix gently, and allow to stand overnight at −30 °C.
4. Recover the 2 M LiCl soluble nucleic acids by centrifugation at 12,000 × *g* for 10 min at 4 °C, and discard carefully the supernatant.
5. Air-dry the resulting pellet for 10 min at room temperature.
6. Resuspend pellet at an appropriate amount (e.g., 100–160 μL) of DEPC water.

3.3 DNase 1 Treatment

1. Dissolve the precipitate in 160 μL DEPC-treated water, add 40 μL 5× DNase I buffer and 1 μL DNase I (10–20 U/ μL), and incubate for 30 min at 37 °C (*see* **Note 5**).
2. Add equal volume (200 μL) of phenol/chloroform (1.1) for extracting the DNase I digest, and centrifuge at 12,000 × *g* for 5 min at 4 °C.

3. Collect aqueous phase in another tube, add 2.5 volumes of absolute ethanol, and stand for 1 h at −20 °C.
4. Recover the DNA-free low-molecular-weight RNA by centrifugation at 12,000 × *g* for 10 min at 4 °C, and air-dry the resulting pellet for 10 min at room temperature.
5. Resuspend the pellet in 100 μL RNase-free DEPC-treated water.

3.4 Two-Dimensional PAGE (2D-PAGE)

1. Low-molecular-weight RNA (e.g., 200 μg/100 μL) is mixed with 0.9 volume of 50% urea solution (90 μL) and 0.1 volume of marker dye solution (10 μL) and heat-denatured at 68 °C for 15 min.
2. Apply the sample to a 7.5% polyacrylamide gel (acrylamide:/ bisacrylamide = 39:1) containing 89 mM Tris, 89 mM boric acid, and 2 mM EDTA (1 × TBE) −8 M urea, and run at 250 V (constant voltage) until the xylene cyanol blue marker dye migrates to bottom of gel (*see* **Note 6**).
3. The returning run was at 350 V for 2 h in 1/8 × TBE. Attach the front glass plate with rubber heater of 80 °C during the second run.
4. After electrophoresis, remove one of gel plates, and place a dry Hybond N+ nylon membranes to transfer by upward capillary action in 20× SSC buffer.
5. Wrap the assembly with vinyl film to prevent drying out, place a bottle containing ca. 1–2 L of water on top, and leave several hours to overnight at 25 °C for contact blotting of RNAs to nylon membrane.
6. Remove the membrane, and excise a gel slice from 3 to 5 cm below from the meniscus, i.e., the region containing small RNAs with the size 15–35 nucleotides.
7. Crush the gel into small pieces, add 1 mL RNA gel elution buffer, and shake gently overnight at 37 °C.
8. Small RNAs are precipitated by adding 2.5 volumes of absolute ethanol, overnight storage at −30 °C, and centrifugation at 12,000 × *g* for 15 min at 4 °C.
9. Air-dry the resulting pellet for 10 min at room temperature and dissolve in 10 μL RNase-free distilled water.

3.5 Northern Blot Hybridization

The hybridization protocol was performed according to Aliyari et al. [19]:

1. Prehybridization and hybridization is carried out in a buffer containing 50% formamide, 10% dextran sulfate, 0.18 M NaCl, 20 mM sodium cacodylate, 0.1% SDS, 500 μg/mL yeast t-RNA, and 25 μg/mL sonicated salmon sperm DNA.

2. After prehybridization for 4–10 h, change the buffer with new one containing RNA probe (1 μL DIG RNA probe/1 mL hybridization solution) at 50–55 °C overnight.
3. After hybridization for 16–24 h at room temperature, wash membranes twice in 2 × SSC for 15 min at room temperature once in 2 × SSC containing 1 μg/mL RNase A for 30 min at room temperature. After that wash membranes twice in 0.1 × SSC and 0.1% SDS for 15 min at 68 °C.
4. Place the membrane in chemiluminescent chamber ChemiDoc (XRS, Bio-Rad), expose for 10 min to 2 h to visualize viroid-specific small RNAs.

3.6 RNA Cleanup

There are three methods in this section: (1) RNeasy MinElute Cleanup Kit, (2) removal of ribosomal RNA by Ribo-Zero rRNA Removal Kit, and (3) ethanol precipitation.

3.6.1 RNeasy MinElute Cleanup Kit

1. Bring the RNA volume to 100 μL with RNase-free water, and mix with 350 μL Buffer RLT (provided with the kit).
2. Add 250 μL of absolute ethanol and mix by pipetting.
3. After injecting 700 μL RNA sample into the micro-adsorption column, centrifuge at 8000 × *g* for 15 s at room temperature, and discard the filtrate.
4. Add 500 μL Buffer RPE (provided with the kit) to the adsorption column to wash the adsorption membrane, centrifuge at 8000 × *g* for 15 s at room temperature, and discard the filtrate.
5. Repeat the washing once as in **step 4** above, and centrifuge at 8000 × *g* for 2 min at room temperature to remove residual Buffer RPE and ethanol.
6. Open the lid of the column, and centrifuge for 5 min at 12000 × *g*.
7. Place the column in a new 1.5 mL centrifuge tube, add 10–40 μL of RNase-free water, and centrifuge for 1 min at 12000 × *g* to obtain clean RNA.

3.6.2 Removal of Ribosomal RNA by Ribo-Zero rRNA Removal Kit

Take 5 μg of clean RNA (no DNA contamination) to remove rRNA by using RNA by Ribo-Zero rRNA removal Kit following the manufacturer's protocol; key steps are as follows:

Magnetic Bead Preparation

1. Gently vortex the beads to obtain a uniform suspension. Each reaction requires 225 μL of magnetic beads. Carefully pipette the beads into a 1.5 mL centrifuge tube to avoid air bubbles.
2. Place the tube with beads on a magnetic stand until the solution is clear (about 1 min), and remove the supernatant by aspiration.

3. Remove the tube from the magnetic stand, and resuspend the beads by slow vortexing with an equal volume (225 μL) of RNase-free water.
4. Place tube on a magnetic separator for 1 min. Aspirate and discard the supernatant.
5. Repeat **steps 3–4** once.
6. Remove and discard all supernatant, and remove from the magnetic stand.
7. Add 65 μL magnetic bead resuspension solution, and vortex to resuspend at room temperature.

RNA Sample Preparation/ Hybridization of Probes

In this step, probes in the removal solution hybridize to rRNA present in the sample; the steps are as follows:

1. For each sample, combine the following volumes in a 0.5 mL microcentrifuge tube. Pipette to mix:
   4 μL Ribo-Zero Reaction Buffer.
   5 μg DNA-free RNA sample.
   10 μL Ribo-Zero Removal Solution.
   RNase-free water to 40 μL of total solution.
2. After thoroughly mixing by pipetting, incubate at 68 °C for 10 min, remove from high-temperature incubator, and continue to incubate for 5 min at room temperature.
3. Add 40 μL RNA sample to a 1.5 mL tube containing 65 μL washed magnetic beads. Immediately pipette to mix.
4. Vortex for 10 s, and incubate at room temperature for 5 min.
5. Vortex the tube again at medium speed for 10 s, and incubate for 5 min at 50 °C.
6. Immediately place on a magnetic stand, with cap open, and wait until the liquid is clear (about 1 min).
7. Transfer 85–90 μL supernatant containing depleted RNA to a fresh 1.5 mL tube.
8. Store at −25 °C to −15 °C overnight.

Concentration of rRNA-Free RNA Using Ethanol Precipitation

The steps of the ethanol precipitation method are as follows:

1. Bring the rRNA free RNA-containing supernatant volume to 200 μL with RNase-free water.
2. Add 20 μL of 3 M NaAc (pH 5.2) and 1 μL of glycogen, and mix gently by vortexing.
3. Add 500 μL of absolute ethanol, mix well, and incubate at −80 °C for 30 min or overnight.
4. Centrifuge at 12,000 × *g* for 20 min at 4 °C, and carefully discard the supernatant without disturbing the pellet.

5. Add 200 μL of 70% ice-cold ethanol.
6. Centrifuge the tube for 5 min at 12,000 × *g* at 4 °C. Discard the supernatant carefully without disturbing the pellet.
7. Repeat **steps 5** and **6** once.
8. Air-dry the pellet for 5 min. Resuspend the RNA pellet in appropriate amount (10–30 μL) of DEPC-treated water.
9. Place RNA on ice and proceed to desired downstream application or store RNA at −80 °C.

3.7 Circular RNA Enrichment

RNase R digestion is one of important steps for detection of viroids because it can digest almost all linear RNA while retaining circular RNA or intron loops [17, 20] (*see* **Note 7**). The reaction system is as follows:

2 μL 10 × reaction buffer: 0.2 M Tris–HCl (pH 8.0), 1 M KCl, 1 mM $MgCl_2$, and 5 μg DNA-free RNA (cleaned RNA by RNA cleanup protocols obtained from earlier steps in Subheading 3.4) 5 U RNase R Up to 20 μL RNase-free water

1. Mix the abovementioned components gently by pipetting, and incubate at 37 °C for 30 min.
2. Integrity of the resulting RNA preparations must be checked before preparing cDNA libraries for deep sequence analysis. Check RNA integrity using the Agilent Technologies 2100 Bioanalyzer. The RNA Integrity Number (RIN) value should be greater than 8 (*see* **Note 8**). Otherwise, electrophoresis can also assess the integrity of total RNA extraction and preparation on a formaldehyde −1% agarose gel and ethidium bromide staining. Two clear bands of 28S and 18S rRNA should be shown presenting the 28S rRNA twice of 18S rRNA with almost no visibility of 5.8S RNA.

3.8 Small RNA and RNA-seq Library Construction

Use the purified and enriched RNA to construct the small RNA library to be sequenced by the Illumina Hiseq platform *2000* [11] (*see* **Note 9**).

3.8.1 First-Strand cDNA Synthesis

1. Prepare the reagents required for reverse transcription reaction according to Table 1.
2. Reagents will be added in the reaction in two steps (Table 1).
3. Add into a PCR reaction tube the 3′-Primer and the RNA target, and bring reaction to its final volume with RNase-free water as described in Table 1, **step 1**.
4. Mix well, and place tube in a thermocycler programmed at 95 °C for 2 min and 65 °C for 5 min.
5. After program is completed, place tube immediately in ice bath, and add the 5× reaction buffer, RNase inhibitor, dNTPs, and the RevertAid M-MuLV as described in Table 1, **step 2**.

Table 1
Reaction composition for first-strand cDNA synthesis

Reaction components	Quantity
Step 1. Initial reaction	
3′-primer(10μM)	1μL
RNA	5μL
RNase-free water	to 12μL
Step2. Following template denaturation and primer annealing	
5 × reaction buffer	4μL
RNase inhibitor	1μL
10 mM dNTPs	2μL
RevertAid M-MuLV	1μL

6. Mix well, and place tube in a thermocycler programmed at 25 °C for 5 min, 42 °C for 60 min, and 70 °C for 5 min.
7. After program is completed, place tube immediately in ice bath, and add 3 μL of 10 × Digestion Buffer, 2 μL of RNase H and 5 μL of RNase-free water.
8. Mix well, and place tube in a thermocycler programmed at 37 °C for 30 min and 65 °C for 10 min.
9. To clean up the first cDNA strand, purify using 54 μL of Ribo-Zero beads (Subheading 3.4, **step 2**).

3.8.2 Second-Strand cDNA Synthesis

1. Add the reagents (Table 2) on ice in order and mix gently by vortexing.
2. Incubate at 15 °C for 2 h. Do not let the temperature rise above 15 °C.
3. Terminate the reaction by adding 5 μL of 0.5 M EDTA, (pH 8.0).
4. Add 1.5 μL of RNase 1 for digestion of RNA. Incubate for 30 min at 37 °C.

3.8.3 Fragment Selection

Fragments can be purified using MinElute Gel Extraction Kit. Extraction of 1 μg or 0.5 μg starting DNA from an agarose gel with a recovery of 80% using the MinElute Gel Extraction Kit. Loading dye is provided for analysis of purified DNA samples using electrophoresis. It contains three marker dyes (bromophenol blue, xylene cyanol, and orange G) that facilitate estimation of DNA migration distance and optimization of agarose gel run time. All centrifugation steps are carried out at 10,000 × *g* in a conventional microcentrifuge at room temperature.

Table 2
Reaction composition for second-strand cDNA synthesis

Reaction components	Quantity
1st cDNA	20μL
5′ primer (100μM)	1μL
10 × NEB buffer 2	3μL
10 mM dNTP mix	1.5μL
Klenow enzyme	3μL
ddH_2O	to 30μL

1. Run agarose gel electrophoresis (with TAE (Tris-acetate/EDTA) or TBE (Tris-borate/EDTA) buffer) and using ethidium bromide dye.
2. Cut the band with 300–400 bp DNA, weigh it, and then add the appropriate amount of Buffer QG (300 μL buffer per 100 mg gel).
3. After dissolving at 50 °C for 10 min, add an equal volume of isopropanol to the mass of the gel.
4. Add all the solution to the adsorption column (maximum volume 800 μL), centrifuge for 1 min, and discard the filtrate.
5. Add 500 μL Buffer QG, centrifuge, discard the filtrate, and centrifuge the empty adsorption column for 1 min.
6. Add 10 μL Buffer EB to the adsorption membrane, let column stand for 1 min at room temperature, and centrifuge to obtain DNA.

3.8.4 PCR Amplification

1. Formulate the following system on ice (Table 3), thaw, and add in following order: water (ddH_2O), buffer, dNTPs, $MgCl_2$, template primers, *Taq* polymerase (Phusion).
2. After gently mixing, centrifuge.
3. Denaturation: the initial denaturation is at 98 °C at 30 s which is sufficient for DNA templates.
4. Second denaturation is performed at 98 °C for 10 s.
5. Annealing is done at 60 °C for 30 s which allows the primers to attach to template.
6. Extension is carried out at 70 °C for 30 s.
7. After 10–20 cycles, follows the final extension at 72 °C for 5 min (Table 4).

Table 3
Demonstration of PCR amplification reaction to be used

Reaction components	Quantity
5 × HF buffer	5μL
2.5 mM dNTP mix	2μL
5′Ion-primer(10μM)	0.5μL
3′Ion-primer(10μM)	0.5μL
cDNA	5μL
Phusion	0.25μL
ddH_2O	to 25μL

Run the PCR reaction after mixing

Table 4
Conditions for PCR reaction steps with temperature and time

Steps	Temperature	Reaction time
Initial denaturation	98 °C	30 s
Denaturation	98 °C	10 s
Annealing	60 °C	30 s
Extension	72 °C	30 s
Go to step 2	Go to step 2	10–15 cycles
Final extension	72 °C	5 min

3.8.5 Library Purification and Sequencing

1. Place all of the following reaction components on ice and thaw: CRL (Custom RNA Ladder).

   HRL (high-resolution ladder).

   Ultrapure water.

   Amplified cDNA construct.

   Novex TBE running buffer (5×).

   Ultrapure ethidium bromide 10 mg/mL.

   10 M NaOAc, pH 5.2.

   10 mM Tris–HC1, pH 8.5.

   Absolute ethanol.

   Glycogen.

   Pellet Paint NF Co-precipitant.

2. Determine the volume of 1× TBE Buffer needed for gel electrophoresis. Dilute the 5× Novex TBE Buffer to 1×.
3. Place 6% Novex TBE gel into the gel electrophoresis unit per manufacturer instructions.

4. Combine 2 μL CRL and 2 μL DNA Loading Dye in a new 1.5 mL microcentrifuge tube. Pipette to mix.
5. Combine 1 μL HRL and 1 μL DNA Loading Dye in a new 1.5 mL microcentrifuge tube. Pipette to mix.
6. Combine all amplified cDNA construct (typically 48–50 μL) and 10 μL DNA Loading Dye in a new 1.5 mL microcentrifuge tube. Pipette to mix.
7. Load two gel lanes with 2 μL CRL/loading dye mixture.
8. Load 1 gel lane with 2 μL HRL/loading dye mixture.
9. Load two gel lanes with 25 μL each of amplified cDNA construct/loading dye mixture.
10. The total load volume is 50 μL.
11. Run the gel for 60 min at 145 V or until the blue front dye leaves the gel.
12. View the gel on a UV transilluminator.
13. Using a razor blade, cut out the bands from the two lanes that correspond approximately to the adapter-ligated constructs derived from the 22 nt and 30 nt small RNA fragments.
14. Place the band of interest into the 0.5 mL gel breaker tube.
15. Centrifuge the nested tubes at 20,000 × *g* for 2 min to move the gel through the holes into the 2 mL tube.
16. Add 200 μL ultrapure water to the gel debris in the 2 mL tube.
17. Rotate or shake for at least 2 h to elute the DNA.
18. Transfer the eluate and gel debris to the top of a 5 μm filter.
19. Centrifuge at 10 s at 600 × *g*.
20. Normalize library concentration to 2 nM using Tris–HCl 10 mM, pH 8.5.
21. Store at −25 °C to −15 °C.

    The library concentration (appropriate conc. 1–100 ng/μL for library preparation) can be determined using a Qubit 2.0 fluorometer and sequencing with the Ion Torrent™ Personal Genome Machine™ (PGM) System.

3.9 Quality Control and Adaptor Trimming and Mapping

1. RNA-sequenced reads in FASTQ format can be uploaded in FASTQC to check if the quality of the reads is good (*see* **Note 10**). If the sample has bad-quality reads, you can use trimmomatic software (http://www.usadellab.org/cms/?page=trimmomatic).
2. There are many efficient tools available for removing adaptor sequences, e.g., FastqMcf (http://code.google.com/p/ea-utils/wiki/FastqMcf),
   Trimmomatic (http://www.usadellab.org/cms/index.php?page=trimmomatic).

HTSeq (http://www.huber.embl.de/users/anders/HTSeq/doc/overview.html Cutadapt (http://www.ark-genomics.org/events-onlinetraining-eu-training-course/adapter-and-quality-trimmingillumina-data).

3. Trimmomatic supports both single- and paired-end reads. To use trimmomatic, first download the .jar file from the link (http://www.usadellab.org/cms/uploads/supplementary/Trimmomatic/Trimmomatic-Src-0.36.zip) (read manual) (*see* **Note 11**).
4. Extract the zip file and add the PATH.
5. Access the jar file using java with the command given below:

```
java -jar trimmomatic-0.30.jar SE <input.fastq> <output.fastq> ILLUMINACLIP:TruSeq3-SE:2:30:10 LEADING:3 TRAILING:3 SLIDINGWINDOW:4:15 MINLEN:36
```

6. For mapping and alignment, use bowtie [22, 23] with default parameters as it is an ultrafast, memory-efficient short read aligner, which is a popular tool for alignment of small RNAs (*see* **Note 12**).

3.10 Working of PFOR

1. The databases and software required for the analysis are: (1) clean reads with the 5′ and 3′ adapters removed, (2) host genome sequence information if it is available in NCBI or special Website supplied in the published papers, and (3) PFOR program [5] (*see* **Note 13**).

   PFOR/PFOR2 program starts working with a simple command line and initiates analysis of small RNA library input file with k-mer values from 17 to 22 (i.e., 17+5) [5].

   Command line:

```
perl PFOR.pl -f input_file -x 5.
Options include:
```

- f <file.fasta> input reads
-o <output_file> sequences of viroid candidates (default: input_file.contigs)
-k <number> initial assembling k-mer (default = 17)
-s <number> minimal length of reads for assembling (default = 18)
-l <number> maximal length of reads for assembling (default = 40)
-n <number> the minimal number of reads for extension (default = 1)
-x <number> the range of test k-mer (i.e., 17+x) (default: +0)
-D <number> the minimal length of candidate sequence (default = 100)
-U <number> the maximal length of candidate sequence (default = 800)

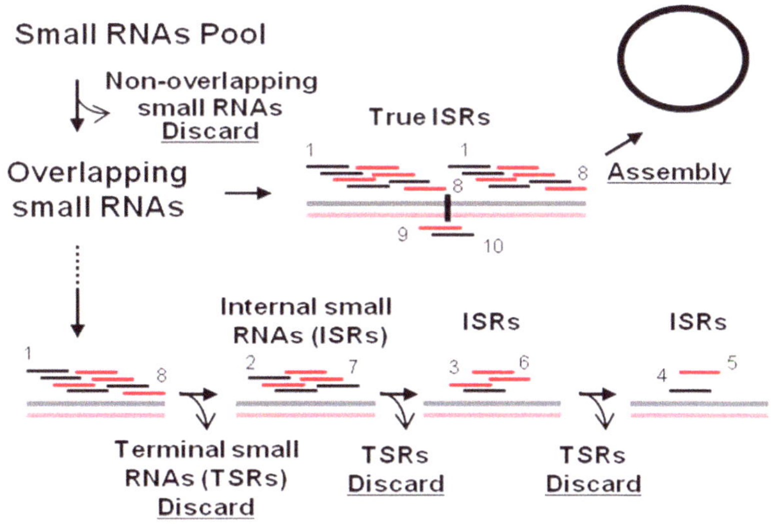

Fig. 1 The key steps in progressive filtering of overlapping small RNAs (PFOR)

2. The basic steps involved in the processing of data by PFOR is the categorization of overlapping sRNAs in a particular library into two groups (Fig. 1). These includes the terminal sRNA (TSR) and the internal sRNA (ISR).
3. The overlapping of the two types of sRNA occurs by two different ways.
4. The terminal sRNA (TSR) overlaps at least one other sRNA in the pool (default k-mer = 17) at only one of the two ends resulting to a contig longer than it.
5. Contrarily, an internal sRNA (ISR) overlaps at least one other sRNA in the pool at both ends of the ISR to yield a contig longer than ISR.
6. All the nonoverlapping sRNAs as well as TSRs detected in initial step are removed by the algorithm and the remaining sRNAs into the pool are categorized as ISRs and TSRs.
7. The process is repeated continuously until the complete elimination of all of the TSRs from the sRNA pool. The remaining obtained sRNA are true ISRs.
8. Finally, the residual true ISRs are then assembled into contigs from the sRNA library.
9. The first phase of PFOR processing involves the elimination of nonoverlapping as well as small RNA overlaps with size less than 17 nt from the pool. This step is helpful in decreasing the complexity of the sRNA library.
10. All of the siRNAs, progressively termed as TSRs, were removed from the pool as they are processed by a Dicer nuclease from a

linear dsRNA. Nevertheless, because of the presence of the small RNAs derived from junction and progressively enriched by PFOR, the overlapping siRNAs processed from a circular RNA or an incomplete or complete head-to-tail direct repeat dsRNA would be classified as true ISRs.

11. Hence, PFOR is designed to discover viroids according to the models suggesting that viroid siRNAs can be processed either from the structural components of the circular viroid RNA or from dsRNA head-to-tail repeats produced as intermediate replicas from the rolling-circle viroid replication [9, 21, 23–25] (*see* **Note 14**).
12. PFOR removes all the overlapping TSRs along with nonoverlapping sRNAs as well as processes the retained overlapping ISR simultaneously in the final genome assembly irrespective of its copy number in the pool.
13. In PFOR, a two-level hash table is built at each iteration process to store all sequences in the pool, whereas in PFOR2, a two-level hash table is only established at the first iteration, and non-ISRs are deleted from the two-level hash table for each subsequent iteration (Fig. 2).
14. The final number of generated virtual sRNAs is dependent on two metrics: sRNA size and overlap size between neighboring sRNAs (step size).
15. Typically, a longer read is cut into contiguous 21-nt sRNAs covering the whole read, in which each sRNA overlapped 20 nt with its 59 and 39 neighboring sRNAs (step size = 1).
16. In principle, SLS-PFOR2 can identify novel viroid circular RNAs in host species that replicate but do not trigger the biogenesis of viroid-specific siRNAs. Therefore, SLS-PFOR2 has the potential to expand the list of both viroids and host species that support viroid infection.

3.11 Evaluation of PFOR2

PFOR2 has the ability to analyze RNA-seq libraries depleted of either ribosomal RNAs or linear RNAs for the discovery of both replicating and non-replicating circular RNAs in diverse range of samples:

1. To determine the efficacy of SLS-PFOR2, use the PSTVd-infected potato rRNA-depleted library.
2. Run the program, yields 774,621 reads longer than 100 nt, among which 83 reads would be derived from PSTVd with a mean length of 160 nt.
3. A total of 92,093 reads longer than 100 nt obtained from the RNase R treated library, with 55 reads from PSTVd.

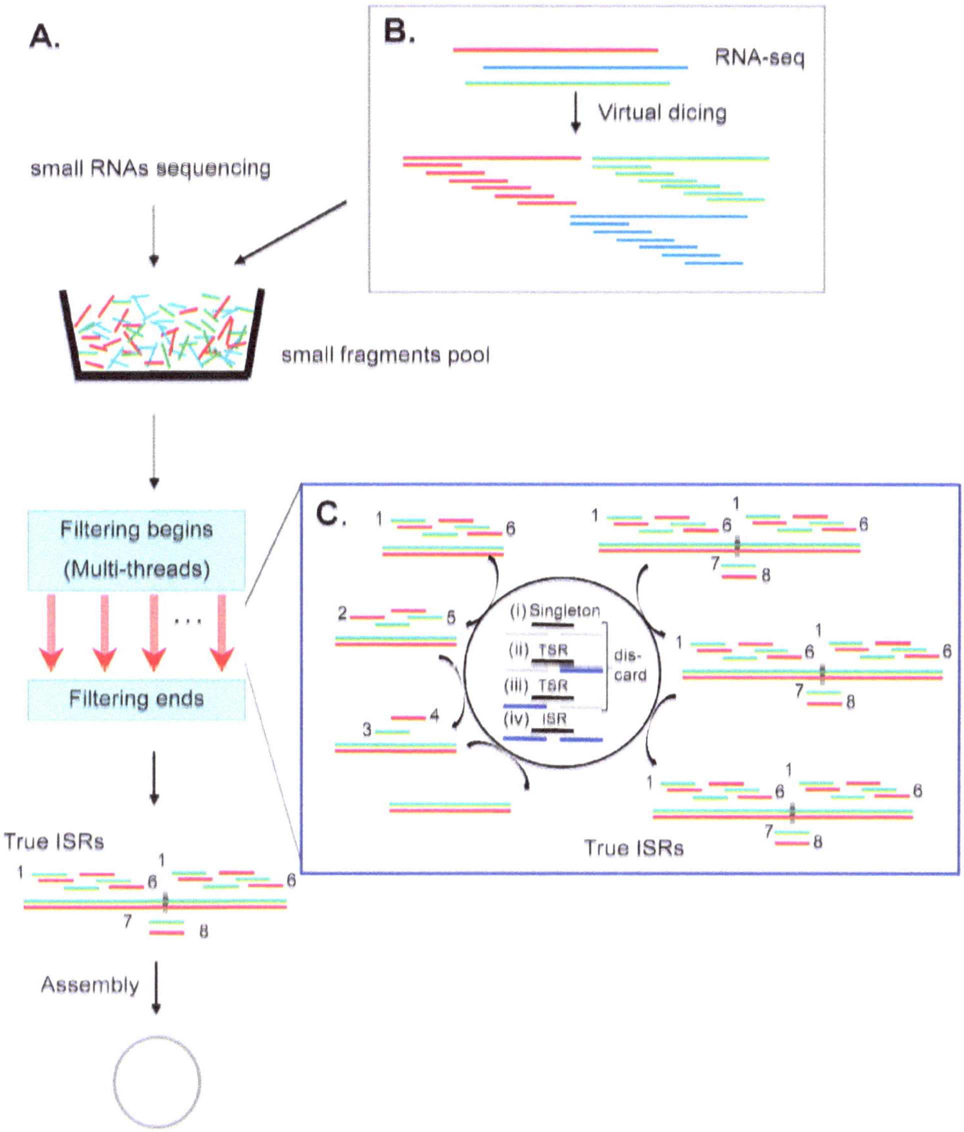

Fig. 2 Illustration of the principles of the PFOR2 and SLS computational programs. (**a**) Flowchart showing the steps followed by PFOR2 for filtering all non-overlapping small RNAs (singleton) and terminal small RNAs (TSRs) with multi-thread processing and the assembly of the remaining true internal small RNAs (ISRs). (**b**) Schematic view of the virtual dicing of longer RNA reads into small overlapping sRNAs by SLS. (**c**) Close-up view of the filtering criterion of TSRs implemented in PFOR2. Each sRNA in the small fragments pool is placed into one of four groups based on the presence or absence of 5′ and 3′ overlapping sRNAS [12]

4. The full-length PSTVd molecule of 354 nt would be readily identified by SLS-PFOR2 from both the rRNA-depleted library (k-mer 19 or 20) and the RNase R library (k-mer 17 or 18) with a running time of 3 h 20 min and 103 h 14 min, respectively, as shown in Table 5 (*see* **Note 15**).

Table 5
PFOR2 Processing ability of RNA-seq library

PFOR2 analysis result	RNA-seq library	
	Ribo-Minus RNA	RNase R treated RNA
Total number of library sequences greater than 100 nt	774,621	92,093
Number of library sequences derived from PSTVd	83	55
Proportion of sequences from PSTVd sources	0.01%	0.06%
Run time for PFOR2 analysis	3 h 20 min	103 h 14 min

3.12 Homology Search for PFOR/PFOR2 Assembled Contigs

The resulting contigs can be queried against the GenBank nt and nr databases using the BLAST program (see manual of BLAST program) [26, 27] (*see* **Note 16**).

3.13 Validation of PFOR/PFOR2 Assembled Contigs

1. In order to verify the results of PFOR and PFOR2 analysis, contigs without any homology with available genomes are further validated by PAGE and northern blot hybridization (*see* Subheadings 3.4 and 3.5 described above) as (*see* **Note 17**).
2. Secondary RNA structure could be predicted by MFold program [28].

4 Notes

1. Linux operating system or workstation is mostly recommended to run the algorithms as various modules are easily available and work efficiently.
2. It should be checked and assured that correct modules are installed on the server to avoid any possible error while running the scripts in Perl.
3. Use gloves and eye protection when using TRIzol reagent. Avoid contact with skin or clothing and use in a chemical fume hood. Avoid breathing the vapor. TRIzol reagent is a ready-to-use reagent for the isolation of total RNA from cells and tissues. During sample homogenization, TRIzol reagent maintains the integrity of the RNA, while disrupting cells and dissolving cell components. Addition of chloroform followed by centrifugation separates the solution into aqueous and organic phases.
4. Chill the RNA on ice before starting the LiCl fractionation, and add the 4 M solution dropwise with constant mixing. A flocculent precipitate containing primarily rRNA and mRNA should

appear only slowly (after 1–2 h). Too rapid precipitation may nonspecifically trap viroid RNA.

5. RNase-free DNase I (e.g., RQ1 DNase) should be used for DNA digestion.
6. It is a very crucial step to apply nucleic acids in gel. When low-molecular-weight RNAs preparation from viroid-infected plants are fractionated on a normal size gel, 10–20 μg RNA (in 10–20 μL heat-denatured sample) per well allows reliable detection of viroid-specific small RNAs.
7. This enzyme is ideally suited to the study of lariat RNA produced by traditional splicing as well as circular RNAs which arise through back-splicing. By removing linear RNAs from RNA extracts, RNase R greatly facilitates the identification of circular species through RNA-sequencing. This enables researchers to probe the landscape of splicing events with greater depth.
8. In order to standardize the process of RNA integrity interpretation, the RNA Integrity Number (RIN) is used to remove individual interpretation in RNA quality control. The RIN software algorithm allows for total RNA, based on a numbering system from 1 to 10, with 1 being the most degraded profile and 10 being the most intact.
9. For high-quality sequencing and better yield, it is strongly recommended to follow Illumina's Hiseq method.
10. Upload fastq reads on FastQC, a quality control tool of fastq files (https://www.bioinformatics.babraham.ac.uk/projects/fastqc/). This will provide base quality as well as other related information about reads, for example, read length, GC distribution, duplicated reads, adapters, or any other overrepresented sequence.
11. Remove overrepresented sequences and adapters before aligning. Institutions and companies that offer sequencing services often provide the services of removing adaptors and return the clean reads to the investigator. The investigator can also provide the list to trim the adaptors.
12. Make sure that you have good-quality RNA-sequencing reads. Poor-quality reads can affect the alignment and accuracy.
13. The method employed in PFOR can effectively deal with the problems created by sequencing inaccuracies or heterogeneity of the viroid population, facilitating the full-length genome assembly from viroid-specific siRNAs produced by the RNA silencing complex of the host. We noted that the use of PFOR with different k-mer values may retain different sets of overlapping siRNAs derived from distinct and closely related

molecules of the same viroid for the final assembly. Thus, identification of viroids by PFOR provides important clues about the population structure of the viroid [5]. Although this approach works efficiently for the discovery of viroids, it may also be applicable to the identification of circular DNA viruses. PFOR analysis of the grapevine small RNA library required ~0.5 GB RAM and 2 min, which recovered GYSVd molecules and the novel grapevine hammerhead viroid-like RNA (GHVd RNA, 372 and 375 nt in length), which shares 96% nucleotide sequence identity with the verified GHVd RNAs. GYSVd of 363 and 365 nt in length were obtained by PFOR, using k-mer of 17 and 18, respectively. The sequence identity was 96% between two recovered GYSVd molecules [5].

14. Asymmetric and symmetric rolling-circle replication models of viroids in the *Pospiviroidae* and *Avsunviroidae* families implicate a monomer circular viroid RNA in one and two polarities, respectively, but both produce plus- and minus-strand multimeric head-to-tail replicative intermediates [6, 9, 29].
15. SLS-PFOR2 is capable of discovering circular RNAs that may or may not trigger in vivo production of specific small silencing RNAs. The application of PFOR2 should facilitate the discovery of novel circular RNAs and expand the list of known host species that can be independently infected by viroids [12].
16. The assembled contigs often include many corresponding to host ribosomal or genomic repeat sequences and can be removed after they are identified by BLAST analysis. If we want the reads to have for example at least 90% identity with the reference sequence in order to be included in the final mapping, set this value to 0.9. The similarity fraction does not apply to the whole read; it relates to the "Length fraction."
17. The circularity of contigs can be confirmed by RT-PCR using two adjacent primers of opposite polarity. If the new potential viroid-like RNA can be folded into rod-like or branched rod-like secondary structure typical of viroids in the families *Pospiviroidae* and *Avsunviroidae*, it can be considered as a candidate new viroid species. Further direct evidence can be obtained by PAGE, northern blot analysis, and finally bioassay to show the autonomous replication in a certain host species. Additionally, search for conserved motifs such as the hammerhead ribozyme of the *Avsunviroidae* family or the central conserved region (CCR) of the *Pospiviroidae* family is as reported previously [12, 30].

Acknowledgments

This project was supported by grants from the National Natural Science Foundation of China (32030087 and 31871927) to Q Wu and from the US-Israel Binational Agricultural Research and Development Fund and the Agricultural Experimental Station of the University of California, Riverside to S.-W.D.

References

1. Zhang YZ, Shi M, Holmes EC (2018) Using metagenomics to characterize an expanding virosphere. Cell 172:1168–1172. https://doi.org/10.1016/j.cell.2018.02.043
2. Geoghegan JL, Holmes EC (2017) Predicting virus emergence amid evolutionary noise. Open Biol 7:170189. https://doi.org/10.1098/Rsob.170189
3. Shi M, Lin XD, Tian JH, Chen LJ, Chen X, Li CX, Qin XC, Li J, Cao JP, Eden JS, Buchmann J, Wang W, Xu JG, Holmes EC, Zhang YZ (2016) Redefining the invertebrate RNA virosphere. Nature 540:539. https://doi.org/10.1038/nature20167
4. Wu Q, Luo Y, Lu R, Lau N, Lai EC, Li WX, Ding SW (2010) Virus discovery by deep sequencing and assembly of virus-derived small silencing RNAs. Proc Natl Acad Sci U S A 107:1606–1611. https://doi.org/10.1073/pnas.0911353107
5. Wu Q, Wang Y, Cao M, Pantaleo V, Burgyan J, Li WX, Ding SW (2012) Homology-independent discovery of replicating pathogenic circular RNAs by deep sequencing and a new computational algorithm. Proc Natl Acad Sci U S A 109:3938–3943. https://doi.org/10.1073/pnas.1117815109
6. Diener TO (2003) Discovering viroids—a personal perspective. Nat Rev Microbiol 1:75–80. https://doi.org/10.1038/nrmicro736
7. Flores R, Hernandez C, de Alba AEM, Daros JA, Di Serio F (2005) Viroids and viroid-host interactions. Annu Rev Phytopathol 43:117–139. https://doi.org/10.1146/annurev.phyto.43.040204.140243
8. Owens RA (2008) Identification of viroids by gel electrophoresis. Curr Protoc Microbiol. Chapter 16:Unit 16G 1 1-16G 1 9. https://doi.org/10.1002/9780471729259.mc16g01s10
9. Ding B (2009) The biology of viroid-host interactions. Annu Rev Phytopathol 47:105–131. https://doi.org/10.1146/annurev-phyto-080508-081927
10. Hu CC, Hsu YH, Lin NS (2009) Satellite RNAs and satellite viruses of plants. Viruses 1:1325–1350. https://doi.org/10.3390/v1031325
11. Guo ZX, Li Y, Ding SW (2019) Small RNA-based antimicrobial immunity. Nat Rev Immunol 19:31–44. https://doi.org/10.1038/s41577-018-0071-x
12. Zhang Z, Qi S, Tang N, Zhang X, Chen S, Zhu P, Ma L, Cheng J, Xu Y, Lu M, Wang H, Ding SW, Li S, Wu Q (2014) Discovery of replicating circular RNAs by RNA-seq and computational algorithms. PLoS Pathog 10: e1004553. https://doi.org/10.1371/journal.ppat.1004553
13. Alexandrescu A (2001) Modern C++ design: generic programming and design patterns applied. MA. Addison-Wesley Professional, Boston
14. Levin JZ, Yassour M, Adiconis X, Nusbaum C, Thompson DA, Friedman N, Gnirke A, Regev A (2010) Comprehensive comparative analysis of strand-specific RNA sequencing methods. Nat Methods 7:709–715. https://doi.org/10.1038/nmeth.1491
15. Tariq MA, Kim HJ, Jejelowo O, Pourmand N (2011) Whole-transcriptome RNAseq analysis from minute amount of total RNA. Nucleic Acids Res 39:e120. https://doi.org/10.1093/nar/gkr547
16. Wu Q, Kim YC, Lu J, Xuan Z, Chen J, Zheng Y, Zhou T, Zhang MQ, Wu CI, Wang SM (2008) Poly A- transcripts expressed in HeLa cells. PLoS One 3:e2803. https://doi.org/10.1371/journal.pone.0002803
17. Suzuki H, Zuo Y, Wang J, Zhang MQ, Malhotra A, Mayeda A (2006) Characterization of RNase R-digested cellular RNA source that consists of lariat and circular RNAs from pre-mRNA splicing. Nucleic Acids Res 34:e63. https://doi.org/10.1093/nar/gkl151
18. Adkar-Purushothama CR, Zhang Z, Li S, Sano T (2015) Analysis and application of viroid specific small RNAs generated by viroid-inducing RNA silencing. Methods Mol Biol

1236:135–170. https://doi.org/10.1007/978-1-4939-1743-3_12

19. Aliyari R, Wu Q, Li HW, Wang XH, Li F, Green LD, Han CS, Li WX, Ding SW (2008) Mechanism of induction and suppression of antiviral immunity directed by virus-derived small RNAs in drosophila. Cell Host Microbe 4:387–397. https://doi.org/10.1016/j.chom.2008.09.001
20. Vincent HA, Deutscher MP (2006) Substrate recognition and catalysis by the exoribonuclease RNase R. J Biol Chem 281:29769–29775. https://doi.org/10.1074/jbc.M606744200
21. Di Serio F, Gisel A, Navarro B, Delgado S, Martinez de Alba AE, Donvito G, Flores R (2009) Deep sequencing of the small RNAs derived from two symptomatic variants of a chloroplastic viroid: implications for their genesis and for pathogenesis. PLoS One 4:e7539. https://doi.org/10.1371/journal.pone.0007539
22. Langmead B, Trapnell C, Pop M, Salzberg SL (2009) Ultrafast and memory-efficient alignment of short DNA sequences to the human genome. Genome Biol 10:R25. https://doi.org/10.1186/gb-2009-10-3-r25
23. Bolduc F, Hoareau C, St-Pierre P, Perreault JP (2010) In-depth sequencing of the siRNAs associated with peach latent mosaic viroid infection. BMC Mol Biol 11:16. https://doi.org/10.1186/1471-2199-11-16
24. Navarro B, Pantaleo V, Gisel A, Moxon S, Dalmay T, Bisztray G, Di Serio F, Burgyan J (2009) Deep sequencing of viroid-derived small RNAs from grapevine provides new insights on the role of RNA silencing in plant-viroid interaction. PLoS One 4:e7686. https://doi.org/10.1371/journal.pone.0007686
25. Wang Y, Shibuya M, Taneda A, Kurauchi T, Senda M, Owens RA, Sano T (2011) Accumulation of potato spindle tuber viroid-specific small RNAs is accompanied by specific changes in gene expression in two tomato cultivars. Virology 413:72–83. https://doi.org/10.1016/j.virol.2011.01.021
26. Korf I, Yandell M, Bedell J (2003) BLAST. O'Reilly & Associates, Sebastopol, CA
27. Altschul SF, Gish W, Miller W, Myers EW, Lipman DJ (1990) Basic local alignment search tool. J Mol Biol 215:403–410. https://doi.org/10.1016/S0022-2836(05)80360-2
28. Zuker M (2003) Mfold web server for nucleic acid folding and hybridization prediction. Nucleic Acids Res 31:3406–3415. https://doi.org/10.1093/nar/gkg595
29. Flores R, Grubb D, Elleuch A, Nohales MA, Delgado S, Gago S (2011) Rolling-circle replication of viroids, viroid-like satellite RNAs and hepatitis delta virus: variations on a theme. RNA Biol 8:200–206. https://doi.org/10.4161/rna.8.2.14238
30. Fadda Z, Daros JA, Fagoaga C, Flores R, Duran-Vila N (2003) Eggplant latent viroid, the candidate type species for a new genus within the family Avsunviroidae (hammerhead viroids). J Virol 77:6528–6532. https://doi.org/10.1128/Jvi.77.11.6528-6532.2003

Chapter 23

An In Silico Detection of a Citrus Viroid from Raw High-Throughput Sequencing Data

Tyler Dang, Andres Espindola, Georgios Vidalakis, and Kitty Cardwell

Abstract

E-probe Diagnostic for Nucleic Acid Analysis (EDNA) is a user-friendly bioinformatic tool that has been adapted for the detection and identification of citrus exocortis viroid (CEVd). Here, we describe the procedures for RNA extraction from citrus tissues, library and sequencing preparation, and the utilization of EDNA Mi-Finder online platform on raw high-throughput sequencing (HTS) data.

Key words EDNA, Microbe-Finder (MiFi), e-probes, Raw HTS reads

1 Introduction

The use of high-throughput sequencing (HTS) as a plant diagnostic tool has been demonstrated and shows potential as a routine diagnostic method [1–3]. However, analyzing HTS data is time-consuming, laborious, and requires a high-level knowledge and strong background in bioinformatics and computer science. To overcome these challenges, the E-probe Diagnostic for Nucleic Acid Analysis (EDNA) was developed as a user-friendly HTS data analysis tool for diagnostic applications [4].

EDNA is an in silico bioinformatic tool that utilizes short curated electronic probes (e-probes) designed from pathogen-specific sequences. The e-probes allow users to detect and identify a single or multiple pathogens of interest from raw HTS datasets generated from different sequencing platforms such as Illumina and Oxford Nanopore [4, 5]. EDNA has been used for the detection of foodborne pathogens such as *E. coli* O157: H7 [6], plant pathogens such as oomycetes (*Phytophthora ramorum* and *Pythium ultimum*), fungi (*Phakopsora pachyrhizi* and *Puccinia graminis*) [7], and viruses (plum pox virus) [4, 5]. A set of e-probes have been developed specifically for the detection and identification of citrus exocortis viroid (CEVd).

Ayala L. N. Rao et al. (eds.), *Viroids: Methods and Protocols*, Methods in Molecular Biology, vol. 2316,
https://doi.org/10.1007/978-1-0716-1464-8_23,

Here we describe a streamlined protocol for the extraction of RNA from citrus tissue, library preparation, sequencing with the Oxford Nanopore MinION platform, and data analysis for the detection of CEVd in citrus tissues using MiFi, a software platform that provides an easy to use EDNA protocols for diagnosis. This protocol can be implemented into research or diagnostic laboratories interested in CEVd detection.

2 Materials

1. TRIzol® reagent.
2. DNA LoBind microcentrifuge tube (Eppendorf, 022431021).
3. Micropipettes (P2, P10, P20, P100, P1000) and tips.
4. Standard microcentrifuge tubes.
5. Ice bucket with ice.
6. RNase-free water.
7. Liquid nitrogen.
8. Mortar and pestle.
9. Vortex mixer.
10. Refrigerated microcentrifuge.
11. Disposable spatula.
12. Laboratory scale with 0.01 g precision.
13. Chloroform.
14. 75% ethanol.
15. 70% ethanol.
16. Nuclease-free water.
17. Isopropanol.
18. Single-edge razor blades.
19. Instrument to quantify nucleic acid such as NanoDrop™, Qubit™ Fluorometer, or Agilent Bioanalyzer system.
20. −80 °C freezer.
21. Refrigerated microcentrifuge.
22. Oxford Nanopore MinIT companion device or high-end desktop computer with Linux Ubuntu 16.04 LTS, Intel i7 CPU or better with least 4 cores, 16 GB of RAM, 1 TB solid-state hard drive, and Nvidia GTX 10 series GPU or better.
23. Oxford Nanopore Technologies MinION (*see* **Note 1**).
24. MinION Flow Cell R9.4.1.
25. cDNA-PCR Sequencing Kit (ONT, SQK-PCS109): VN primers, Strand-Switching Primers, cDNA primers, Elution

Buffer, Rapid Adapter, Sequencing Buffer, Sequencing Tether, and Loading Beads.

26. Flow Cell Priming Kit (EXP-FLP002): Flush Buffer and Flush Tether.
27. Guppy software: https://community.nanoporetech.com/downloads.
28. 7-Zip software: https://www.7-zip.org/.
29. Mi-Finder: http://www.mi-finder.com/.
30. Standard 0.2 mL PCR tube.
31. Thermal cycler.
32. Incubator.
33. Agencourt AMPure XP beads.
34. Magnetic separation rack.
35. 10 mM dNTP solution (NEB, N0447).
36. LongAmp Taq 2× Master Mix (NEB, M0287).
37. Maxima H Minus Reverse Transcriptase with 5× RT Buffer (ThermoFisher, EP0751).
38. RNaseOUT™, 40 U/ μL (ThermoFisher, 10777019).
39. Exonuclease I (NEB, M0293).

3 Methods

3.1 Tissue Preparation and RNA Extraction

1. Collect stem samples around the canopy of the tree from the latest fully matured growth (1-year-old hardened tissue).
2. Use a razor blade to peel the bark from the stems.
3. Add liquid nitrogen, and grind the tissue into a fine powder with a mortar and pestle.
4. Transfer 100 mg of pulverized tissue with disposable spatula into a standard microcentrifuge with 1000 μL of TRIzol® reagent.
5. Vortex for 20 s and incubate for 10 min at 4 °C.
6. Centrifuge the lysate for 5 min at 12,000 × *g* at 4 °C, and transfer the supernatant to a new microcentrifuge tube.
7. Add 200 μL of chloroform and incubate at room temperature for 2–3 min.
8. Centrifuge the sample for 15 min at 12,000 × *g* at 4 °C.
9. Transfer the clear aqueous (top layer) to a new microcentrifuge tube, avoid touching the interphase or organic layer.
10. Add 500 μL isopropanol and incubate at room temperature for 10 min.

11. Centrifuge for 10 min at 12,000 × *g* at 4 °C and discard the supernatant.
12. Resuspend the pellet with 1 mL of 75% ethanol.
13. Vortex the sample briefly, and then centrifuge for 5 min at 7500 × *g* at 4 °C.
14. Discard the supernatant.
15. Air-dry the RNA pellet for at least 30 min and no longer than 1 h (*see* **Note 2**).
16. Resuspend the pellet in 100 μL of RNase-free water.
17. Incubate the samples in a water bath at 55–60 °C for 10 min.
18. Assess the RNA quantity and quality with NanoDrop™ instrument, Qubit™ Fluorometer, or Agilent Bioanalyzer.
19. Aliquot the RNA equally into four low DNA LoBind microcentrifuge tubes, and store at −80 °C until ready for use.

3.2 Library Preparation (from Oxford Nanopore)

1. Transfer approximately 50 ng total RNA into a DNA LoBind microcentrifuge tube.
2. Adjust the volume to up to 9 μL with nuclease-free water.
3. Mix by flicking the tube to avoid unwanted shearing, and briefly spin down in a microcentrifuge at room temperature.
4. Prepare the reaction: 9 μL of nuclease-free water, 1 μL of 10 mM dNTP, 1 μL VNP at 2 μM, and 50 ng of total RNA.
5. Mix the reaction by gently flicking and briefly spin down in a microcentrifuge.
6. Incubate at 65 °C for 5 min and then immediately cool on ice or freezer block.
7. Prepare strand-switching buffer: 4 μL 5× RT Buffer, 1 μL RNaseOUT, 1 μL nuclease-free water, and 2 μL stand-switching primer at 10 μM.
8. Mix the buffer by gently flicking and briefly spin down in a microcentrifuge.
9. Add the strand-switching mixture to the snap-cooled, annealed mRNA, mix by flicking the tube and spin down.
10. Incubate at 42 °C for 2 min.
11. Add 1 μL of Maxima H Minus Reverse Transcriptase. The final volume is 20 μL.
12. Mix by gently flicking and briefly spin down in a microcentrifuge.
13. Incubate the sample in a thermal cycler for reverse transcription: 42 °C for 90 min, 85 °C for 5 min, and Hold at 4 °C.
14. Prepare the reaction: 25 μL 2× LongAmp Taq Master Mix, 1.5 μL cDNA primer, 18.5 μL nuclease-free water, and 5 μL of reverse-transcribed RNA sample.

15. Amplify the reaction using the following thermal cycler conditions: initial denaturation at 95 °C for 30 s, denaturation 95 °C for 15 s, annealing 62 °C for 15 s, extension 65 °C for 50 s (10–18 cycles), final extension 65 °C for 6 min, and hold at 4 °C.
16. Add 1 μL of Exonuclease I to each PCR reaction.
17. Incubate the reaction at 37 °C for 15 min, then 80 °C for 15 min.
18. Pool PCR reactions into a DNA LoBind tube (total 204 μL).
19. Resuspend the AMpure XP beads by vortexing (*see* **Note 3**).
20. Add 160 μL of AMpure XP beads and mix by pipetting.
21. Incubate on a rotator mixer for 5 min at room temperature.
22. Briefly spin down in a microcentrifuge, and put the tube on a magnetic separation rack.
23. Wash the sample with 200 μL of 70% ethanol without disrupting the pellet.
24. Remove the ethanol with a pipette.
25. Repeat **steps 23** and **24**.
26. Briefly spin down in a microcentrifuge, and place the tube on a magnetic separation rack. Pipette any residual ethanol and allow to air-dry for a maximum of 30 s.
27. Remove the tube from the magnetic separation rack, resuspend the pellet in 12 μL of Elution Buffer, and incubate at room temperature for 10 min.
28. Pellet beads on the magnetic separation rack until the elute is clear, and pipette 12 μL of the elute into a DNA LoBind tube without disturbing the pellet.
29. Save 1 μL cDNA to evaluate the quality and quantity if necessary.
30. Prepare 100–200 fmol of cDNA, and adjust the volume to up to 12 μL in Elution Buffer.
31. Add 1 μL of Rapid Adapter to the amplified cDNA.
32. Mix by pipetting and briefly spin down in a microcentrifuge.
33. Incubate the reaction for 5 min at room temperature, and briefly spin down in a microcentrifuge.
34. Store the library on ice until ready to load.

3.3 Priming and Loading cDNA on the MinION Flow Cell R9.4.1

1. Thaw and then mix all components thoroughly with a vortex before use.
2. Prepare the flow cell priming mix: add 30 μL of Flush Tether directly to the Flush Buffer, and mix by pipetting up and down three to four times.

3. Prepare library for loading: add 37.5 μL Sequencing Buffer, 25.5 μL Loading Beads, and 12 μL of cDNA library (*see* **Notes 3**).
4. Open the MinIon device, and slide the flow cell priming port cover clockwise to expose the priming port.
5. Gently drawback 20–30 μL of buffer with P1000 micropipette from the priming port, avoid introducing air bubbles (*see* **Note 4**).
6. Gently load 800 μL priming mix into the flow cell via priming port, and avoid introducing air bubbles into the flow cells with P1000 pipette. Allow the flow cell to sit for 5 min (*see* **Note 5**).
7. Lift the SpotON sample port cover on the flow cell, and gently load 200 μL with P1000 via priming port without introducing air bubbles.
8. Mix the prepared library gently by pipetting up and down three to four times prior to loading.
9. Load 75 μL of samples to the flow cell via the SpotON sample port in a dropwise fashion.
10. Place SpotON sample port cover, close the priming port, and replace the MinION lid.
11. Open the MinKNOW GUI Program.
12. Choose the flow cell types and then check the "Available" box.
13. Click "New Experiment." Create an experiment number and enter a sample ID. Select the proper kit used to prepare the library. Turn off basecalling. Use the default parameters on under "Run Options." Indicate a proper output path and ensure FAST5 file format has been chosen. Turn off custom scripts (*see* **Note 6**).
14. Click "Start run."
15. The MinIon will begin sequencing and collecting data. The user can run the sequencing machine until all flow cells are exhausted, which can take up to 48 h. Alternatively, the user can stop the sequencer earlier by estimating the required sequencing depth using the tool provided in **Note** 7.

3.4 Data Processing and Analysis with MiFi

1. Run the post-sequencing basecalling for FAST5 files using the Guppy software by Oxford Nanopore (*see* **Note 6**).
2. Compress FASTQ files (.gz) using 7-Zip software (*see* **Note 8**).
3. Create an account on Mi-Finder website (*see* **Note 9**).
4. Upload the compressed sequence file (.gz) under the "Metagenomes" tab. Click "Metagenome Upload" and input the study name, sample name, and sequencing library type, and choose the file to upload. The upload time will depend on the file size.

5. Under the "E-probe Libraries" tab, search and select the citrus control gene and the CEVd E-probe, and add them to a new list in the drop-down window.
6. To submit a job in "MiDetect," provide the name of the job, and go to the Citrus Library of e-probes. Select the CEVd e-probe from the list, select metagenome file to analyze, and hit Run.
7. An email will be sent to the user once the job is complete. The results can be viewed under the "Jobs" tab. Select "Finish" under the "Status" column.
8. The MiFi software will generate results based on a probability algorithm or a simple *t*-test for the targeted citrus gene control and viroid. Samples with CEVd *p*-value <0.5 are considered positive for the viroid.

4 Notes

1. The protocol described is specific with MinION sequencer by Oxford Nanopore Technologies; however, the Microbe-Finder (MiFi) is not limited to sequencing data generated by the MinION. Users can also use metagenomic data compiled from other sequencing platforms (Illumina, PacBio) combined with MiFi software for diagnostics. The data does not have to be cleaned or assembled, because the search query e-probes will find and match the fragments of sequence of the pathogen if they are present.
2. Avoid over drying the RNA; it will make the RNA pellet more difficult to dissolve into solution as it can affect downstream applications.
3. Place the AMpure Xp beads at room temperature for at least 30 min before use. In addition, the AMpure XP beads and Loading Beads should be mixed or vortexed periodically to prevent the beads from settling to the bottom of the container.
4. The flow cells are extremely fragile and should be handled with care when removing the fluid from the priming port. Use a P1000 micropipette and place the tip into the priming port. Slowly rotate the ring clockwise to increase the volume to aspirate approximately 20–30 μL of buffer.
5. Load the priming mix with care. Use a P1000 and aspirate 810 μL of the mix. Place the tip on to the priming port, and slowly turn the ring counterclockwise to dispense 800 μL of the priming mix. This will reduce the chance of introducing air bubbles and damaging the flow cells.

6. The MinKNOW GUI Program can perform live basecalling simultaneously with the sequencing runs to generate the FASTQ files; however, this process is not recommended at this time because of potential data loss. It is recommended to first generate FAST5 output files and then perform the basecalling to convert the files from FAST5 to FASTQ with Guppy software, (https://community.nanoporetech.com/downloads). The basecalling does not require high-performance computing servers and can be performed quickly on a desktop computer equipped with a Nvidia GTX 10 series or higher GPU. Do not perform the basecalling with the desktop CPU; it could potentially take upward of a month to complete the analysis. Refer to the Oxford Nanopore community forum for detailed protocol and walkthrough guides to perform basecalling with a GPU. Alternatively, users can order the MinIT companion device to the MinION that is pre-configured for data acquisition and basecalling without the need for high-end desktop computers, hence simplifying the analysis.
7. The tool provided allows the user to estimate how much sequencing depth is required to detect the target of interest based on the information provided by the user. https://sequence-calculator.shinyapps.io/sequence_calculator_shinyapp/.
8. 7-Zip is a free open-source compression software that can be downloaded at https://www.7-zip.org/. Compressed files will the data to be uploaded more quickly to the Microbe-Finder website.
9. Follow the instructions to create an account at http://www.mi-finder.com/.

Acknowledgments

This research was supported by the Citrus Research Board (project # 5300-179), the USDA National Institute of Food and Agriculture, Hatch project 1020106, and the National Clean Plant Network (12-8100- 1544-CA and 14-8130-0419-CA). We would like to acknowledge the assistance of the dedicated personnel of the Citrus Clonal Protection Program, Oklahoma State University (OSU) Technical Business Development Program, USDA-ARS NGCR, and EPCC, Beltsville, MD. Mention of trade names or commercial products in this publication is solely for the purpose of providing specific information and does not imply recommendation or endorsement by the University Of California (UC) or Oklahoma State University. UC is an equal opportunity provider and employer:

1. CRB: https://www.citrusresearch.org
2. USDA NIFA: https://nifa.usda.gov
3. NCPN: http://nationalcleanplantnetwork.org

References

1. Al Rwahnih M, Daubert S, Golino D et al (2015) Comparison of next-generation sequencing versus biological indexing for the optimal detection of viral pathogens in grapevine. Phytopathology 105(6):758–763
2. Rott M, Xiang Y, Boyes I et al (2017) Application of next generation sequencing for diagnostic testing of tree fruit viruses and viroids. Plant Dis 101:1489–1499
3. Jones S, Baizan-Edge A, MacFarlane S et al (2017) Viral diagnostics in plants using next generation sequencing: computational analysis in practice. Front Plant Sci 8:1770
4. Stobbe AH, Daniels J, Espindola AS et al (2013) E-probe diagnostic nucleic acid analysis (EDNA): a theoretical approach for handling of next generation sequencing data for diagnostics. J Microbiol Methods 94:356–366
5. Stobbe AH, Schneider WL, Hoyt PR et al (2014) Screening metagenomic data for viruses using the e-probe diagnostic nucleic acid assay. Phytopathology 104:1125–1129
6. Blagden T, Schneider W, Melcher U et al (2016) Adaptation and validation of E-probe diagnostic nucleic acid analysis for detection of *Escherichia coli* O157:H7 in metagenomic data from complex food matrices. J Food Prot 79:574–581
7. Espindola A, Schneider W, Hoyt PR et al (2015) A new approach for detecting fungal and oomycete plant pathogens in next generation sequencing metagenome data utilising electronic probes. Int J Data Min Bioinform 12:115–128

Part V

RNA Silencing Techniques

Chapter 24

Detection of Viroid RNA and vd-siRNA in *N. benthamiana* Plants: Northern Blot Analyses for Viroid and vd-siRNAs

Konstantina Katsarou, Nikoleta Kryovrysanaki, and Kriton Kalantidis

Abstract

Viroids are considered the most minimalistic group of pathogens. Despite their presumed inability to encode for proteins, viroids induce several diseases in plants of primary economic importance. The production of viroid derived siRNAs (vd-siRNAs) of 21–24 nt, accompanies viroid infections in plants and results from the activation of the RNA silencing mechanism and specifically the function of Dicer endonucleases. A comprehensive set of experiments for the study and thorough analysis of viroid-infected plants has been developed. Here we present a detailed experimental plan including optimized protocols for plant infection by agroinfiltration, RNA extraction, and northern blot hybridization for the detection of both viroid genomic RNA and vd-siRNAs.

Key words Viroid, Agroinfiltration, RNA extraction, Northern blot, vd-siRNAs

1 Introduction

Viroids are small circular RNA pathogens with the capacity of infecting economically important crops, among many other plants. Their size varies from 246 to 401 nucleotides (nt) depending on the type and isolate of the viroid [1, 2]. Since the discovery of RNA silencing, it was accepted that viroids could be targeted by the Dicer-like (DCL) proteins, resulting in the production of small RNAs (sRNAs) [3, 4]. The exact role of each DCL in viroid infection was later determined following the production of plants expressing hairpins against each of the four DCL transcripts expressed in *Nicotiana benthamiana* (*N. benthamiana*) [5–7]. Each DCL protein is responsible for the production of a specific sRNA size class. DCL1 and DCL4 are responsible for the production of 21 nt small RNAs, DCL2 for the production of 22 nt, and DCL3 of 24 nt. In order to identify vd-sRNA populations upon viroid infection, NbDCLi plants were infected via agroinfiltration with *Potato spindle tuber viroid* (PSTVd), and 3 weeks post-infection (wpi), tissue was collected, RNA was extracted, and

Ayala L. N. Rao et al. (eds.), *Viroids: Methods and Protocols*, Methods in Molecular Biology, vol. 2316,
https://doi.org/10.1007/978-1-0716-1464-8_24,

northern blots were performed. DCL4 was identified as the nuclease of preference targeted by viroids during infection after it was shown that the combined action of DCL2 and DCL3 are able to targets viroids more efficiently [6]. This is in contrast to what is observed with viruses, where DCL4 together with DCL2 are recognized as the major antiviral proteins of *N. benthamiana*. The miRNA pathway seems also to be involved in viroid infectivity since when DCL1 and SERRATE, another important protein of this pathway, are downregulated in both *N. benthamiana* and *N. tabacum* plants, viroids accumulate lower titers [8].

For the analysis of all the abovementioned experiments, agroinfiltration, RNA extraction, and northern blots were essential. Agroinfiltration is a very popular technique in tobacco. It is a simple and efficient method for delivering transgenes into host cells for transient expression. It is mostly used for the rapid production of recombinant proteins for research or industrial purposes [9, 10]. There is also an increasing tendency to use this technique to introduce pathogens into plants, mostly due to the speed and stability of the produced infection [9–15]. The method is based on the ability of *Agrobacterium tumefaciens* (*A. tumefaciens*), a well-studied plant pathogen, to transfer part of its tumor-inducing (Ti) plasmid, called transfer DNA (T-DNA), into wounded plant cells. The tumor-inducing genes in the T-DNA can be replaced by the gene of interest, rendering modified Ti plasmids the most useful vectors for gene delivery [11]. Various reports have described efforts to optimize the conditions and efficiency of agroinfiltration methods, including *Agrobacterium* strain, agrobacterial cell density, concentration of acetosyringone, and expression vectors in various plant species [16–18]. The protocol presented here is optimized for efficient expression of viroids in *N. benthamiana* plants to produce stable, efficient, and highly reproducible infections [5–8].

Three weeks post-infiltration, young systemic leaves are collected, ground to a fine powder, and RNA extraction performed. Earlier RNA extraction techniques utilized guanidine thiocyanate and hot phenol. Under acidic conditions, total RNA remained in the upper aqueous phase, while most of the DNA and proteins remained either in the interphase or in the lower organic phase [19, 20]. Often, RNA precipitation was then achieved using ultracentrifugation and thus was time-consuming [19, 20]. Later, Chomczynski et al. developed an improved RNA extraction protocol using guanidine thiocyanate and phenol in a single step. In addition, RNA recovery was achieved using simple centrifugations and not ultracentrifugation allowing the simultaneous analysis of a great number of biological samples in approximately 4 h [21, 22]. The RNA extraction protocol described in this chapter is based on protocols described by Chomczynski et al. and is optimized for total RNA extraction including sRNAs. The methods have been used successfully across laboratories for a variety of downstream applications including detection of viroid RNAs and vd-siRNAs [5–8, 21–23].

The analysis of viroid RNA and vd-siRNA is performed by northern blot analysis. Northern blot was first described in 1977 as a modification of the southern blot technique and soon after was used for virus and viroid detection [24, 25]. The technique is based on the size separation of RNAs via electrophoresis, followed by transfer to a blotting membrane using either capillary or electro-transfer. This technique is considered the gold standard for the detection of both the viroid genome and vd-siRNAs. The major advantage of northern blots is the qualitative analysis, since the exact size can be calculated and the specificity of the probe verified. Also, the RNA undergoes little manipulation since no amplification or enzymatic reaction is needed. On the other hand, this technique is not suitable for the detection of very low viroid titers or the identification of new viroids in an unknown infection. Additionally, a northern blot only shows the steady state of the accumulation of RNAs at a specific moment. It is not suitable for the measurement of the viroid replication rate or RNA stability.

In this chapter, we describe a detailed protocol for the analysis of both viroid RNA and vd-siRNAs. We describe agroinfiltration, RNA extraction, northern blot using DIG-labeled probe for the detection of viroid genomic RNA, and finally, northern blot using ^{32}P-labeled probe for the detection of vd-siRNAs. A flowchart showing each of the steps is presented in Fig. 1.

2 Materials

Prepare all solutions using distilled water. Take all necessary precautions for dangerous and toxic products used in this protocol. Get informed about the disposal regulations of dangerous substances in your department and especially isotope handling.

2.1 Agroinfiltration

2.1.1 Material-Equipment

1. Plant material: Use 5- to 6-week-old *N. benthamiana* plants with around six leaves. The stage of plant material is critical for various downstream applications and symptom development.
2. *A. tumefaciens* strain hosting a viroid expression construct. Materials and the procedure described are suitable for use with agrobacterial strains C58C1, LBA4404, or GV3101.
3. Antibiotics appropriate to the *Agrobacterium* strain and viroid RNA-expressing plasmid. Usually, kanamycin and rifampicin are used.
4. Centrifuge for 15 mL or 50 mL tubes.
5. 15 mL or 50 mL tubes.

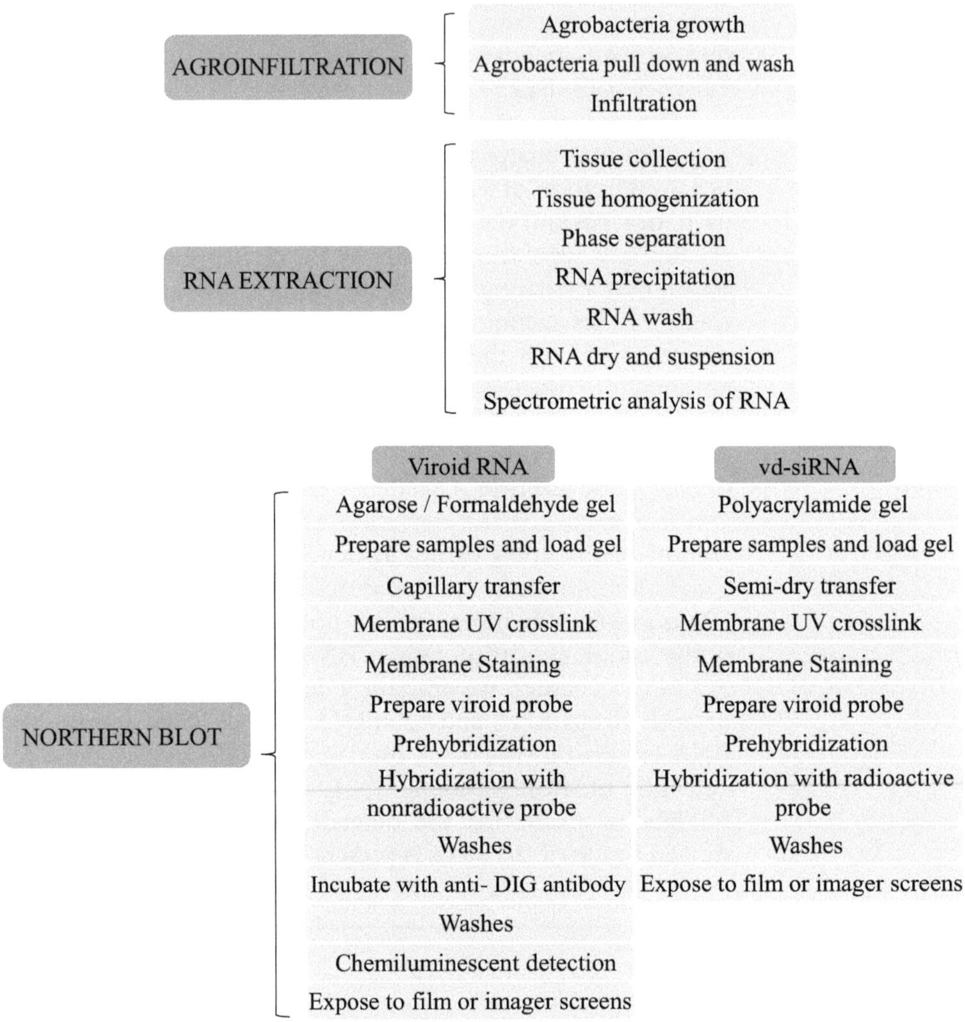

Fig. 1 Flowchart showing the steps for a successful viroid infection, followed by RNA extraction and northern blot analysis for both viroid RNAs and vd-siRNAs

6. Spectrophotometer.
7. Syringes.
8. Needles.

2.1.2 Solutions

1. Luria-Bertani (LB) medium with appropriate antibiotics: 0.5% w/v yeast extract, 1% w/v sodium chloride, and 1% w/v tryptone. For 1 L LB, add 800 mL water in 1 L beaker. Weigh 5 g yeast extract, 10 g sodium chloride, and 10 g tryptone, and transfer to the beaker. Mix and adjust to pH 7.0 using NaOH. Adjust to 1 L with water, autoclave, and store at room temperature.

2. Murashige and Skoog medium (MS): Macroelements ($CaCl_2$, 332.02 mg/L, KH_2PO_4, 170 mg/L, KNO_3, 1900 mg/L, $MgSO_4$, 180.54 mg/L, NH_4NO_3, 1650 mg/L), microelements ($CoCl_2{\cdot}6H_2O$, 0.025 mg/L, $CuSO_4{\cdot}5H_2O$, 0.025 mg/L, FeNaEDTA, 36.70 mg/L, H_3BO_3, 6.20 mg/L, KI, 0.83 mg/L, $MnSO_4{\cdot}H_2O$,16.90 mg/L, $Na_2MoO_4{\cdot}2H_2O$, 0.25 mg/L, $ZnSO_4{\cdot}7H_2O$, 8.60 mg/L), vitamins (glycine, 2.00 mg/L, myo-inositol, 100.00 mg/L, nicotinic acid, 0.50 mg/L, pyridoxine HCl, 0.50 mg/L, thiamine HCl, 0.10 mg/L). It is possible to prepare all the above substances separately in 100× concentration and then dilute to 1× working concentration. There are ready mixes for MS salts that contain all the above elements and can be used according to manufacturer instructions. Adjust pH to 5.6–5.8 with 1 M KOH. Autoclave and store at 4 °C.
3. MES pH 5.6: For 50 mL 1 M MES stock solution, weigh 9.76 g MES in 20 mL distilled water. Mix thoroughly until the powder dissolves, and adjust pH to 5.6 with 5 N NaOH. Add water to a final volume of 50 mL, and sterilize with 0.2 μM filter. Split MES into 1.5 mL Eppendorf tubes and store at −20 °C. Before use, let the MES thaw on ice.
4. Acetosyringone: For 50 mL 200 mM stock solution, weigh 9.81 g acetosyringone in 30 mL DMSO. Mix thoroughly until the powder dissolves, and add water to a final volume of 50 mL. Sterilize the solution with 0.2 μM filters. Prepare aliquots of 100 μL and store at −20 °C. Before use, let the acetosyringone thaw on ice.
5. MMA: MS, 10 mM MES pH 5.6, 200 μM acetosyringone. For 10 mL MMA, mix in a 15 mL falcon, 10 mL MS, 10 μL MES 1 M, and 200 μL acetosyringone 200 mM (*see* **Note 1**).
6. Magnesium chloride ($MgCl_2$): For 500 mL 1 M $MgCl_2$ stock solution, weigh 47.6 g $MgCl_2$ in 400 mL distilled water. Mix thoroughly, and bring the final volume to 500 mL using a volumetric cylinder. Autoclave and store at room temperature. Use at a final concentration of 10 mM.

2.2 RNA Extraction

2.2.1 Material-Equipment

1. Liquid nitrogen.
2. Pre-chilled mortar, pestle, forceps, spatula, and 15 mL tubes.
3. Oven or water bath.
4. Stirrer.
5. 1.5 mL tube.
6. Centrifuge for 1.5 ml tube.
7. Small-volume spectrophotometer (e.g., NanoDrop-ThermoFisher, Waltham, Massachusetts, USA).

2.2.2 Solutions

1. Tris–HCl buffer, pH 8.0: For 1 L 0.5 M Tris–HCl, add 500 mL distilled water to a 1 L beaker. Weigh 60.57 g Tris–HCl and transfer to the beaker. Mix and adjust pH with HCl. Add water to 1 L. Store at 4 °C.
2. Saturated phenol, pH 8.0 (*see* **Note 2**): Place the bottle of crystal phenol at 68 °C until crystals are melted. Place the dark glass bottle, which will be used for the phenol also at 68 °C to warm. For 100 mL saturated phenol, add 100 mg 8-hydroxyquinoline and stir to dissolve (*see* **Note 3**). Equilibrate the phenol with an equal volume of 0.5 M Tris–HCl pH 8.0, and stir for 15 min. Stop stirring and leave for 15 min at RT, until two distinct phases are formed. The lower organic phase corresponds to the phenol and the upper phase to the equilibrating 0.5 M Tris–HCl pH 8.0. Remove and discard the upper layer without touching the lower phase. Repeat the addition of 0.5 M Tris–HCl pH 8.0 twice. Add an equal volume of 0.1 M Tris–HCl pH 8.0 to the phenol. Stir for 15 min, and after the separation of the two phases, remove the upper one with a glass pipette. Repeat the step using 0.1 M Tris–HCl pH 8.0 until the aqueous phase is ~pH 7.8 (measure with pH paper). Once the phenol is ready, remove the magnet, leave a ~1 cm layer of 0.1 M Tris–HCl pH 8.0 over the phenol, and store at 4 °C for 1–6 months. Test the pH periodically. Phenol quantity will be reduced during the preparation therefore, start with at least 2.5× the final volume of phenol wanted.
3. RNA extraction buffer: Phenol in saturated buffer 38%, guanidine thiocyanate 0.8 M, ammonium thiocyanate 0.4 M, sodium acetate 0.1 M, pH 5.0, and glycerol 5 M. For 100 mL RNA extraction buffer, add the following reagents in order: 38 mL saturated phenol to a 100 mL dark glass bottle. Weigh 9.453 g guanidine thiocyanate and 3.04 g ammonium thiocyanate, and transfer in the glass bottle. With a glass pipette, take 3.34 mL sodium acetate (3 M stock solution, pH 5.0) and 5 mL glycerol, and transfer to the glass bottle. Add distilled water and leave the solution stirring until it become transparent. Store in the dark at 4 °C.
4. Sodium salts: Sodium citrate 0.8 M, sodium chloride 1.2 M. For 50 mL of salts, add about 20 mL water to a 100 mL beaker. Weigh 11.76 g sodium citrate and 5.5 g sodium chloride, and transfer to the beaker. Mix and make up to 50 mL with water in a cylinder. Store at room temperature.
5. Chloroform.
6. Ice-cold isopropanol.
7. Ice-cold 70%w/v EtOH.

2.3 Formaldehyde Denaturing Agarose Gel

2.3.1 Material-Equipment

1. Horizontal electrophoresis apparatus.
2. Nylon membrane 0.45 or 0.2 μm.
3. Whatman 3MM paper.
4. UV cross-linker.
5. Scanner or smartphone camera.

2.3.2 Solutions

1. 0.5 M EDTA: Dissolve 186.12 g disodium ethylenediamine-tetraacetate-$2H_2O$ (EDTA) in 800 mL distilled water. Adjust pH to 8.0 using NaOH pellets and add water to a final volume of 1 L. Sterilize and store at room temperature.
2. 10× MOPS: 0.2 M MOPS pH 7, 50 mM sodium acetate, and 10 mM EDTA pH 8.0. Add 41.8 g MOPS (4-morpholinopropanesulfonic acid), 16.6 mL 3 M sodium acetate and 20 mL of 0.5 M EDTA to a final volume of 1 L. Adjust pH to 7.0 using NaOH (*see* **Note 4**). Store at 4 °C.
3. Deionized formamide: Transfer 100 mL of formamide in a beaker and add 5 g of ion exchange resin. Stir for 30 min at room temperature and filter the beads using Whatman filter. Store at −20 °C (*see* **Note 5**).
4. 5× loading buffer A: Add 720 μL formaldehyde, 3084 μL formamide, 80 μL 0.5 M EDTA, 40 μL 1% bromophenol blue (BPB), 2 mL glycerol, and 4 mL 10× MOPS, and water to a final volume of 10 mL.
5. Running buffer: Add 882 mL distilled water, 100 mL 10× MOPS, and 18 mL formaldehyde for a final volume of 1 L (*see* **Note 6**).
6. 20× SSC: 3 M NaCl, 300 mM trisodium citrate dehydrate. Add 175.3 g of NaCl and 88.23 g of sodium citrate to 800 mL of distilled water. Adjust pH to 7.0 using HCl, and then add water to a final concentration of 1 L.
7. Formaldehyde 37% w/v.
8. Ethidium bromide 1 mg/mL.
9. Agarose.

2.4 DIG Labeling

2.4.1 Material-Equipment

1. Pasteur pippette.
2. Plasmid containing the viroid sequence of choice under the control of a prokaryotic promoter (T7, T3, or SP6).
3. DIG RNA labeling mix (Roche).
4. Anti-digoxigenin AP Fab Fragments (Roche).
5. CDP-Star (Roche).
6. Small-volume spectrophotometer.

7. Hybridization glass tubes with screw cap. Size depends on the size of the membrane. Usually, for one membrane use 14.5 × 4 cm and for two membranes 30 × 4 cm.
8. Hybridization oven with rotating glass-tube attachment.
9. 1.5 mL tubes.
10. 50 mL tubes.
11. Forceps.

2.4.2 Solutions

1. Stainning buffer: 0.3 M sodium acetate/0.03% methylene blue. Add 5 mL of 3 M sodium acetate, 0.015 g of methylene blue, and distilled water to a final volume of 50 mL. Prepare fresh.
2. 20% SDS: Dissolve 200 g of sodium dodecyl sulfate in 900 mL of distilled water. Heat at 68 °C to dissolve. Adjust pH to 7.2 using HCl and add distilled water to 1 L. Store at room temperature (*see* **Note 7**).
3. 50× Denhardt's: 1% Ficoll, 1% polyvinylpyrrolidone, and 1% BSA. Prepare stock buffer of all components, and then mixed them to the desired volume. Store at −20 °C.
4. tRNA: Dilute 100 mg of ribonucleic acid in 15 mL distilled water in a 50 mL tube, and leave it to rotate in an oven overnight at 50 °C. Perform phenol/chloroform extraction to clean/purify, with precipitation using two volumes of 100% ethanol and 3M sodium acetate pH 5.2 (1/10 of the volume). Wash the pellet with 70% ethanol and resuspend in 10 mL distilled water. Determine the concentration using a spetrophotometer. Store at −20 °C.
5. Prehybridization buffer: 5× SSC, 1% SDS, 1× Denhardt's solution, 0.25 mg/mL t-RNA, and 50% formamide. Add 5 mL of 20× SSC, 1 mL of 20% SDS, 400 μL of 50× Denhardt's solution, and 500 μL of 10 mg/mL t-RNA. Add distilled water to a final volume of 20 mL. Heat prehybridization buffer to the approproate hybridization temperature before use.
6. Appropriate restriction enzyme to cut and linearize the viroid RNA-expressing plasmid.
7. 50 mM DTT.
8. RNAse inhibitor (40 u/μL).
9. T7, T3, or SP6 polymerase depending on the promoter present in the viroid containing plasmid.
10. Agarose.
11. 1× Maleic acid: Add 11.6 g of maleic acid and 8.76 g of NaCl in 800 mL water. Bring the solution to pH 7.5 using NaOH and adjust the volume to 1 L.

12. Washing buffer: 1× Maleic acid, 0.3% Tween-20. Add 300 μL Tween-20 in 100 mL of 1× Maleic acid.
13. Blocking buffer: 1× Maleic acid, 0.3% Tween-20, and 3% milk powder. Add 1.5 g milk powder in a 50 mL tube and adjust volume with washing buffer.
14. Detection buffer: 0.1 M Tris pH 9.5, 0.1 M NaCl. Add 2 mL Tris pH 9.5 and 400 μL NaCl in 20 mL of distilled water.

2.5 Polyacrylamide Gel

2.5.1 Material-Equipment

1. Parafilm.
2. Plastic volumetric tube.
3. Vertical electrophoresis apparatus.
4. Semidry blotting apparatus.
5. Nylon membrane 0.2 μm.
6. Whatman 3MM paper.
7. Pasteur pipettes.

2.5.2 Buffers

1. Acrylamide/bisacrylamide 38:2: Add 38 g of acrylamide and 2 g of bisacrylamide in 80 mL distilled water, and stir overnight. When fully dissolved, add water to a final volume of 100 mL, and store in the dark at 4 °C.
2. 10× TBE: Dissolve 108 g Tris and 55 g boric acid in 900 mL distilled water. Add 40 mL 0.5 M EDTA pH 8.0, and adjust the volume to 1 L. Store at room temperature (*see* **Note 8**).
3. 10% APS: Dissolve 1 g of ammonium persulfate in 8 mL distilled water. Adjust volume to 10 mL, make aliquots, and store at −20 °C.
4. Loading buffer B: 80% formamide, 10 mM EDTA pH 8.0, 1 mg/mL BPB, and 1 mg/mL xylene. For 10 mL, add 8 mL formamide, 200 μL 0.5 M EDTA, 10 mg BPB, and 10 mg xylene.
5. TEMED (tetramethylethylenediamine).
6. Urea.

2.6 Solutions for Radioactive Work

2.6.1 Material-Equipment

1. An area or room restricted for radioactive work, with specific regulations depending on each country and institution, is necessary. Before starting, undertake appropriate training.
2. Hybridization glass cylinders with screw caps. Size depends on the size of the Nylon membrane used. Usually, for a single Nylon membrane, use 14.5 × 4 cm hybridization glass cylinders, whereas for two Nylon membranes, we suggest the 30 × 4 cm hybridization glass cylinders.
3. Hybridization oven with rotating glass-tube attachment.
4. Pasteur pipettes.
5. Sephadex G25 superfine beads.

6. Micro Bio-Spin chromatography column (Bio-Rad).
7. Thermocycler.
8. Primers specific to the viroid.
9. Horizontal electrophoresis apparatus.
10. 1.5 mL tubes.
11. 2 mL tubes.
12. Forceps.
13. Small-volume spectrophotometer (e.g., NanoDrop).

2.6.2 Solutions

1. Prehybridization buffer: 5× SSC, 7% SDS, 20 mM Na_2HPO_4 pH 7.2, and 1× Denhardt's. Add 5 mL 20× SSC, 7 mL 20% SDS, 400 μL 1 M Na_2HPO_4, and 400 μL 50× Denhardt's in a final volume of 20 mL.
2. 2.5× Random priming buffer: 125 mM Tris–HCl pH 6.8, 12.5 mM $MgCl_2$, and 25 mM β-mercaptoethanol. Add 1.25 mL of 1 M Tris–HCl pH 6.8, 125 μL of 1 M $MgCl_2$, 250 μL β-mercaptoethanol and adjust volume to 10 mL.
3. Probe buffer: 1 μM random primers, 1 μM dTTP, 1 μM dGTP, 1 μM dATP, and random priming buffer 1×. Add 1 μL 50 μM random primers, 1 μL 50 μM dTTP, 1 μL 50 μM dGTP, 1 μL 50 μM dATP, and 20 μL random priming buffer in a final volume of 40 μL.
4. Polymerase and buffer for PCR.

3 Methods

All procedures are carried at room temperature unless specified otherwise. Safety issues are to be taken into consideration.

3.1 Agroinfiltration

3.1.1 Agrobacteria Growth

1. A single colony of *Agrobacterium tumefaciens* carrying a plasmid expressing a longer than unit-length viroid sequence under the control of a 35S promoter is inoculated into LB medium with antibiotics and grown overnight at 28 °C with continuous shaking (225 rpm).
2. The following day, inoculate the appropriate volume of LB medium with 1/1000 of the overnight grown culture (*see* **Note 9**), and incubate at 28 °C with continuous shaking (225 rpm).
3. Measure the OD of the new subculture at 600 nm. Usually, an OD_{600} of 0.4 to 0.7 is used but do not exceed an OD_{600} of 1.

3.1.2 Preparation of Agrobacteria for Inoculation

1. Transfer the culture into an appropriate tube, and centrifuge for 15 min at 950 × *g* at 4 °C.
2. Carefully remove the supernatant.
3. Resuspend the pelleted agrobacteria in an equal volume of MMA solution (*see* **Note 10**).
4. Incubate for 1–3 h at 28 °C with continuous shaking (225 rpm).
5. Centrifuge agrobacteria for 15 min at 950 × *g* at 4 °C.
6. Carefully take out the supernatant.
7. Resuspend the pellet in half volume with 10 mM $MgCl_2$.
8. Centrifuge for 15 min at 950 × *g* and 4 °C.
9. Take out the supernatant.
10. Repeat **steps 7–9** twice.
11. Resuspend in same volume with 10 mM $MgCl_2$.
12. Measure OD_{600} and dilute the bacteria to a final OD_{600} of 0.2. Use this culture for plant infiltration.

3.1.3 Infiltration

1. Use two young leaves of each plant to be infiltrated (*see* **Note 11**).
2. With a needle, make a very delicate hole on the abaxial surface (bottom part) of the leaf, avoiding the central veins of the leaves.
3. Perform the infiltration with 1 mL needless syringe by pressing the syringe gently over the hole and exert a counter-pressure with a finger on the other side of the leaf. Successful infiltration is often observed as a spreading "wetting" area in the leaf with a darker green color. It is recommended to infect two entire leaves in each plant to achieve a homogenous infection.
4. Immediately post-agroinfiltration water the plants, and maintain them in the greenhouse or plant growth chamber with a 16-h light/8-h dark photoperiod, at stable incubation temperature (*see* **Note 12**).

3.2 RNA Extraction

All steps are carried out in the fume hood because of the volatile and caustic phenol. Get informed about the safety regulations in your department regarding phenol-containing reagents. Always use gloves and eye protection. Avoid contact with skin or clothing as well as breathing vapor.

3.2.1 General Recommendation When Working with RNA

1. Wear gloves at all times.
2. Clean all the equipment before use to avoid the presence of RNAses (incubation with a solution of dishwashing liquid (1%) for >4 h is usually enough).
3. Use autoclaved and RNAse-free plastics and buffers.

3.2.2 Tissue Collection

The collection of upper, systemic leaves of an agroinfiltrated plant is usually performed 3 weeks post-inoculation (*see* **Note 13**). Collected tissue is then snap-frozen in liquid nitrogen.

3.2.3 Tissue Homogenization

1. Pre-chill mortar and pestle, and grind tissue in liquid nitrogen for 3–5 min until the tissue is turned into a fine powder (*see* **Note 14**).
2. Transfer powder to pre-chilled 15 mL tubes and store at −80 °C.
3. Weigh 100 mg pulverized tissue into 1.5 mL pre-chilled tubes.

3.2.4 Phase Separation

1. Add 1 mL RNA extraction buffer in 10:1 ratio (buffer/tissue).
2. Mix samples vigorously by vortex (10 s), and incubate the samples at room temperature for 5 min. Cell lysis and stabilization of nucleic acids occur in this step.
3. Centrifuge samples at 13,400 × *g* for 15 min at 4 °C so that the aqueous phase, including nucleic acids, separates from cell debris.
4. Transfer the supernatant carefully using a pipette into a new 1.5 mL tube, avoiding the cell debris.
5. Add 0.2 volumes of chloroform. Mix the samples vigorously by vortex (10 s), and incubate at room temperature for 10 min.
6. Centrifuge samples at 13,400 × *g* for 10 min at 4 °C.
7. Two phases and interphase are formed: the colorless upper aqueous phase, the interphase, and the lower organic phase. Under the acidic conditions used, total RNA remains in the upper aqueous phase, while most of the DNA and proteins remain either in the interphase or in the lower organic phase containing the phenol and chloroform. Transfer the supernatant carefully into a new 1.5 mL tube, avoiding the interphase and lower phase.

3.2.5 RNA Precipitation

From this point on, two different protocols can be used depending on the type of RNAs desired for isolation. For total (large) RNAs, the protocol is shorter and the RNAs isolated cleaner, but no small RNAs are present in the precipitate (Subheading 3.2.6). To isolate both sRNAs and large (total) RNAs, Subheading 3.2.7 should be followed.

3.2.6 Total (Large) RNA Precipitation

1. Add 250 μL of cold isopropanol.
2. Add 250 μL of sodium salts.
3. Mix by gentle inversion.
4. Incubate samples for 10 min at room temperature.
5. Centrifuge samples at 13,400 × *g* for 15 min at 4 °C. Hydrophobic RNA is separating from the aqueous phase, and a transparent to white pellet is formed.
6. Remove the supernatant carefully, by inverting the tube, trying to keep the pellet of precipitated RNA intact.

3.2.7 Total RNA Precipitation with Small RNA Enrichment

1. Add 500 μL of cold isopropanol.
2. Mix by gentle inversion.
3. Incubate samples for 2 h at −80 °C (*see* **Note 15**).
4. Let the samples thaw on ice.
5. Centrifuge at 13,400 × *g* for 30 min at 4 °C.
6. Remove the supernatant carefully trying to keep the pellet of precipitated RNA intact.

3.2.8 RNA Washing

1. Discard supernatant without disturbing the pellet.
2. Wash with 0.5 mL ice-cold 70% ethanol by gentle inversion. In this step, any residual salt is removed from the pelleted RNA.
3. Centrifuge at 13,400 × *g* for 5 min at 4 °C.

3.2.9 RNA Drying and Resuspension

1. Discard the ethanol from the wash step and air-dry pellets for 5–15 min in a hood, until no visible ethanol remains. Do not exceed 15 min as excess drying makes resuspension of the RNA difficult.
2. Resuspend the pellet in 50 μL distilled water by pipetting up and down a few times. If RNAs are not dissolved, heat them at 65 °C in a water bath for 10 min, then place on ice for 3 min. Pipette up and down a few more times to dissolve the pellet. Store diluted RNA at −20 °C if they are to be used soon; if not store at −80 °C. Thaw on ice before use.

3.2.10 Spectrometric Analysis of RNA

The concentration and purity of RNA is obtained by post spectrophotometric readings at wavelengths of 260 nm and 280 nm. 1 μL of each sample is measured in a spectrophotometer (e.g. NanoDrop). Approximately, up to 75 μg RNA can be obtained from 100 mg of leaf tissue from *N. benthamiana* plants.

3.3 Northern Blot for PSTVd

3.3.1 Denaturating Formaldehyde Agarose/Gel (1.4%)

1. Wash with soap all the equipment used for running an agarose gel (Tank, tray, etc.) (*see* **Note 16**).
2. For a 100 mL gel: Add 88.2 mL distilled water and 1.4 g of agarose in a glass flask, and weigh it before boiling in the microwave until agarose melts.
3. Weigh it again, and add distilled water to regain the same weight to retain the desired agarose concentration.
4. Add 10 mL 10× MOPS.
5. Add 1.8 mL formaldehyde.
6. Add 7 μL ethidium bromide 1 mg/mL.
7. Pour into a caster (*see* **Note 17**).
8. When gel is ready, put it into the tank together with running buffer to cover the gel entirely.

3.3.2 Prepare Samples

1. Combine 5 μg of RNA with distilled water in a final volume of 10 μL. Keep RNA samples always on ice.
2. Add 10 μL of loading buffer A to each RNA sample from **step 1**.
3. Boil samples for 5 min.
4. Place samples immediately on ice for 3 min.
5. Spin the tubes briefly.
6. Load samples into the gel.
7. Run the gel at 100 V for around 1 h. Run times vary depending on the apparatus (due to the distance between electrodes). Typically electrpohoresis is stopped when the BPB dye has migrated for 2/3 of the gel.

3.3.3 Capillary Transfer of RNAs

Capillary transfer is the gold standard for large RNA transfer since semidry transfer involves higher temperatures, which may melt agarose gels.

3.3.4 Prepare the Gel

1. After running, wash the gel with water (2 × 5 min) to remove any remaining formaldehyde.
2. Take a picture of the gel under UV light. Avoid long exposure times as this may affect the RNA.
3. Incubate the gel in 10× SSC for 15 min.
4. In the meantime, prepare three pieces of Whatman 3MM paper the size of the gel, as well as one piece at 20 × 20 cm and one piece at 20 × 57 cm.
5. Prepare the Nylon membrane (0.45 or 0.2 μm) the size of the gel.

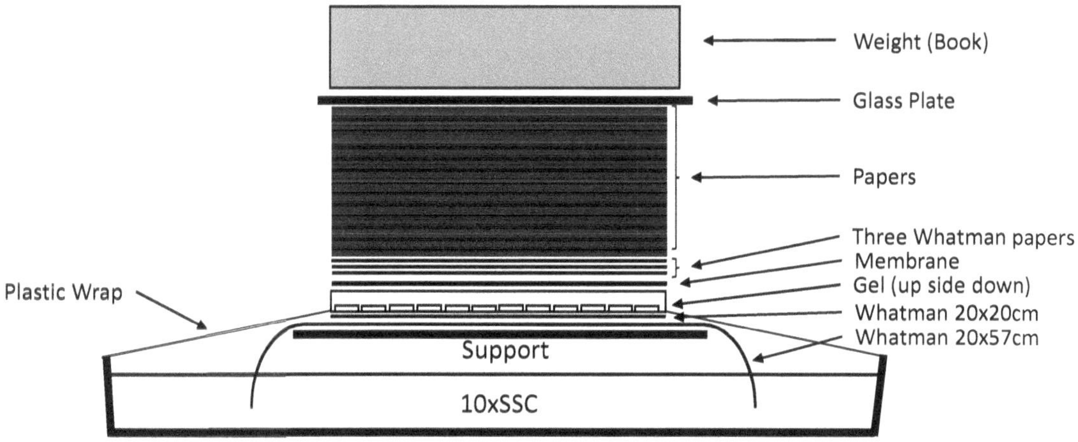

Fig. 2 Scheme representing the assembly of the capillary transfer during a northern blot

3.3.5 Preparation of the Bridge

1. Wash with dishwashing soap (contains SDS) all the glass components prior to assembly.
2. Use a glass tray 40 × 25 cm and a piece of glass of 20 × 25 cm as a support on the top. Put the 20 × 57 cm Whatman 3MM paper piece on top of the support. Use 10× SSC to wet the paper, and shape it using a Pasteur pipette as a rolling pin to make it flat. Bend the sides of the paper to create the shape of a reverse "U" (*see* Fig. 2).
3. Place the 20 × 20 cm Whatman 3MM paper piece of on the top of the first piece of Whatman 3MM, and use 10× SSC to wet the surface. Remove air bubbles using the Pasteur pipette and pay attention not to leave any air bubbles between sheets and membranes as transfer will not occur in those parts.

3.3.6 Transfer Mounting

1. Place the gel onto the Whatman 3MM papers, with the upper side facing down. Use the Pasteur pipette to remove any air bubbles.
2. Soak the Nylon membrane in 2× SSC and place it in the top of the gel. Remove any air bubbles (*see* **Note 18**).
3. Wet in 2× SSC one Whatman 3MM paper in the size of the gel, place it on top of the Nylon membrane, and remove any trapped air bubbles. Repeat the same with the other two Whatman 3MM papers.
4. Place plastic wrap around the gel to connect it with the end of the glass tank, so that liquid evaporation is minimized.
5. Add absorbent papers on the top of the Whatman 3MM papers, to arrive at the height of 9–10 cm.
6. Use a glass or plastic plate on the top of absorbent papers as a support, and add a 2 to 3 kilos weight on top (we usually use a thick catalog).

7. Leave overnight at room temperature.
8. The following day, remove the absorbent paper. The "good" side of the membrane is the one in contact with the gel. Mark the membrane using a pencil in one corner, and let it dry (good side facing up) for at least 1–2 h; the extent of humidity on the membrane affects cross-linking of the membrane (see next step). Membranes that have retained different humidity require different cross-linking regimes (*see* **Note 19**).

3.3.7 Membrane Staining

1. UV cross-link the membrane using a specialized apparatus according to manufacturer suggestions (usually 120 mJ/cm^2).
2. Wet the membrane with 2× SSC for 2 min.
3. Add Staining buffer for 3 min. RNAs should appear as dark blue bands.
4. Wash with distilled water multiple times until the desired contrast is achieved.
5. Place the membrane into a plastic bag and scan (or take a picture with a camera on your smartphone). The Nylon membrane can dry relatively fast. The scanning procedure must be carried out in a short time. When dried, the Nylon membrane produces "stripe shapes" which affect the picture quality. Thoroughly dried Nylon membranes can be stored between Whatman 3MM papers at room temperature for months if necessary.

3.3.8 Prepare Viroid Probe for DIG labeling

1. Use a plasmid containing the viroid of choice under a prokaryotic promoter, and cut with an appropriate restriction enzyme to linearize it. Usually, this type of plasmids contains a promoter at both ends of the insert (here the viroid); thus it is important to choose the restriction enzyme to use according to the orientation of the viroid sequence. In addition, it is preferable to use enzymes that leave blunt ends or 5′ overhangs since enzymes that produce 3′ overhangs usually result in aberrant transcripts.
2. Perform in vitro transcription according to manufacturer recommendations. Usually, 1 μg of a linearized plasmid is used, together with rNTPs (containing DIG labelled rUTPs) and RNAse inhibitor. Incubation occurs between 37 °C and 42 °C, depending on the manufacturer, for up to 2 h.
3. Run a 1.5% agarose gel to verify the success of in vitro transcription.
4. Measure the concentration of RNA in a small-volume spectrophotometer.
5. Prepare aliquots and store at −20 °C.
6. Defrost an aliquot just before use.

7. Add the appropriate volume (between 50 and 100 ng) to 90 μL of distilled water.
8. Boil probe for 5 min.
9. Place the probe on ice for 3 min.
10. Quick spin and keep on ice (see below).

3.3.9 DIG Protocol for Viroids

1. Wet membrane with 2× SSC.
2. Roll the membrane around a Pasteur pipette, and use a forceps to place it inside a hybridization glass tube.
3. Add 10 mL of prehybridization buffer (*see* **Note 20**).
4. Put the cylinder in an oven with rotation at 65 °C for 1 h. During this, prepare the probe to use (see above).
5. Remove the prehybridization buffer and add 20 mL of fresh buffer (*see* **Note 21**).
6. Add the probe prepared as mentioned above, and place the hybridization tube back in the oven at 65 °C with rotation. Leave overnight (*see* **Note 22**).
7. Remove hybridization buffer with probe (*see* **Note 23**).
8. Wash membrane with 10 mL 2× SSC for 5 min in 65 °C with rotation.
9. Wash membrane twice with 10 mL 2× SSC + 0.2% SDS for 30 min at 65 °C with rotation.
10. Wash membrane with 10 mL 2× SSC + 0.2% SDS for 15 min at 65 °C with rotation.
11. Open the door of the oven to lower the temperature for the next wash. Wash membrane with 10 mL washing buffer for 10 min at room temperature with rotation.
12. Add 20 mL blocking buffer for 30 min at room temperature with rotation.
13. Add anti-DIG antibody to the buffer using the dilution proposed by the manufacturer. For this, add anti-DIG antibody in 200 μL of blocking buffer, and after gently mixing, add it in the hybridization glass tube without touching the Nylon membrane to avoid possible staining.
14. Place the hybridization tube in the rotating oven at room temperature for 30 min.
15. Wash twice with 10 mL washing buffer for 15 min at room temperature with rotation.
16. Wash with 10 mL of 1× Maleic acid for 10 min at room temperature with rotation.

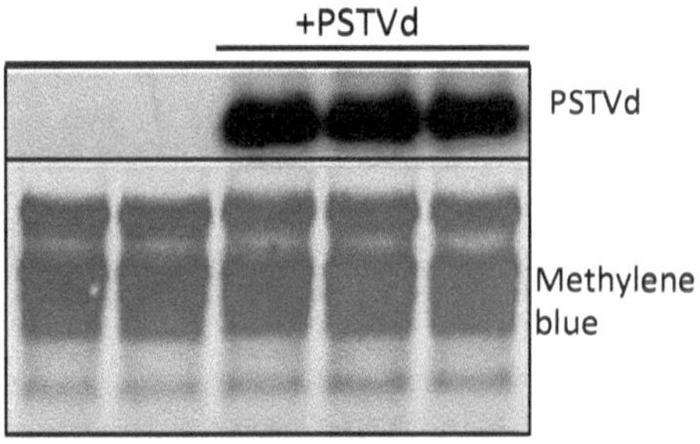

Fig. 3 Representative picture of a northern blot for the detection of viroid RNA. As a loading control the picture of methylene blue staining of RNAs is shown

17. Wash membrane with 15 mL of detection buffer for 10 min at room temperature with rotation.
18. Develop with CDP-Star as described by the manufacturer.

3.3.10 Anticipated Results

In Figure 3 a typical northern blot for PSTVd viroid is presented. The probe used here was specific for this viroid. Therefore, only one band is observed. If more bands are observed, check the following points:

1. The sequence of the probe.
2. The orientation of the probe.
3. The loading buffer A.
4. The hybridization temperature.

3.4 Northern Blot for vd-siRNAs of PSTVd

3.4.1 17% Polyacrylamide Gel

1. For a 75 mL gel add 31.9 mL of acrylamide/bisacrylamide 38:2 and 31.5 g urea (8 M final concentration) in a plastic volumetric tube, and seal with parafilm.
2. Heat it at 45 °C until completely dissolved. Invert from time to time to accelerate the dilution.
3. Add 7.5 mL of 10× TBE and water to adjust to 75 mL.
4. Add 312 μL of 10% APS.
5. Add 37.5 μL TEMED.
6. Quickly pour the gel avoiding air bubble formation. The apparatus glasses are 24.5 × 20 × 0.4 cm (*see* **Note 24**).
7. When gel is ready, perform a pre-run of the gel at 35 mA for 30 min using 1× TBE as a running buffer. This is an essential step as it improves the running temperature and thus the band separation.

3.4.2 Prepare Samples

1. 25 μg of total RNA are diluted in 30 μL of distilled water.
2. Add 30 μL of loading buffer B.
3. Boil samples for 5 min.
4. Put samples on ice for 3 min.

5. Quick spin.
6. Load the samples in the gel using Hamilton syringe or long pipette tips.
7. Run the gel at 40 mA until the xylene dye runs out of the gel. It is recommended to add a metal plate on the one side of the glass during running to keep the temperature homogenous (*see* **Note 25**).
8. After running, let the gel cool down since glass plates can break when opening.
9. After releasing from the glass plates, cut unnecessary parts and also one corner of the gel to mark the direction of the gel.
10. Transfer the gel into a piece of tulle (or cheesecloth), in order to continue without risking the integrity of the gel. The tulle may decrease the quality of images.

3.4.3 Semidry Transfer of RNAs

For small RNAs, semidry transfer is preferable since it is faster.

3.4.4 Preparation of the Gel

1. With the help of the tulle, transfer the gel in a large tray, and add cold 1× TBE enough to completely cover it, and leave it gently shaking for 5 min.
2. Add 20 μL EtBr (10 mg/mL) in the cold 1× TBE and shake for 15 min.
3. Use the tulle to move the gel from the tank to take a photo of the gel under UV. A typical image is presented in Fig. 4. Small RNAs cannot be visualized on the gel, but large RNAs help evaluate the loading of the gel.
4. Return the gel with the help of the tulle in an empty tank, and wash the gel in cold 1× TBE in a shaker for 10 min.
5. In the meantime, cut Whatman 3MM paper and Nylon membrane at the size of the gel (for this protocol, the size of the gel is 18 × 17 cm).

3.4.5 Mounting for Semidry Transfer

1. Use one by one, five Whatman 3MM papers soaked into cold 1× TBE. Each time use a Pasteur pipette as a rolling pin to remove air bubbles.
2. Use absorbent paper to wipe all liquid around the Whatman 3MM papers.
3. Place the gel on the top of Whatman 3MM papers.
4. Soak the membrane in cold 1× TBE and put it above the gel. Use a Pasteur pipette to remove air bubbles.
5. Add one by one, as in **step 1**, 5 Whatman 3MM papers.
6. Use standard absorbent paper to recover all liquid around the Whatman 3MM papers.

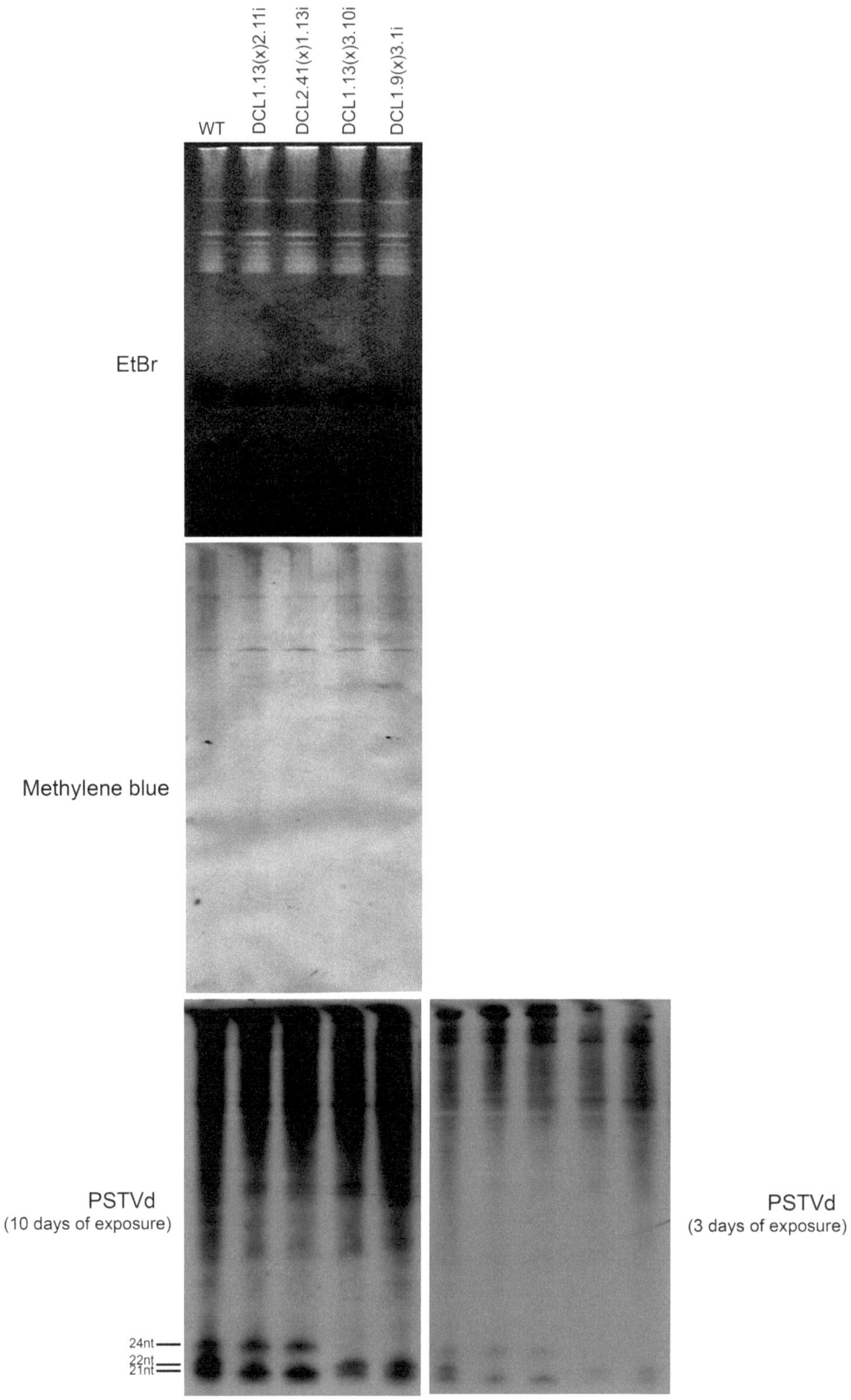

Fig. 4 Representative picture of a northern blot for the detection of vd-siRNAs

7. Run the apparatus according to manufacturer instructions (for this gel size at 615 mA) but in the cold room, to avoid temperature increase.
8. After the semidry transfer has been completed, remove the gel and the Nylon membrane.
9. The "good side" of the membrane is the one touching the gel. Let the Nylon membrane dry for 1–2 h, with the good side upward.
10. In the meantime, put the gel in a tank and cover it with distilled water.
11. Add 20 μL EtBr (10 mg/mL) and shake for 15 min.
12. Take a picture of the gel under UV light. In this way, an evaluation of the efficiency of transfer is achieved.

3.4.6 Membrane Staining

UV cross-linking and staining of the Nylon membrane is carried out as described in Subheading 3.3.7.

3.4.7 Prepare Viroid Template for Radioactive Probe

Probes for the described procedure are PCR probes. Therefore, a plasmid containing the viroid sequence of choice is necessary. It is recommended to have a PCR product between 200 and 800 nt.

1. Perform a PCR amplifying the viroid.
2. Verify by gel electrophoresis the presence of a product of the expected size.
3. Cut the band and recover the corresponding DNA fragment.
4. Quantify the product either by running 1 μL of recovered PCR product in a gel or by quantification using a small-volume spectrophotometer.
5. Dilute the PCR product to 50 ng/μL, and make aliquots.
6. Boil for 5 min.
7. Put on ice for 3 min. Store at −20 °C

3.4.8 Radioactivity to Detect vd-siRNAs

Even though there are some reports about nonradioactive probes for the detection of small RNAs, radiolabeled probes are still the method of choice since they produce much less background noise. Therefore, part of the described procedure must be performed in radioactivity restricted area or room, and, thus, there are general and specific rules according to each institution that need to be followed regarding handling, storage, and safely discarding radioactive ^{32}P. Before starting, get informed accordingly.

1. Wet the Nylon membrane in 1× TBE.
2. Roll the Nylon membrane around a Pasteur pipette, and place the membrane inside the hybridization glass tube.
3. Add 10 mL of prehybridization buffer and incubate for 1 h in a rotating oven at 42 °C (for small hybridization glass tube).

4. Use 40 μL of probe buffer, and add 2 μL of probe, prepared as in Subheading 3.4.7.
5. From this part on, everything must be performed in a radioactivity area. Add 4 μL of 5 u/μL Klenow (20 u in total) and 4 μL of ^{32}P-alpha-dCTP.
6. Leave the probe for 1 h at 37 °C.
7. Just before using, prepare the column. Add 700 μL of Sephadex G25 beads in a micro Bio-Spin chromatography column, and put the column in a 2 mL tube.
8. Centrifuge at 700 × *g* for 2 min.
9. Place the column in a new 2 mL tube.
10. Add the radioactive mix on the top.
11. Centrifuge at 700 × *g* for 2 min.
12. Discard the column into radioactive waste, and keep the tube with the probe.
13. Heat the probe at 95 °C for 5 min in a heat block.
14. Place the tube on ice for 3 min.
15. Place the probe directly in the hybridization glass tubes, and incubate in the oven at 42 °C rotating overnight.
16. Discard the probe into an appropriate radioactive liquid waste.
17. Wash Nylon membrane with 10 mL 2× SSC for 5 min in a rotating oven at 42 °C.
18. Wash Nylon membrane twice with 10 mL 2× SSC + 0.2% SDS for 30 min in a rotating oven at 42 °C.
19. Wash Nylon membrane with 10 mL 2× SSC + 0.1% SDS for 10 min at 42 °C with rotation.
20. Put the Nylon membrane in a plastic bag and seal. Do not let the membrane dry out.
21. Put the sealed Nylon membrane in a cassette.
22. Expose membrane to film or phospho-imaging screen in a darkroom, and place in a cassette.
23. Store the cassette at −80 °C.
24. Develop a film after 24 h to check if additional washing of the membrane is necessary. This can be estimated depending on the background observed. If the background is too high, place membrane back to hybridization tube, and repeat **steps 19–23** (*see* **Note 26**).

3.4.9 Anticipated Results

In Figure 4, a typical northern blot for vd-siRNAs of PSTVd is presented. In this particular case, we used DCLi crossed lines. DCL1.9i and DCL1.13i correspond to plant lines with decreased levels of DCL1 transcript, DCL2.11i and DCL2.41i lines have plants suppressed for DCL2, and DCL3.10i and DCL3.1i are plant lines with decreased DCL3 levels. F1-crossed plants were infected with the protocol described above. As it is obvious when DCL2 or DCL3 trancripts are decreased, a significant reduction of the 22nt and 24nt vd-siRNA population respectively is observed. This type of analysis is essential when the involvement of a specific DCL is in question. Since each plant Dicer produces a specific class of sRNAs, their detection by northern analysis signifies the activity of the respective DCL protein.

3.4.10 Stripping of the Membrane

If the re-probing of a membrane is needed, it is not recommended to strip membranes prepared via this described protocol as there is a possibility of losing a significant portion of the bound sRNA target. In this case, it may be preferable to wait until the signal from the previous probe has faded away and repeat radioactive probing as described in Subheading 3.4.7–8. It is evident that the original targets, however, cannot be reidentified with this method.

If it is necessary to perform a probing for an abundant target like U6 (usually used as a loading control), we propose to perform stripping as following:

1. Put the membrane in a hybridization glass tube.
2. Add 10 mL of stripping buffer preheated at 95 °C (quantity sufficient for one membrane in a small hybridization glass tube).
3. Heat at 85 °C for 15 min (*see* **Note 27**).
4. Remove the stripping buffer and wash for 5 min with 2× SSC.
5. Test with the Geiger counter that the Nylon membrane is not radioactive. If the membrane is still emitting signal, repeat **steps 2–4**.
6. Place the Nylon membrane in a cassette.
7. Add a film or phospho-imaging screen in a darkroom to test for any signal left. If there is no signal, membrane can be reused or stored.

4 Notes

1. MMA solution is to be prepared immediately before use under sterile conditions.
2. All solutions containing phenol should be prepared in a hood with appropriate protection measures.
3. 8-hydroxyquinoline protects the phenol from oxidation. It gives the solution a yellowish color.
4. It is not recommended to autoclave MOPS. However, it is possible to do so if it is to be kept for a long period. Post-autoclave, the solution usually turns yellow. Nevertheless, this does not affect the efficiency of MOPS or its pH.

5. Formamide is to be handled with caution since it is toxic.
6. Formaldehyde can be replaced with distilled water if necessary, but it is preferable to run gels in the presence of formaldehyde to increase denaturation of the RNAs.
7. If SDS crystals precipitate due to cold temperatures, warm the solution at 37 °C to redissolve.
8. If kept for a long time, 10× TBE solution may precipitate. Make a dilution of 5× or 1× to store.
9. The quantity of agrobacterial culture to be used depends on the number of leaves to be infiltrated. Five mL of initial culture is sufficient for about 25 leaves.
10. MMA contains acetosyringone, which induces agrobacterial infectivity resulting in more efficient agroinfiltration.
11. Use two young leaves but avoid cotyledons. Plants should not be watered the day of performing agroinfiltration as the infiltrated material will pass easier through the opened plant stomata.
12. The temperature during agroinfiltration procedure as well as the days after is crucial for the experiment. High temperature decreases the efficiency of this technique. We propose temperatures between 22 °C and 25 °C.
13. Infection is visible in a northern blot even at 1 wpi, but infection cannot be detected in most leaves at this time point. Using the described procedure, we propose to collect tissue at 3 wpi. If a comparison between samples is necessary, always collect exactly the same leaves, and increase the number of plants analyzed to avoid collection bias.
14. Samples should remain well frozen during the grinding procedure. Therefore liquid nitrogen has to be added in regular intervals while pulverizing the sample. Clean mortar and pestle thoroughly between samples to avoid cross-contamination.
15. Increased incubation in low temperature enhances the precipitation of small RNAs but also increases salt precipitation. This is the reason why no extra salt is added in this step. In cases where very low RNA quantity is predicted, longer incubation at −80 °C is recommended.
16. It is best to keep aside an electrophoresis apparatus only for RNA experiments although overnight incubation in dishwashing liquid would inactivate/remove any remaining RNAse activity.
17. This protocol is adapted for a 12 × 14 cm tray. Adjust the gel volume to your tray of choice.
18. It is better to soak the membrane together with the protective papers to avoid any contact with impurities. Uncover the membrane when thoroughly wet. Handle gently with the use of forceps.

19. Absorbent papers can be reused for at least two to three times if dried to a horizontal surface. Nevertheless, at least some fresh absorbent papers must be used immediately after Whatman 3MM papers to obtain maximum and homogenous absorption and, as a result, homogenous transfer.
20. For small hybridization glass tubes containing one Nylon membrane at least 10 mL of appropriate buffer should be used, whereas for large hybridization glass tubes with two Nylon membranes 20 mL should be used.
21. Prehybridization buffer can be reused at least three times.
22. When adding the probe, avoid touching the membrane, since this may leave stains on the membrane that will then interfere with the signal.
23. Hybridization buffer with the probe can be reused at least three times. Upon reuse, preheat the buffer at 65 °C.
24. The gel may need up to 1 h to solidify. If necessary, the gel can be kept at room temperature covered with wet paper to avoid drying out for up to 24 h.
25. Polyacrylamide gels for small RNAs need to run at around 55 °C, which is usually achieved by applying high voltage. Thick glasses are preferred to prevent crashing.
26. Hybridized wet membranes can be stored up to 1 year, if sealed and kept relatively wet at room temperature.
27. The time of stripping depends on the signal observed on the membrane.

Acknowledgments

The authors would like to acknowledge support of this work by project "PlantUP" [MIS MIS5002803] which is implemented under the Action "Reinforcement of the Research and Innovation Infrastructure," funded by the Operational Programme "Competitiveness, Entrepreneurship and Innovation" (NSRF 2014-2020) and co-financed by Greece and the European Union (European Regional Development Fund). This work was also supported by the grant "Emblematic Action for Research in the Cretan Agrofood sector: Four Institutions, Four References" (2018ΣE01300000) held by the General Secretary for Research and Technology of Greece.

References

1. Flores R, Di Serio F, Navarro B, Duran-Vila N, Owens R (2011) Viroids and viroid diseases of plants. Wiley-Blackwell, Hoboken, New Jersey, pp 307–341

2. Katsarou K, Rao ALN, Tsagris M, Kalantidis K (2015) Infectious long non-coding RNAs. Biochimie 117:37–47
3. Papaefthimiou I, Hamilton A, Denti M, Baulcombe D, Tsagris M, Tabler M (2001) Replicating potato spindle tuber viroid RNA is accompanied by short RNA fragments that are characteristic of post-transcriptional gene silencing. Nucleic Acids Res 29 (11):2395–2400
4. Itaya A, Folimonov A, Matsuda Y, Nelson RS, Ding B (2001) Potato spindle tuber viroid as inducer of RNA silencing in infected tomato. Mol Plant Microbe Interact 14 (11):1332–1334
5. Dadami E, Boutla A, Vrettos N, Tzortzakaki S, Karakasilioti I, Kalantidis K (2013) DICER-LIKE 4 but not DICER-LIKE 2 may have a positive effect on potato spindle tuber viroid accumulation in Nicotiana benthamiana. Mol Plant 6(1):232–234
6. Katsarou K, Mavrothalassiti E, Dermauw W, Van Leeuwen T, Kalantidis K (2016) Combined activity of DCL2 and DCL3 is crucial in the defense against potato spindle tuber viroid. PLoS Pathogens 12:1–24
7. Katsarou K, Mitta E, Bardani E, Oulas A, Dadami E, Kalantidis K (2018) DCL-suppressed Nicotiana benthamiana plants : valuable tools in research and biotechnology. Mol Plant Pathol 20(3):432–446
8. Kryovrysanaki N, Alexiadis A, Grigoriadou AM, Katsarou K, Kalantidis K (2019) SERRATE, a miRNA biogenesis factor, affects viroid infection in *Nicotiana benthamiana* and *Nicotiana tabacum*. Virology 528:164–175
9. Chen Q, Lai H, Hurtado J, Stahnke J, Leuzinger K, Dent M (2014) Agroinfiltration as an effective ans scalable strategy of gene delivery for production of pharmaceutical proteins. Adv Tech Biol Med 1(1):1–21
10. Leuzinger K et al (2013) Efficient agroinfiltration of plants for high-level transcient expression of recombinant proteins. J Vis Exp 77:50521
11. Rivera AL, Gómez-Lim M, Fernández F, Loske AM (2012) Physical methods for genetic plant transformation. Phys Life Rev 9(3):308–345
12. Del Toro F, Tenllado F, Chung BN, Canto T (2014) A procedure for the transient expression of genes by agroinfiltration above the permissive threshold to study temperature-sensitive processes in plant-pathogen interactions. Mol Plant Pathol 15(8):848–857
13. Chaturvedi S, Jung B, Gupta S, Anvari B, Rao ALN (2012) Simple and robust in vivo and in vitro approach for studying virus assembly. J Vis Exp 61:1–5
14. Annamalai P, Rao ALN (2006) Delivery and expression of functional viral RNA genomes in planta by Agroinfiltration. Curr Protoc Microbiol, Chapter 16:16B.2.1-16B.2.15
15. Pamela RC, Kanneganti T-D, Huitema E, Kamoun S (2006) In planta expression of oomycete and fungal genes. In: Plant-pathogen interactions. Blackwell, Oxford, pp 35–43
16. Yamamoto T et al (2018) Improvement of the transient expression system for production of recombinant proteins in plants. Sci Rep 8 (1):1–10
17. Wroblewski T, Tomczak A, Michelmore R (2005) Optimization of agrobacterium-mediated transient assays of gene expression in lettuce, tomato and Arabidopsis. Plant Biotechnol J 3(2):259–273
18. Norkunas K, Harding R, Dale J, Dugdale B (2018) Improving agroinfiltration-based transient gene expression in Nicotiana benthamiana. Plant Methods 14(1):1–14
19. Feramisco JR, Helfman DM, Smart JE, Burridge K, Thomas GP (1982) In: Maniatis T, Fritsch EF, Sambrook J (eds) Molecular cloning. Cold Spring Harbor Laboratory, Cold Spring Harbor, NY, pp 194–195
20. Chirgwin JM, Przybyla AE, MacDonald RJ, Rutter WJ (1979) Isolation of biologically active ribonucleic acid from sources enriched in ribonuclease. Biochemistry 18 (24):5294–5299
21. Chomczynski P, Sacchi N (1987) Single-step method of RNA isolation by acid guanidinium thiocyanate-phenol-chloroform extraction. Anal Biochem 162(1):156–159
22. Chomczynski P, Sacchi N (2006) The single-step method of RNA isolation by acid guanidinium thiocyanate-phenol-chloroform extraction : twenty-something years on. Nat Protoc 1(2):581–585
23. Chomczynski P, Mackey K (1995) Short technical reports. Modification of the TRI reagent procedure for isolation of RNA from polysaccharide- and proteoglycan-rich sources. Biotechniques 19(6):942–945. PMID: 8747660.
24. Alwine JC, Kemp DJ, Stark GR (1977) Method for detection of specific RNAs in agarose gels by transfer to diazobenzyloxymethyl-paper and hybridization with DNA probes. Proc Natl Acad Sci 74(12):5350–5354
25. Gucek T, Trdan S, Jakse J, Javornik B, Matousek J, Radisek S (2017) Diagnostic techniques for viroids. Plant Pathol 66:339–358

Part VI

Splicing Techniques

Chapter 25

Production of Recombinant RNA in *Escherichia coli* Using Eggplant Latent Viroid as a Scaffold

Beltrán Ortolá and José-Antonio Daròs

Abstract

Similar to viruses, viroids can also be engineered and transformed into useful biotechnological tools. We describe here a viroid-based system to produce large amounts of recombinant RNA in *Escherichia coli*. A precursor of eggplant latent viroid (ELVd), with the RNA of interest inserted between positions U245 and U246, is co-expressed in *E. coli* along the chloroplastic isoform of the eggplant tRNA ligase, the enzyme that mediates the circularization of this viroid in the infected plants. In the bacterial cells, the chimeric ELVd-RNA-of-interest precursor self-cleaves through the embedded hammerhead ribozymes, and the monomer is recognized and circularized by the co-expressed tRNA ligase. The resulting circular RNA, likely bound to the tRNA ligase, accumulates to a high concentration in the bacterial cells.

Key words Recombinant RNA, Circular RNA, tRNA ligase, *Escherichia coli*, Hammerhead ribozyme, RNA ligation, Ribonucleoprotein complex

1 Introduction

Despite viroids consist of small circular RNA molecules (246–401 nt), which do not code for any protein, when these infection agents of higher plants manage to enter the appropriate host, they are able to recruit factors and structures to replicate their genomes and to invade the whole plant [1]. In addition, they manage to evade the defensive mechanisms that the host plants oppose to infection. Definitively, viroid RNAs pack mighty biological properties in minimal sequence space. As we gain insight into viroid molecular biology, these unique, infectious agents will keep supplying a wealth of parts, elements, structures, and strategies for designs in biotechnology and synthetic biology. As an example, we have recently developed a viroid-based system to produce large amounts of recombinant RNA in *Escherichia coli* [2, 3]. In contrast to DNA or proteins, and possibly as a consequence of their intrinsically low half-lives, production or large amounts of recombinant RNAs is not easy. Yet, these types of

Ayala L. N. Rao et al. (eds.), *Viroids: Methods and Protocols*, Methods in Molecular Biology, vol. 2316,
https://doi.org/10.1007/978-1-0716-1464-8_25,

molecules are demanded in large amounts for research and biotechnological and pharmaceutical applications [4–6]. One of the most successful strategies to produce large amounts of recombinant RNAs is based on expression in *E. coli* of the RNA of interest inserted into a stable RNA scaffold, such as a tRNA or an rRNA, which provides stability to the whole ensemble [7, 8]. We found that eggplant latent viroid (ELVd; family *Avsunviroidae*) is an excellent scaffold to produce large amounts of recombinant RNAs in *E. coli*, particularly if co-expressed with the chloroplastic isoform of eggplant tRNA ligase, the enzyme that mediates circularization of this viroid in the infected plant [9].

In our system, an ELVd precursor with the RNA of interest inserted between position U245 and U246 is constitutively co-expressed along with the eggplant tRNA ligase in *E. coli* (Fig. 1). The RNA precursor self-cleaves through the viroid hammerhead ribozymes producing a chimeric ELVd-RNA-of-interest linear monomer with 5′-hydroxyl and 2′,3′-cyclic phosphodiester termini. Both the ELVd moiety and the terminal groups of the chimeric RNA are recognized by the co-expressed eggplant tRNA ligase that catalyzes the circularization of the molecule (Fig. 1). This way, a circular form of the chimeric RNA accumulates at high concentration in the *E. coli* cells based on (1) the stability of the highly base-paired viroid scaffold, (2) the circularity that provides resistance to exonucleases, and (3) the likely formation of a ribonucleoprotein complex with the tRNA ligase that also contributes to stabilize the bound RNA. Using the method described below, multi-milligram amounts of ELVd and ELVd-derivatives fused to RNAs of interest can be quickly produced per liter of *E. coli* culture in normal laboratory conditions. Figure 2 shows the two-dimensional polyacrylamide gel electrophoresis (2D-PAGE) separation of total RNAs from *E. coli* that expressed empty ELVd, and ELVd-derivatives fused to the fluorescent aptamer Spinach [10] and the streptavidin binding aptamer [11]. The circular forms of the recombinant RNAs are visible after the ethidium bromide staining of the gels both after the first (Fig. 2a) and second separation (Fig. 2b). Note that after the 2D-PAGE, the circular forms of the recombinant RNAs are entirely separated from the bulk of linear RNAs that migrate in a faster diagonal (Fig. 2b), which allows purification of these molecules to homogeneity. Compactness of the viroid scaffold that likely folds as an independent domain in the chimeric RNA must also contribute to the functionality of the RNA of interest, as exemplified here by the green fluorescence exhibited by the *E. coli* that express ELVd-Spinach when the 3′5-difluoro-4-hydroxybenzylidene (DFHBI) fluorophore is added to the culture (Fig. 3).

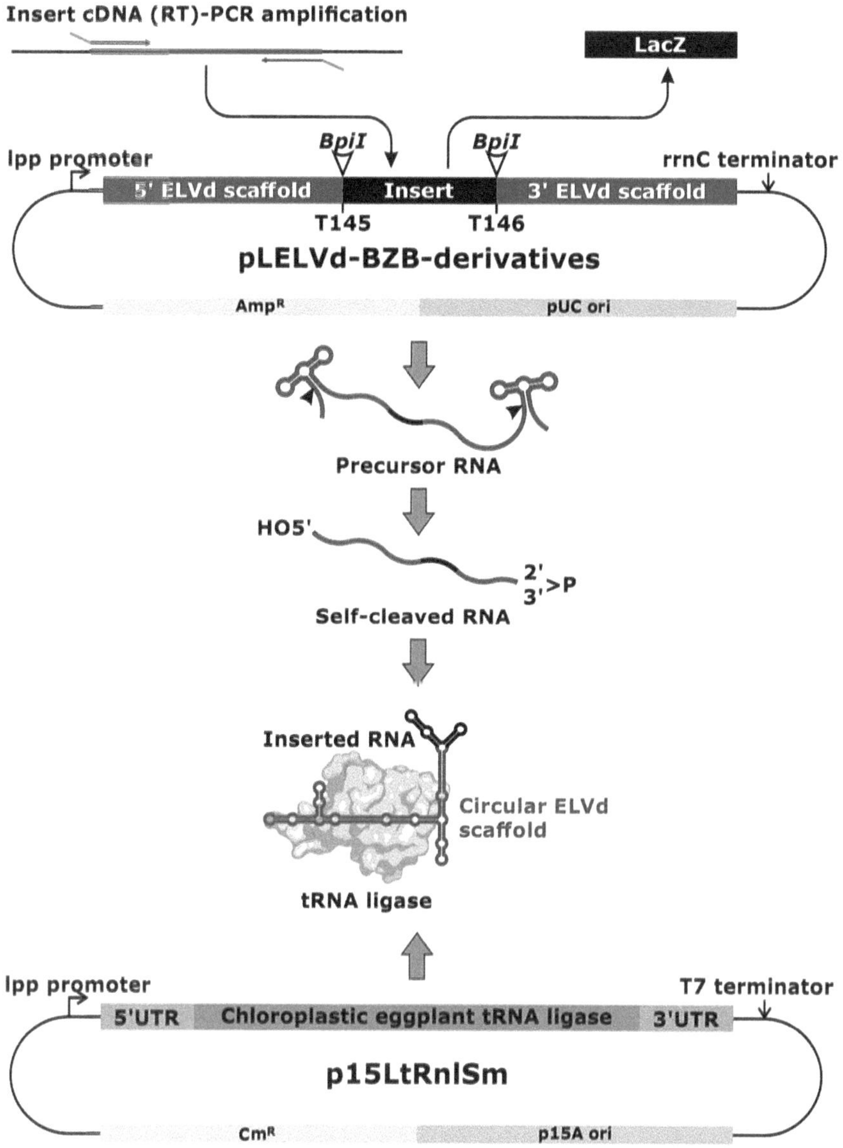

Fig. 1 Graphical representation of the ELVd-based system to produce recombinant RNAs in *E. coli*. The cDNA corresponding to the RNA of interest replaces the LacZ blue-white reporter in expression plasmid pLELVd-BZB. The expression *E. coli* strain is co-transformed with the pLELVd-BZB-derivative and p15LtRnlSm. In the bacteria, a precursor of the chimeric RNA is transcribed and self-cleaved (black arrowheads) to a monomer by the flanking ribozymes. The chloroplastic isoform of eggplant tRNA ligase that is expressed from p15LtRnlSm circularizes and remains bound to the chimeric ELVd-RNA-of-interest monomer. pUC ori, pUC replication origin; Amp^R, ampicillin resistance gene; lpp, *E. coli* murine lipoprotein promoter; rrnC, *E. coli* rRNA terminator; p15A ori, p15A replication origin; Cm^R, chloramphenicol resistance gene; T7 terminator, phage T7 transcription terminator

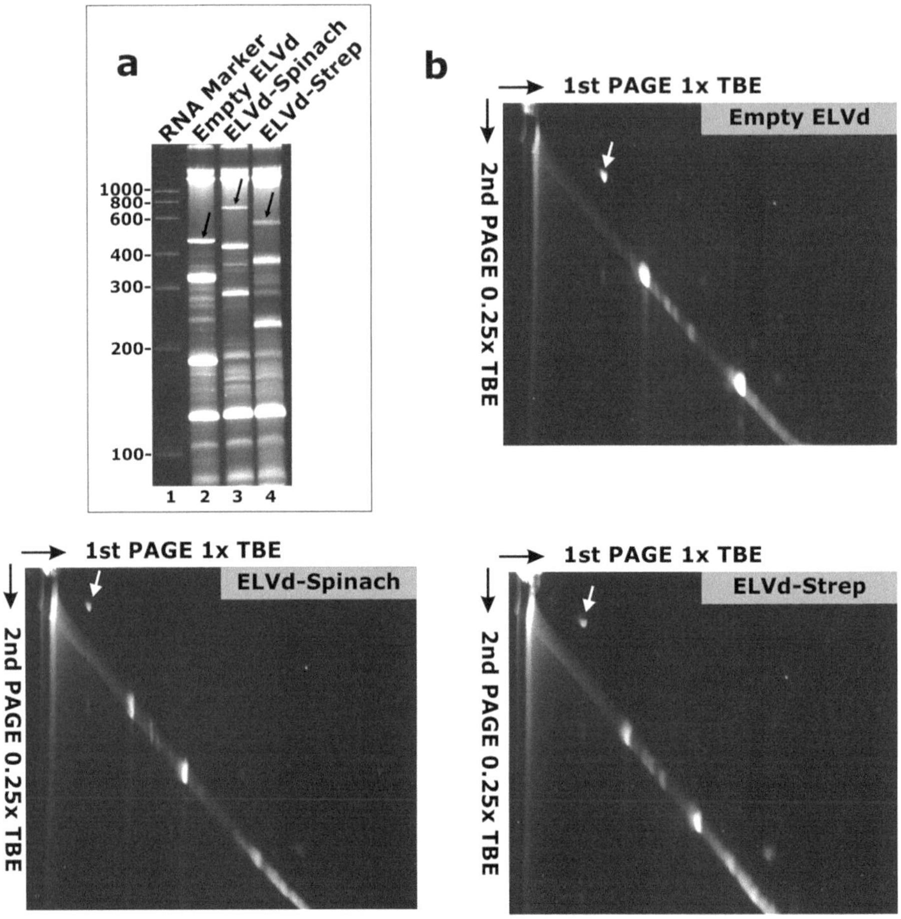

Fig. 2 Analysis by 2D-PAGE of recombinant RNAs produced in *E. coli* using the ELVd-based system. (**a**) RNA aliquots were separated by denaturing PAGE in ×1 TBE and the gel stained with ethidium bromide. Lane 1, RNA markers (with sizes in nt on the left); lanes 2 to 4, RNAs from *E. coli* clones that express the empty ELVd (333 nt; lane 2) and the ELVd fused to the Spinach (333 + 98 nt; lane 3) and streptavidin binding aptamer (333 + 42 nt; lane 4). The circular forms of the recombinant RNAs are indicated by black arrows. (**b**) The whole lanes from the first separation (**a**) were cut and transversely laid on top of a second denaturing gel in ×0.25 TBE. After electrophoresis the three gels were stained with ethidium bromide. The circular forms of the recombinant RNAs are indicated by white arrows. In the 2D-PAGEs the directions of RNA migration are indicated by black arrows. Gels were loaded with RNA aliquots corresponding to 0.8 mL of *E. coli* cultures

2 Materials

Prepare all solutions using ultrapure water and analytical grade reagents. Follow all waste disposal regulations.

2.1 Bacterial Strains

1. Use *E. coli* DH5α or another general cloning strain to clone plasmids.
2. Use *E. coli* BL21 or any other BL21-derivative to produce recombinant RNA.

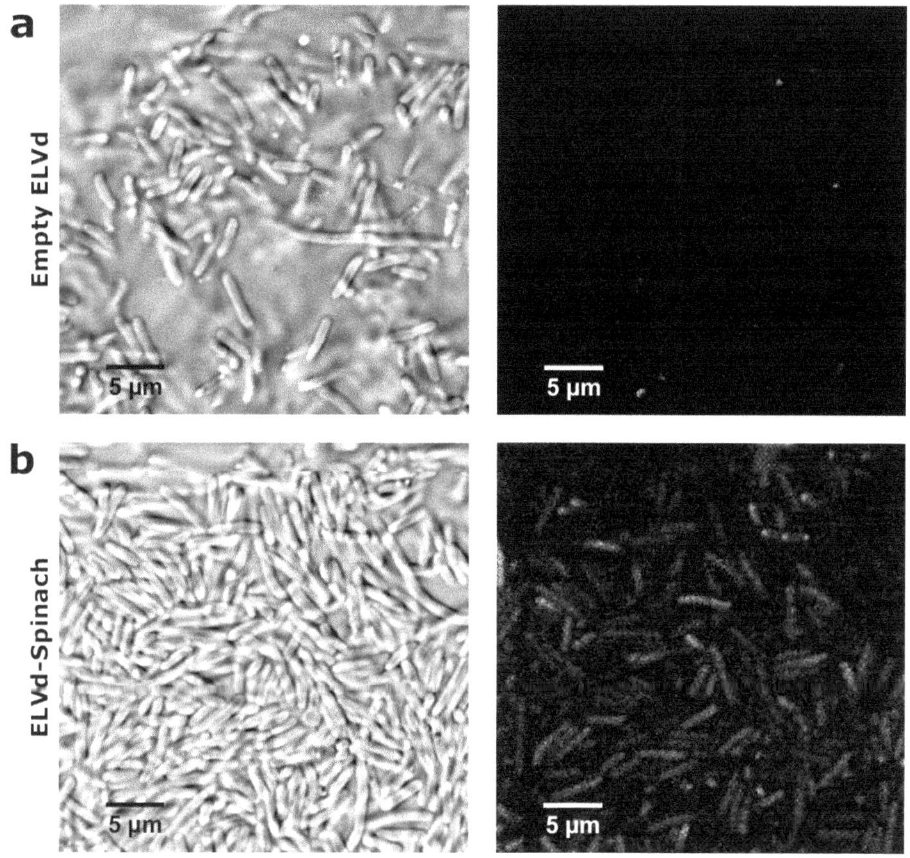

Fig. 3 Pictures of *E. coli* that express (**a**) the empty ELVd and (**b**) the chimeric ELVd-Spinach RNA. Liquid cultures were grown for 16 h and DFHBI added to 200 μM to an aliquot of cells. Incubation at 37 °C was continued for 1 h and bacteria finally concentrated by centrifugation. Images were taken with a Zeiss LSM 780 confocal microscope through a 63×/1.40 oil objective, under bright field (left panels) or under an argon laser of 488 nm coupled with the dichroic beam splitter MBS 488 nm and filtering the emission between 463 and 580 nm (right panels). Scale bars: 5 μm

2.2 Culture Media

1. Luria-Bertani (LB) medium: 1% tryptone, 0.5% yeast extract, and 1% NaCl. Weigh 10 g of tryptone, 5 g of yeast extract, and 10 g of NaCl. Dissolve in water and bring it to 1 L. Transfer to a bottle and sterilize by autoclaving. For solid LB media, add 15 g of agar (1.5%) before sterilization.
2. Super optimal broth with catabolite repression (SOC): 20 g/L tryptone, 5 g/L yeast extract, 0.5 g/L NaCl, 2.5 mM KCl, 10 mM $MgCl_2$, 20 mM glucose, and pH 7.0. Prepare 1 M $MgCl_2$ stock solution by dissolving 20.3 g $MgCl_2$ ($\cdot 6H_2O$) in water. Bring to 100 mL and sterilize by autoclaving. Prepare 1 M KCl stock solution by dissolving 7.4 g KCl in water. Bring to 100 mL and sterilize by autoclaving. Prepare 1 M glucose by dissolving 18 g glucose in water. Bring to 100 mL and sterilize by filtration. Weigh 20 g of tryptone, 5 g of yeast extract and

0.5 g of NaCl, and dissolve in water. Add 2.5 mL 1 M KCl and 10 mL 1 M $MgCl_2$. Adjust the pH to 7.0 with NaOH and bring to 980 mL. Transfer to a bottle and sterilize by autoclaving. After cooling, add 20 mL of sterile 1 M glucose.

3. Terrific broth (TB): 1.2% tryptone, 2.4% yeast extract, 0.4% glycerol, 17 mM KH_2PO_4, and 72 mM K_2HPO_4. Weigh 12 g of tryptone, 24 g of yeast extract, and 5.8 g of 87% glycerol (density 1.228 g/mL), and dissolve in 900 mL of water. Weigh 2.3 g of KH_2PO_4 and 16.4 g of K_2HPO_4 ($\cdot 3H_2O$), and dissolve in 100 mL of water. Transfer to a bottle and sterilize by autoclaving. Mix 9:1 parts of each solution before use.
4. Ampicillin 50 mg/mL: weigh 1 g of ampicillin and dissolve in water. Bring to 20 mL. Sterilize by filtration and aliquot. Store aliquots at −20 °C.
5. Chloramphenicol 34 mg/mL: weigh 340 mg of chloramphenicol and dissolve in ethanol. Bring to 10 mL and store at −20 °C.
6. 5-bromo-4-chloro-3-indolyl-β-D-galactopyranoside (X-gal) 50 mg/mL: weigh 0.25 g of X-gal and dissolve in 5 mL of *N*, *N*-dimethylformamide. Prepare 0.5 mL aliquots in tubes protected from light and store at −20 °C.

2.3 Plasmids

1. Plasmids pLELVd-BZB and p15LtRnlSm are described in [2].

2.4 Oligonucleotides

1. Forward primer for PCR amplification of the cDNA corresponding to the RNA of interest, 5′- TCTCCCCTCC CAGGTACTATCCCCTTXXXXXXXXXXXXXXXXXXXXX - 3′, X represents nucleotides homologous to the 5′ end of the RNA of interest.
2. Reverse primer for PCR amplification of the cDNA corresponding to the RNA of interest, 5′- CCCTCCTAGG GAACACATCCTTGAXXXXXXXXXXXXXXXXXXXXX -3′, X represents nucleotides homologous to the 3′ end of the RNA of interest.
3. Forward sequencing primer 5′- CCTTTTTCAATATTATT GAAGC-3′.
4. Reverse sequencing primer 5′- GATGCTCGT CAGGGGGGCGGAG-3′.

2.5 Enzymes

1. Thermostable DNA polymerase with proofreading activity. We use Phusion high-fidelity DNA polymerase (Thermo Scientific).
2. Restriction enzyme *Bpi*I (Thermo Scientific) or isoschizomer *Bbs*I (New England Biolabs).
3. Gibson assembly enzyme mix. We use NEBuilder HiFi DNA assembly master mix (New England Biolabs).

2.6 Recombinant RNA Extraction Purification

1. TE buffer: 10 mM Tris(hydroxymethyl)aminomethane (Tris), and 1 mM ethylene diamine tetraacetic acid (EDTA), pH 8.0. Prepare 1 M Tris–HCl, pH 8.0. Weigh 60.6 g of Tris and dissolve in water. Adjust pH to 8.0 with HCl and bring it to 500 mL. Sterilize by autoclaving and store at room temperature. Prepare 0.5 M EDTA, pH 8.0. Weigh 18.6 g of disodium EDTA ($\cdot 2H_2O$) and add 75 mL of water. While stirring the mix, adjust pH to 8.0 with 10 M NaOH. Notice that disodium EDTA would not dissolve until pH is close to 8. Bring the solution to 100 mL, sterilize by autoclaving, and store at room temperature. Finally, mix 1 mL of 1 M Tris–HCl, pH 8.0, and 200 μL of 0.5 M EDTA, pH 8.0, and bring to 100 mL with water. Transfer to a bottle and keep at room temperature.
2. 1:1 (v/v) buffered phenol/chloroform, pH 8.0. Work in a hood when preparing and handling this reagent. Prepare water-saturated phenol mixing 400 mL of 90% phenol and 120 mL of water. Mix and transfer to an amber bottle. Store at 4 °C. Mix 250 mL of water-saturated phenol and 50 mL of 1 M Tris–HCl, pH 8.0. Stir the mix for 5 min and settle to separate the phases. Remove most of the upper aqueous phase. Add 250 mL of chloroform. Stir the mix for 5 min and settle to separate the phases. Again remove most of the upper aqueous phase (*see* **Note 1**).
3. 3 M sodium acetate, pH 5.5. Weigh 40.8 g of sodium acetate ($\cdot 3H_2O$) and dissolve in water. Adjust pH to 5.5 with glacial acetic acid and bring to 100 mL. Sterilize by autoclaving and store at room temperature.
4. 70% ethanol. Pour 73 mL of 96% ethanol and add water to 100 mL. Mix, transfer to a bottle and store at −20 °C.
5. Chromatography buffer: 50 mM Tris–HCl, pH 6.5, 150 mM NaCl, and 0.2 mM EDTA. Prepare 0.5 M Tris–HCl, pH 6.5. Weigh 6.06 g of Tris, dissolve in water, adjust to pH 6.5 with HCl, and bring to 100 mL. Sterilize by autoclaving and store at room temperature. Weigh 0.88 g NaCl, add 10 mL of 0.5 M Tris–HCl, pH 6.5, 40 μL 0.5 M EDTA, pH 8.0, and water, and dissolve. Bring to 100 mL. Sterilize by autoclaving and store at room temperature.
6. Elution buffer: 50 mM Tris–HCl, pH 6.5, 1 M NaCl, 0.2 mM EDTA. Weigh 2.92 g NaCl, add 5 mL of 0.5 M Tris–HCl, pH 6.5, 20 μL of 0.5 M EDTA, pH 8.0, and water, and dissolve. Bring to 50 mL. Sterilize by autoclaving and store at room temperature.
7. Polyacrylamide elution buffer (PEB): 0.1 M Tris–HCl, pH 9.0, 0.1 M 2-mercaptoethanol, 10 mM EDTA, and 1% sodium dodecyl sulfate (SDS). Prepare 10% SDS. Weigh 10 g of SDS

and dissolve in water. Bring to 100 mL. Store at room temperature. Prepare 1 M Tris–HCl, pH 9.0. Weigh 60.6 g of Tris and dissolve in water. Adjust the pH to 9.0 with HCl and bring to 500 mL. In a graduated tube, mix 1 mL of 1 M Tris–HCl, pH 9.0, 200 μL of 0.5 M EDTA, pH 8.0, and 71 μL of 2-mercaptoethanol. Add water to 9 mL. Finally, add 1 mL of 10% SDS. Mix and store at room temperature.

3 Methods

3.1 Plasmid Construction

1. Amplify the cDNA corresponding to the RNA of interest by PCR (*see* **Note 2**), using a high-fidelity DNA polymerase and the sequence-specific primers.
2. Digest 100 ng of plasmid pLELVd-BZB using restriction enzyme *Bpi*I or equivalent.
3. Separate the PCR products and the digestion products by electrophoresis in an agarose gel and elute the DNAs migrating at the expected positions using silica gel columns. *Bpi*I digestion of pLELVd-BZB produces two DNA fragments of 2046 and 528 base pairs (bp). Elute the slowest DNA. The fastest DNA corresponds to the released polylinker that includes a LacZ blue-white reporter.
4. After elution from the agarose gel, mix the amplified cDNA and the digested plasmid at a molar ratio 3:1. Add one volume of NEBuilder HiFi DNA assembly master mix, and incubate at 50 °C for 1 h. Purify the reaction products using a silica gel column and use them to electroporate *E. coli* DH5α.
5. Using 1 mm cuvettes, electroporate cells at 1500 V, and quickly add 1 mL of SOC medium. Recover bacteria from the cuvette and incubate them with shaking for 1 h at 37 °C.
6. Spread the electroporated bacteria on a solid LB plate containing 50 μg/mL ampicillin and X-gal (*see* **Note 3**). Grow overnight at 37 °C.
7. Pick several *E. coli* colonies of white color and inoculate liquid LB media containing 50 μg/mL ampicillin. Grow overnight at 37 °C with shaking.
8. Miniprep the plasmids and analyze them by electrophoresis in an agarose gel. Use empty pLELVd-BZB as a control. Select one plasmid with the expected electrophoretic migration according to the inserted cDNA (*see* **Note 4**).
9. Confirm the right sequence of the whole expression cassette by sequencing the selected plasmid (*see* primers in Subheading 2.4).

3.2 Transformation of the E. coli Expression Strain

1. Electroporate competent *E. coli* BL21 cells with the pLELVd-BZB-derivative that contains the cDNA corresponding to the RNA of interest and p15LtRnlSm (*see* **Note 5**). Repeat the Subheading 3.1, **steps 5** and **6**, but spread electroporated bacteria in LB plates containing 50 μg/mL ampicillin and 34 μg/mL chloramphenicol.
2. Incubate the plates overnight at 37 °C.

3.3 Analysis of Recombinant RNA Production

Before getting engaged in a large-scale production and purification process of the recombinant RNA, we recommend an analytical assay to select a good *E. coli* colony and to learn about the yield and expression time-course of a particular RNA of interest.

1. Pick *E. coli* colonies using sterile toothpicks, and inoculate 10 mL of TB, containing 50 μg/mL ampicillin and 34 μg/mL chloramphenicol, in 50 mL tubes. Incubate at 37 °C with vigorous shaking (225 revolutions per minute, rpm).
2. Take 2 mL aliquots of the cultures at the planned time points, and pellet the bacteria by centrifugation at 15,000 × *g* for 2 min. Discard the supernatant, and resuspend the bacteria in 50 μL of TE buffer by vortexing. Freeze the bacteria until all aliquots are collected.
3. Break the cells by adding one volume (50 μL) of a 1:1 mixture of phenol/chloroform and vigorous vortexing. Centrifuge at 15,000 × *g* for 5 min, and recover the aqueous phase with the total bacterial RNA. This preparation can be directly subjected to analysis or stored at −20 °C until use.
4. Routinely, analyze aliquots of total RNA by denaturing polyacrylamide gel electrophoresis (PAGE) in 5% polyacrylamide gels containing 8 M urea in Tris-boric acid-EDTA (TBE) buffer.
5. In some cases, it is beneficial to analyze the RNAs by two-dimensional PAGE (2D-PAGE) (*see* the Chapter 8 in this book).
6. Visualize the RNA in the gel by staining with ethidium bromide.

3.4 Large-Scale Production of Recombinant RNA

1. Pick an *E. coli* colony using a sterile toothpick, and inoculate a 2 L baffled Erlenmeyer flask containing 250 mL of TB, including 50 μg/mL ampicillin and 34 μg/mL chloramphenicol.
2. Incubate the culture at 37 °C with vigorous shaking at 225 rpm between 12 and 16 h (*see* **Note 6**).

3.5 Extraction of Recombinant RNA

1. Pour the culture in a 250 mL, centrifuge bottle, and pellet the cells by centrifugation at 12,000 × *g* for 10 min. Discard supernatant. Add approximately 45 mL of water and vortex

vigorously to resuspend bacteria. Transfer the cells to a 50 mL tube and fill with water.

2. Sediment the cells again by centrifuging at 7500 × *g* for 15 min. Discard supernatant, add 10 mL of chromatography buffer, and vortex vigorously to resuspend bacteria. Cells can be directly subjected to purification or stored at −20 °C until use.
3. Break the cells by adding one volume (10 mL) of 1:1 phenol/chloroform and vortexing the mix intensely.
4. Centrifuge at 7500 × *g* for 10 min to separate the aqueous and organic phases. Recover the aqueous phase and transfer to a new tube.
5. Re-extract the aqueous phase with 10 mL of chloroform. Vortex, centrifuge at 7500 × *g* for 10 min, and recover the aqueous phase. The RNA preparation can be subjected to chromatographic purification or stored at −20 °C to proceed in any other moment.

3.6 Chromatographic Purification of Recombinant RNA

Further purification of the RNA preparation can be done by anion exchange chromatography. For this purpose, we use an ÄKTA-prime plus liquid chromatography system (GE Healthcare), operated at a 1 mL/min flow rate.

1. Equilibrate a 1 mL DEAE-Sepharose column with 10 mL of chromatography buffer.
2. Filter the RNA extract with a 0.45 μm syringe filter and load it into the column.
3. Wash the column with 10 mL of chromatography buffer.
4. Elute the RNA with 20 mL of elution buffer. Collect 1 mL aliquots while monitoring the absorbance at 260 nm.
5. To store and reuse the column, wash it with 10 mL of water, followed by 10 mL of 20% ethanol, and store at 4 °C.
6. Store the fractions in which RNA eluted at −20 °C.

3.7 2D-PAGE Separation of the Circular Form of the Recombinant RNA

If required, the circular form of the recombinant RNA can be purified to homogeneity by 2D-PAGE.

1. Precipitate the RNA present in the selected chromatographic fraction or fractions by adding 0.1 volumes of 3 M sodium acetate, pH 5.5, and 1 volume of isopropanol at −20 °C. Incubate overnight at −20 °C.
2. Centrifuge at 12,000 × *g* for 15 min. Discard the supernatant, and wash the sediment carefully with cold (4 °C) 70% ethanol. Air-dry the sediment and dissolve in water.

3. Separate the RNA by denaturing PAGE in a 5% polyacrylamide gel containing 8 M urea in TBE buffer (*see* the Chapter 8 in this book).
4. Cut the whole lane of the polyacrylamide gel that contains the preparation of recombinant RNA, and transversely load on top of a second 5% polyacrylamide gel containing 8 M urea in ×0.25 TBE buffer.
5. Continue separation in a second denaturing PAGE under low ionic strength (*see* the Chapter 8 in this book).
6. Stain the RNA in the gel with ethidium bromide, and visualize it using a UV transilluminator.
7. In the second PAGE, the circular form of the recombinant RNA will be delayed with respect to the linear RNAs. Cut the gel fragment that contains the RNA of interest and freeze.

3.8 Elution of the Circular Form of the Recombinant RNA from the Polyacrylamide Gel

1. Crush the gel fragment in a mortar and add 5 mL of 1:1 phenol/chloroform. Transfer to a 15 mL centrifuge tube. Wash the mortar twice with 2.5 mL of PEB and transfer it to the same tube.
2. Centrifuge at 7500 × *g* for 15 min and recover the aqueous phase.
3. Re-extract the organic phase with 1.5 mL of PEB. Vortex, centrifuge at 7500 × *g* for 15 min, and recover the aqueous phase (*see* **Note 7**). Combine the two aqueous phases, precipitate the RNA as explained in Subheading 3.7, **step 1** (*see* **Note 8**), and resuspend in water or an appropriate buffer for the intended application.

4 Notes

1. The 1:1 (v/v) phenol/chloroform mix can be stored in a bottle with some aqueous phase on top. When using, make sure the lower organic phase is taken.
2. Perform a reverse transcription reaction before PCR amplification if starting from an RNA preparation.
3. Fifteen minutes before plating the electroporated cells spread 30 μL of 50 mg/mL X-gal on the plates.
4. When analyzing the migration of recombinant plasmids, keep in mind that parental pLELVd-BZB includes a 528-bp polylinker with the LacZ reporter, which is replaced by the cDNA of interest during cloning. A restriction analysis may also help to select the right recombinant plasmid.

5. Although *E. coli* can be sequentially transformed with both plasmids, we obtain optimal results co-electroporating bacteria in a single step with approximately 100 ng of each plasmid.
6. Although a time window to harvest the cells is indicated, recombinant RNA is transiently accumulated in *E. coli*, and the best moment to harvest bacteria for each particular experimental condition may require a preliminary time-course experiment to be determined.
7. Do not re-extract the aqueous phase with chloroform because this causes massive precipitation of SDS.
8. To remove remains of polyacrylamide, before precipitation, spin the combined aqueous phases at 7500 × *g* for 15 min, and take the supernatant. Discard the polyacrylamide that remains in the sediment at the bottom of the tube. Additionally, the RNA preparation can also be loaded in a clarification column to remove polyacrylamide remains further.

Acknowledgments

This work was supported by grants BIO2017-83184-R and BIO2017-91865-EXP from Ministerio de Ciencia e Innovación (Spain), co-financed by European Regional Development Fund (European Commission). BO is the recipient of a predoctoral contract from Universitat Politècnica de València (PAID-01-17).

References

1. Daròs J-A (2016) Viroids: small noncoding infectious RNAs with the remarkable ability of autonomous replication. In: Current research topics in plant virology. Springer International Publishing, Cham, pp 295–322
2. Daròs JA, Aragonés V, Cordero T (2018) A viroid-derived system to produce large amounts of recombinant RNA in *Escherichia coli*. Sci Rep 8:1904. https://doi.org/10.1038/s41598-018-20314-3
3. Cordero T, Aragonés V, Daròs J-A (2018, 2018) Large-scale production of recombinant RNAs on a circular scaffold using a viroid-derived system in *Escherichia coli*. J Vis Exp. https://doi.org/10.3791/58472
4. Dowdy SF (2017) Overcoming cellular barriers for RNA therapeutics. Nat Biotechnol 35:222–229. https://doi.org/10.1038/nbt.3802
5. Jang S, Jang S, Yang J et al (2018) RNA-based dynamic genetic controllers: development strategies and applications. Curr Opin Biotechnol 53:1–11. https://doi.org/10.1016/j.copbio.2017.10.005
6. Ohno H, Akamine S, Saito H (2018) RNA nanostructures and scaffolds for biotechnology applications. Curr Opin Biotechnol 58:53–61. https://doi.org/10.1016/j.copbio.2018.11.006
7. Ponchon L, Beauvais G, Nonin-Lecomte S, Dardel F (2009) A generic protocol for the expression and purification of recombinant RNA in *Escherichia coli* using a tRNA scaffold. Nat Protoc 4:947–959. https://doi.org/10.1038/nprot.2009.67
8. Nelissen FHT, Leunissen EHP, van de Laar L et al (2012) Fast production of homogeneous recombinant RNA--towards large-scale

production of RNA. Nucleic Acids Res 40: e102. https://doi.org/10.1093/nar/gks292

9. Nohales M-Á, Molina-Serrano D, Flores R, Daròs J-A (2012) Involvement of the chloroplastic isoform of tRNA ligase in the replication of viroids belonging to the family Avsunviroidae. J Virol 86:8269–8276. https://doi.org/10.1128/JVI.00629-12

10. Paige JS, Wu KY, Jaffrey SR (2011) RNA mimics of green fluorescent protein. Science 333:642–646. https://doi.org/10.1126/science.1207339

11. Walker SC, Good PD, Gipson TA, Engelke DR (2011) The dual use of RNA aptamer sequences for affinity purification and localization studies of RNAs and RNA-protein complexes. Methods Mol Biol 714:423–444. https://doi.org/10.1007/978-1-61779-005-8_26

Part VII

Structural Analysis

Chapter 26

Predicting the Structure of a Viroid: Structure, Structure Distribution, Consensus Structure, and Structure Drawing

Gerhard Steger

Abstract

Viroids are small non-coding RNAs that require a special sequence and structure to be replicated and transported by the host machinery. Many of these features can be predicted and later experimentally verified. Here, we will present workflows to predict viroid structures and draw the predicted structures in a pleasing and descriptive way using recently developed software.

Key words Structure of minimum free energy, Partition folding, Consensus structure, Sequence and structure drawing, `RNAfold`, `RNAalifold`, `ConStruct`, VARNA, R2R

1 Introduction

Viroids are non-coding RNAs (ncRNA) but—in contrast to other ncRNAs—must be transcribed in a rolling-circle mechanism and systemically transported by the host machinery. Viroids are infectious RNAs and can induce disease in a wide variety of plants. Thus viroids have to present to the host enzymes primary and secondary structures that force these enzymes to act on viroids as if they were host substrates.

Analysis of viroid structures was and is a major topic [96]. Electron microscopy [83] and sequencing [30] showed that the viroid genome is circular; analytical ultracentrifugation demonstrated the rod-like structure of members of *Pospiviroidae* [83]; enzymatic and chemical mapping showed structural details [7, 24, 30, 53, 54]; and analysis of structure formation by fast and slow temperature-jump [37, 48], optically registered denaturation curves, and temperature-gradient gel electrophoresis [80] as well as mutational analysis [57, 89] demonstrated the biological relevance of thermodynamically metastable structures [78]. Most, if not all of these experimental analyses were accompanied by prediction of structures for interpretation of the experimental results: early methods were

Ayala L. N. Rao et al. (eds.), *Viroids: Methods and Protocols*, Methods in Molecular Biology, vol. 2316, https://doi.org/10.1007/978-1-0716-1464-8_26,

base-pair dot plots [101]; then the first version of `Mfold` [123] was extended to circular sequences and to the prediction of suboptimal structures [97]; prediction of kinetically controlled folding detailed metastable viroid structures occurring during synthesis [85].

Today a wide variety of structure prediction programs and structure visualization tools are available. In the following sections, we will demonstrate prediction of the thermodynamically optimal secondary structure with minimum of free energy (mfe) of a single sequence, prediction of the complete ensemble of structures possible for a single sequence (partition function folding), and prediction of a consensus sequence and structure for a set of homologous sequences, and demonstrate a few alternatives for structure drawing. Each section will start with some background on the used method. Figure parts on gray background show command lines in Linux for a workflow to produce the shown results. We will explain program options necessary to obtain the shown output; all programs have many further options for which we refer to manual pages and documentation of the programs and web services (see Table 1). All examples use viroid sequences and demonstrate viroid features known from the literature.

The described workflows present only a few selected programs for secondary structure prediction and structure drawing, but introduced neither 3D structure prediction [76, 95, 100, 103, 117, 118, 120], nor interaction of two RNAs, nor kinetic folding [18, 19, 85, 99, 112] to the reader. For these topics we refer to the mentioned literature.

For the presented programs we mentioned a few alternatives, but many more are available. The selected ones, however, are well documented and were sufficient to solve our own requests in the context of viroid structure prediction and drawing. We hope that they will also be helpful in your projects.

2 Materials

The computing effort and memory usage for most programs predicting RNA secondary structure (see Table 1) scales with n^3 and n^2, respectively, for a sequence of length n. That is, a standard laptop is sufficient for calculations of sequences with the length of mature viroid RNAs, but be aware that doubling the sequence length will result in a factor of 8 in computing time.

Most of the mentioned programs run on Linux, macOS, and Windows; exceptions are `ConStruct` and `R2R`, which run only on Linux.

A text editor is necessary for modification of sequence and alignment files and `R2R` scripts. It should be able to handle very long text lines, capable to search using regular expressions, and to modify selected columns. A few specialized editors are available

Table 1
Web addresses

Package or Program	Address	Reference
Download		
RNAstructure	https://rna.urmc.rochester.edu/RNAstructure.html	[63]
ViennaRNA package[a,b]	https://www.tbi.univie.ac.at/RNA/	[55]
RNAshapes	https://bibiserv.cebitec.uni-bielefeld.de/rnashapes	[43]
MAFFT	https://mafft.cbrc.jp/alignment/software/	[45, 44]
R2R	http://www.bioinf.uni-leipzig.de/~zasha/R2R/	[107]
ConStruct	http://www.biophys.uni-duesseldorf.de/html/local/construct3/index.html	[111]
VARNA	http://varna.lri.fr/	[13]
SQUID library	http://http://eddylab.org/software.html	
EMBOSS	http://emboss.sourceforge.net/	[10]
ClustalW/ClustalX	http://www.clustal.org/	[49]
Web services		
Mfold	http://unafold.rna.albany.edu/?q=mfold	[122]
unafold	http://unafold.rna.albany.edu/?q=DINAMelt	[62]
RNAstructure	https://rna.urmc.rochester.edu/RNAstructureWeb/	[67]
ViennaRNA Web Services	http://rna.tbi.univie.ac.at/	[32]
RNAfold	http://rna.tbi.univie.ac.at/cgi-bin/RNAWebSuite/RNAfold.cgi	
RNAeval	http://rna.tbi.univie.ac.at//cgi-bin/RNAWebSuite/RNAeval.cgi	
RNAalifold	http://rna.tbi.univie.ac.at/cgi-bin/RNAWebSuite/RNAalifold.cgi	
RNAshapes	https://bibiserv.cebitec.uni-bielefeld.de/rnashapes?id=rnashapes_view_download	[43]
MAFFT	https://mafft.cbrc.jp/alignment/server/	[45]
RNA structure logo	https://rth.dk/resources/slogo/	[26]
Sequence logo	http://weblogo.threeplusone.com/	[11]
EMBOSS	http://emboss.sourceforge.net/	[10]

[a]The perl programs mountain.pl, relplot.pl, and rotate_ss.pl, which are used in example Fig.2, are located in ViennaRNA-2.3.5/src/Utils/
[b]RNAplot, used in the example of Sect.3.3.4, is part of the ViennaRNA package

which allow for a user to modify RNA sequence and structure alignments: Emacs users might have a look at RALEE, an RNA alignment editor in Emacs [29]; ConStruct [111] (see below) and 4SALE [90] are graphical tools for RNA alignment and structure editing. With StructureEditor, a part of the RNAstructure package [63], a user can modify a single sequence and structure and view them in different formats.

For refinement of the drawings (Sect. 3.3), which usually are in encapsulated postscript (eps) or portable document (pdf) format, any vector graphics editor (e.g., Adobe Illustrator™ or inkscape [75]) is suitable. Converters from eps to pdf format are available [6, 71].

2.1 Installation and Web Usage

The programs and packages mentioned in Table 1 are available as binary packages or source code. Little experience is required for a direct installation or even compilation. While one may prefer the web versions to avoid dealing with the command line interface of locally installed programs, the user has to be mindful of necessary switches and options also available or at least partially available via the web interfaces. The major disadvantage of web programs becomes obvious when one has to repeat the same actions (e.g., upload of sequences, selecting options, download of results), which is far easier from a command line.

3 Methods

3.1 The Structure of Minimum Free Energy Is One of the Myriad of Structures from the Structure Distribution

Calculation of the free energy ΔG_T^0 of a structure s at temperature T uses the nearest-neighbor model [65]; that is, it sums up the free energy contributions of all structural elements of s:

$$\begin{aligned}\Delta G_T^0(s) &= \sum_{\text{element}\in s} \Delta G^0_{T,\text{element}} = \sum_{\text{stack}\in s} \Delta G^0_{T,\text{stack}} + \sum_{\text{loop}\in s} \Delta G^0_{T,\text{loop}} \\ &= \sum \left(\Delta H^0_{\text{stack}} - T\cdot \Delta S^0_{\text{stack}}\right) + \sum \left(\Delta H^0_{\text{loop}} - T\cdot \Delta S^0_{\text{loop}}\right)\end{aligned}$$

Energetic contributions of stacking basepairs are favorable ($\Delta G^0_{T,\text{stack}} < 0$) while formation of loops is unfavorable ($\Delta G^0_{T,\text{loop}} > 0$) in most cases. The most widely used parameter set contains only $\Delta G^0_{37^\circ\text{C}}$ values and was determined by optical denaturation curves of small oligonucleotides [65, 66, 102] and evaluated by a two-state model. The errors in parameter estimation were lowest at 37°C, which then is the default temperature for all structure prediction programs. An older parameter set contains ΔH^0 and ΔS^0 values and thus allows to calculate ΔG_T^0 values at any temperature, however, with a lower accuracy then the more recent parameter set. Alternative parameter sets are available that are for example obtained with an improved evaluation method [94] or based on an optimization scheme [1]. For a given structure, the details of the energetic contributions can be viewed for example by RNAeval or at the Mfold web server (on the result page at "Thermodynamic details").

All such thermodynamic parameters for RNA structure formation were determined in 1 M NaCl. This is not so far from the ionic conditions in cells except when specific interactions with divalent cations play a role, which for example is the case for pseudoknots.

The energetic difference between a linear molecule and the otherwise identical circular molecule depends on the point of linearization and might be up to several kcal/mol. Note that a $\Delta\Delta G_T^0$

difference of 1 kcal/mol is equivalent to a factor of ~5 in the equilibrium constant

$$K_T = \exp\left(-\Delta\Delta G_T^0/(R \cdot T)\right)$$

with gas constant $R = 1.987$ cal/(K·mol) = 8.314 J/(K·mol) and $T = (273.15 + 37)$ K. That is, calculation for a linear versus a circular sequence might result in quite different structures.

The structure with a minimum of free energy s_0 or s_{mfe} is the thermodynamically optimal structure; it is, however, only the best of a huge amount of alternative structures, which grows exponentially $\sim 1.85^n$ with the sequence length n [39, 105]. All these possible structures s_i of a single sequence coexist in solution with concentrations dependent on their free energies $\Delta G_T^0(s_i)$; that is, each structure is presented as a fraction

$$f_{s_i} = \frac{\exp\left(-\frac{\Delta G_T^0(s_i)}{RT}\right)}{Z}$$

based on the partition function Z (from German "Zustandssumme," sum of states) for the ensemble of all possible structures

$$Z = \sum_{\text{all structures } s_i} \exp\left(-\frac{\Delta G_T^0(s_i)}{RT}\right)$$

Structures quite different from s_0 with identical or very close energies might exist for a single sequence. Thus, one should not assume that an RNA folds into a single, static structure.

The standard procedure to find the mfe structure uses dynamic programming [106, 123]. Dynamic programming can also yield the partition function [68]; this is now implemented in several programs like `RNAfold` [38] of the ViennaRNA package [55], `partition` of the RNAstructure package [79], `DINAMelt` [61], and `unafold` [62]. A good visualization of the partition folding is a dotplot in which all possible basepairs are given as dots with area proportional to the probability of the pair to be part of any structure (e.g., see Fig. 2b and c).

The relative concentration of the mfe structure may be very low when there are many structural alternatives. Furthermore, the free energy of the ensemble is lower (more negative) than that of the mfe structure because of the entropy caused by the number k of different structures:

$$\begin{aligned}\Delta G_T^0(\text{ensemble}) &= -RT\ln Z \\ &= \Delta G_T^0(s_0) - RT\ln\left[1 + \sum_{i=1}^{k} \exp\left(-\frac{\Delta G_T^0(s_i) - \Delta G_T^0(s_0)}{RT}\right)\right] \\ &\le \Delta G_T^0(s_0)\end{aligned}$$

Alternatives to mfe prediction are calculation of a structure with maximum expected accuracy (MEA) [16, 55, 59] or a centroid structure [14, 35]. Both algorithms rely on computing the partition function of a sequence and then select basepairs belonging to the "optimal" structure. The MEA algorithm maximizes the number of basepairs (BP) between nucleotides at positions i and j having a high pairing probability $p_{i,j}$ and the number of nucleotides with a high probability to be single-stranded (SS):

$$\text{Expected accuracy}(s) = \sum_{i,j \in \mathrm{BP}} \gamma \cdot 2 \cdot p_{i,j} + \sum_{k \in \mathrm{SS}} p(k) \quad (1)$$

Thus the MEA structure contains only basepairs and single-stranded nucleotides that have a higher chance to be correctly predicted. The parameter γ is a weighting factor: increasing γ will lead to a MEA structure containing only basepairs of high probability resulting in higher sensitivity; lowering γ encourages the increase of single-stranded regions resulting in higher specificity [16]. The centroid structure consists only of basepairs that have $p_{i,j} > 0.5$.

`RNAsubopt` is able to list all possible structures in a certain energy window above the mfe, but the number of these structures still grows exponentially.

`Mfold` is able to predict for every possible basepair the minimum energy structure containing this basepair [97, 121], which is a number of structures far below the number of all possible structures. Usually `Mfold` shows only a "representative" set of suboptimal structures; that is, the distance d between two shown structures is such that for every basepair i, j of one structure, there is at least one basepair k, l of the other satisfying $|i-k| \leq d$ and $|j-l| \leq d$ [121].

A different approach is used by `RNAshapes` [23, 42, 43]: it searches for "shape" classes that are defined by their arrangement of helices, but lengths of loops and helices are not relevant. Five different abstraction levels of shapes are defined: the most detailed level shows all helices (as pairs of square brackets) and all unpaired regions (as underscores); in the most abstract level nested helices (stem-loop structures) are depicted by a single bracket pair and unpaired regions are not shown. The shape representative ("shrep") is the mfe structure inside a shape class. Unfortunately this program is not able to handle circular RNAs.

3.1.1 mfe Structure

The programs of the ViennaRNA package need as input either a sequence file in Vienna or FASTA format or can directly read from standard input. A Vienna file consists of a header line starting with '>' followed by a sequence name. This name can be of any length; it is used for naming of output files but is then truncated at the first space. In the next line follows the sequence. A sequence consists of A, C, G, T, U in upper or lower case; a T is automatically replaced

```
1 $ readseq -a -f20 -oPSTVd.vie NC_002030.gb
2 $ more PSTVd.vie
3 > NC_002030
4 cggaactaaactcgtggttcctgtggttcacacctgacctcctgagcagaa ...
5 $ RNAfold -c < PSTVd.vie > PSTVd_rnafold_c.log
6 $ more PSTVd_rnafold_c.log
7 > NC_002030
8 cggaacuaaacucgugguuccugugguucacaccugaccuccugagcagaa ...
9 ..(((((((.((.((((((((...((((.((((((..(((((.((((((.. ... (-160.30)
10 $ gv NC_002030_ss.ps
11 $ rotate_ss.pl -a 90 -m NC_002030_ss.ps > NC_002030_ss_m.ps
12 $ epstopdf NC_002030_ss_m.ps
13 $ pdfcrop NC_002030_ss_m.pdf NC_002030_ss_m.pdf
```

(a)

(b)

Fig. 1 Prediction of mfe structure by `RNAfold`. (**a**) Commands to produce the structure shown in (**b**); for a description of the commands see text. (**b**) Drawing of the structure predicted by the commands in (**a**)

by a U; all other characters are handled as nucleotides that are not allowed to basepair. A file might contain more than a single sequence. An example of a file with two sequences might look as follows; here the sequence lines are truncated after 60 nts but have a length of 361 and 360 nts, respectively:

```
> PTVSEQC
GGGUGUUUAGCCCUUGGAACCGCAGUUGGUUCCUCGGAACUAAACUCGUGGUUCCUGUGG...
> AY360446
ggguguuuagcccuuggaaccgcaguugguuccucggaacuaaacucgugguuccugugg...
```

Perform the following steps to predict a mfe structure with `RNAfold`; the step numbers correspond to the line numbers of Fig. 1a. Download the PSTVd sequence with accession number NC_002030 in `GenBank` format from `GenBank` [3] and name the file `NC_002030.gb`. `GenBank` format has the advantage of including the full annotation of the sequence; in the Vienna or FASTA format only the accession number can be kept.

1 Re-format the `GenBank` file to Vienna format using `readseq` with the following options:
   `-a`: Convert all sequences in the `GenBank` file to the output file
   `-f`: Give the output format; '20' is Vienna format
   `-o`: Write output to the given file name

   Note that `readseq` requires parameters without a space separating option and value.
   Further programs for conversion of sequence formats are, for example, `sreformat` from the `SQUID` package [17] or `Seqret` from the EMBOSS package [10].

2 View readseq's output file using more; in this case the file only contains a header line (3) and the sequence in one line (4). Note that that the sequence contains 'T's but not 'U's; the conversion will be done automatically by RNAfold.

5 Predict the optimal secondary structure using RNAfold:
-c: A lower case c forces RNAfold to handle the sequence as circular.
<: Read sequence file PSTVd.vie via standard input (STDIN).
>: Redirect standard output (STDOUT) to file PSTVd_rnafold_c.log.

Without the redirections by '<' and '>' RNAfold asks on the terminal for just a sequence (most easily given by 'copy&paste'), prints the output to the terminal, and asks for the next sequence; termination is done by a single '@' instead of the next sequence:

```
$ RNAfold -c

Input string (upper or lower case); @ to quit
....,....1....,....2....,....3....,....4....,....5....,....6... ...
cggaactaaactcgtggttcctgtggttcacacctgacctcctgagcagaaaagaaaaaagaa ...
length = 359

cggaacuaaacucgugguuccugugguucacaccugaccuccugagcagaaaagaaaaaagaa ...
..((((((.((.((((((((...((((.((((((..(((((.((((((..((((....(((( ...
 minimum free energy = -160.30 kcal/mol

Input string (upper or lower case); @ to quit
....,....1....,....2....,....3....,....4....,....5....,....6... ...
@
```

6 View content of the log file; the output consists of the input file (lines 7, 8) with 'T's converted to 'U's and, in line 9, the predicted structure in dot-bracket notation; at the end of this line the $\Delta G^0_{37°C}$ in kcal/mol is given in round brackets. In dot-bracket format, pairs of opening and closing parenthesis denote basepairs and dots mark unpaired nucleotides.

10 View with gv (ghostview) the drawing of the secondary structure in eps format produced by RNAfold; the file name consists of the header of the input file extended with '_ss.ps'. The file name is rna.ps if input was not from a file.The standard structure-drawing mode of RNAfold (and RNAplot, Sect. 3.3.4) is counter-clockwise without a mark for 5' or 3' ends. The drawing does not reflect the circular sequence; that is, nucleotides at first and last position are not connected.

11 Rotate the drawing by an angle of 90° (counterclockwise) by the option '-a 90' and mirror it by option '-m' to yield a more conventional representation. The used program rotate_ss.

`pl` belongs to the utilities of the ViennaRNA package. Output is redirected ('>') to file `NC_002030_ss_m.ps`.

12 Convert the final drawing from eps to pdf format; the output filename is just the input filename with '.ps' substituted by '.pdf'.

13 Crop the pdf; that is, remove surrounding white margins from the drawing to yield Fig. 1b; the output filename is `<infile>-crop.pdf` if no second filename for the output is given.

3.1.2 Partition Folding

Execute the following steps (Fig. 2a) to perform a partition folding with `RNAfold`. The input sequence file is identical to that shown in Fig. 1a line 3–4.

1 Perform the partition folding and predict the mfe and MEA structures using `RNAfold` with the following options:

`-c`:	A lower case `c` forces `RNAfold` to handle the sequence as circular.
`-p`:	Perform the partition folding.
`--MEA[=`γ`]`:	Predict the structure with maximum expected accuracy; note the double minus signs in front of this "long" option. The default parameter is $\gamma = 1$ (see Eq. (1)).
<:	Read sequence file `PSTVd.vie` via standard input (STDIN).
>:	Redirect standard output (STDOUT) to file `PSTVd_rnafold_c_pf.log`.

This command generates a dotplot file '`NC_002030_dp.ps`' showing the mfe structure and the structure ensemble (Fig. 2b), a structure file '`NC_002030_ss.ps`', and a log file '`NC_002030_c_pf.log`' (lines 3–9); filenames are derived from the header (line 3).

2 View content of the log file; it consists of the following lines:

3–4 Header and sequence as given in the input file (Fig. 1a lines 3–4)

5 The mfe structure in dot-bracket notation; at the end of line the $\Delta G^0_{37°C}$(mfe) in kcal/mol is given in round brackets. This line is identical to the mfe-only prediction (Fig. 1a line 9).

6 The "mean" structure from partition folding; at the end of line the $\Delta G^0_{37°C}$(ensemble) in kcal/mol in square brackets. The possible structure symbols are

`( , )`	for a basepair i, j with $p_{i,j} > 0.66$
`{ , }`	for a basepair i, j with $0.33 < p_{i,j} \leq 0.66$
`|`	for a strongly paired nucleotide without a preferred partner
`,`	for a weakly paired nucleotide
`.`	for an essentially unpaired nucleotide

```
1 $ RNAfold -d -p --MEA < PSTVd.vie > PSTVd_rnafold_d_pf.log
2 $ more PSTVd_rnafold_d_pf.log
3 > NC_002030
4 cggaacuaaacucgugguuccugugguucacaccugaccuccugagcagaa ...
5 ..(((((((.((.((((((((...((((.((((((..(((((.((((((.. ... (-160.30)
6 ..((((((((((.((((((((,..((((.((((((..(((((.((((((.. ... {-166.29}
7 ..((((((.(((.((((((((...((((.((((((..(((((.((((((.. ... {-158.30 d=22.89}
8 ..((((((.(((.((((((((...((((.((((((..(((((.((((((.. ... {-159.30 MEA=324.40}
9  frequency of mfe structure in ensemble 0.000957651; ensemble diversity 29.41
10 $ gv NC_002030_dp.ps
11 $ mountain.pl NC_002030_dp.ps | \
12       xmgrace -pipe -printfile NC_002030_dp_mountain.ps
13 $ rotate_ss.pl -a 90 -m NC_002030_ss.ps > NC_002030_ss_m.ps
14 $ relplot.pl NC_002030_ss_m.ps NC_002030_dp.ps > NC_002030_e_rss.ps
15 $ relplot.pl -p NC_002030_ss_m.ps NC_002030_dp.ps > NC_002030_p_rss.ps
16 $ relplot.pl -a NC_002030_ss_m.ps NC_002030_dp.ps > NC_002030_a_rss.ps
```

(a)

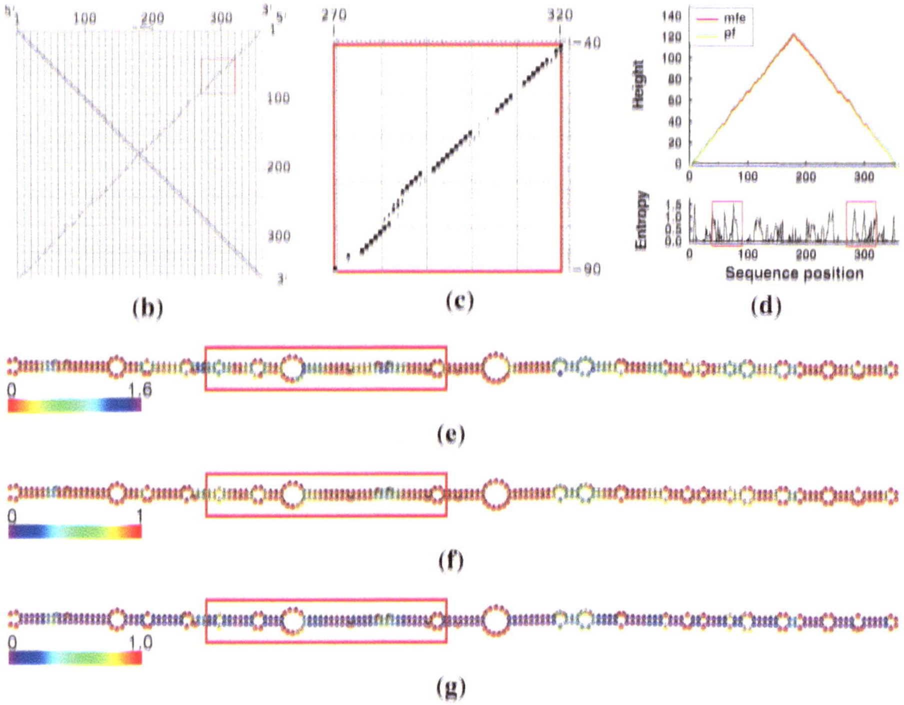

Fig. 2 Prediction of optimal structure and basepair probabilities by `RNAfold`. (**a**) Commands to produce the structure representations shown in (**b**)–(**g**). (**b**) Dotplot `NC_002030_dp.ps`; the axes labels were added manually; the area in the red box is shown enlarged in (**c**) and boxed in (**d**)–(**g**). (**d**) Mountain plot `NC_002030_dp_mountain.ps`. Secondary structure plots `NC_002030_e_rss.ps`, `NC_002030_p_rss.ps`, and `NC_002030_a_rss.ps` with annotation by positional entropy (**e**), pair probability (**f**), and accessibility (**g**), respectively; below each structure is shown the respective color legend

7. The centroid structure: at the end of line the $\Delta G^{0}_{37\,°C}$(centroid) in kcal/mol and its distance to the ensemble structure in curly brackets.

8 The MEA structure: at the end of line the $\Delta G^0_{37°C}$(MEA) in kcal/mol and its distance to the ensemble structure in curly brackets.

9 A line showing the frequency of the mfe structure in the ensemble, and the mean distance $\langle d \rangle = \Sigma_{i,j} p_{i,j} \cdot (1 - p_{i,j})$ between the structures in the ensemble.

10 View the dotplot (Fig. 2b) using gv (ghostview).The axes of the dotplot are labeled with the sequence from 5′ to 3′ (left to right, top to bottom, respectively). The lower left triangle shows the basepairs of the mfe structure with dots of area 1. The top right triangle shows the basepairs predicted by partition folding with dots of area proportional to their pairing probability; pairs with $p_{i,j} < 1 \times 10^{-6}$ are not included in the plot.The mfe and partition folding structure of PSTVd are nearly identical at a first glance; however, the enlarged section (nts 270–320 and 40–90; Fig. 2c) shows many alternative pairings.

11/12 The program `mountain.pl`, which belongs to the utilities of the ViennaRNA package, extracts basepair probabilities from the dotplot file '`NC_002030_dp.ps`'. A mountain plot [41] shows paired regions as slopes, unpaired nucleotides on plateaus, and hairpin loops at peak positions. The program writes three sets of data to STDOUT, which can be used as x, y coordinates for plotting (Fig. 2d). The first set is extracted from the mfe structure part of the dotplot (height $y_k = \Sigma_{i<k<j} p_{i,j}$; $p_{i,j} \in \{0, 1\}$), the second set is derived from the pairing probabilities of the partition folding part of the dotplot ($y_k = \Sigma_{i<k<j} p_{i,j}$; $0 \leq p_{i,j} \leq 1$), and the third set gives the positional entropy $S_i = -\sum_j p_{i,j} \ln(p_{i,j}) - \sum_j (1 - p_{i,j}) \ln(1 - p_{i,j})$, which provides information on the well-definedness of the structure [114]. Regions of low positional entropy are highly basepaired and the reliability of pairing is high, while regions of high positional entropy have structural alternatives. The output is viewed here by piping the data to `xmgrace`, a plotting program with a graphical user interface.

With PSTVd, the mountains of the mfe and the partition folding structure are very close; that is, only minor regions of the structure have weak alternative pairings. An easier insight into the positional entropy then by the simple plot (Fig. 2d bottom) is given by a corresponding color-coding of the secondary structure (line 14; Fig. 2e).

13 Rotate the secondary structure drawing by an angle of 90° (counterclockwise) by the option '`-a 90`' and mirror it by option '`-m`'; output is redirected ('>') to file `NC_002030_ss_m.ps`.

14–16 `replot.pl`, which belongs to the utilities of the ViennaRNA package, is used to annotate the secondary structure drawing by positional entropy (line 14; Fig. 2e), pair probabilities (line 15; Fig. 2f), and accessibility (line 16; Fig. 2g). The color code of positional entropy shows nucleotides with low entropy in red; that is, a nucleotide shown in red is either highly basepaired to one other nucleotide or it is reliably unpaired. A high accessibility is equivalent to a high probability of being unpaired.

For comparison of the free energy and other values computed by `RNAfold` in the example of Fig. 2, we assembled the corresponding values for 267 different PSTVd sequences from GenBank and for 1000 random sequences (Table 2). The random sequences were generated by dinucleotide shuffling of a PSTVd (AC NC_002030); the identical dinucleotide content should preserve the chance for similar pairing energies that are based on base stacking [20, 81]. Obviously, PSTVd (and most other viroids) have a special sequence to allow for extraordinary structural features: all free energy values—irrespective of the calculation method—are by far above that of the random sequences. For the PSTVd sequences, even ΔG^0(MEA) and ΔG^0(centroid) are very close to ΔG^0(mfe);

Table 2
Comparison of prediction results for random and PSTVd sequences. Thousand random sequences with length and dinucleotide composition of PSTVd (AC NC_002030) were generated using `shuffle` [17]. The 267 different PSTVd sequences were selected to be full-length. Predictions were done by `RNAfold -c -p --MEA`. Free energy values are given in kcal/mol

	Random sequences mean$\pm\sigma$	**PSTVd sequences mean$\pm\sigma$**
$\Delta G^0_{37°C}$(mfe)	-127.47 ± 6.05	-159.46 ± 3.59
$\Delta G^0_{37°C}$(ensemble)	-134.00 ± 5.70	-165.57 ± 3.34
$\Delta G^0_{37°C}$(centroid)	-99.14 ± 22.11	-156.22 ± 5.43
$\Delta G^0_{37°C}$(MEA)	-116.02 ± 13.34	-158.04 ± 4.32
$\langle d\rangle$(centroid)	57.56 ± 17.49	23.76 ± 3.86
$\langle d\rangle$(MEA)	270.39 ± 27.75	323.07 ± 5.50
Diversity ensemble	85.83 ± 24.97	35.47 ± 4.73
f(mfe)	$(1.01\pm3.18)\times10^{-4}$	$(6.53\pm5.18)\times10^{-5}$

that is, already most basepairs of the mfe structure have a pairing probability above 0.5 and are quite reliable. This is also obvious from the color-annotated structures in Fig. 2e and f: the positional entropy is low and the pair probability is high for most helices; only a few helices have higher entropy and lower pair probabilities due to alternative pairings (e.g., possible shifts from a pair i, j to $i, j \pm 1$). The frequency of the mfe structure, however, is low, but diversity of the ensemble structures is also quite low. Consequently, alternatives to the mfe structure are present, but these alternatives have a "shape" very similar to the mfe structure.

`RNAfold` offers a lot of further options not used in the example of Figs. 1 and 2. For example, `RNAfold` allows a user to force a wanted structure by applying constraints. One possibility is to give to `RNAfold` "hard" constraints as an additional string of sequence length consisting of the following symbols:

`( ,)` : matching brackets force the two nucleotides to pair.
`x`: the nucleotide is unpaired.
`|`: the nucleotide is paired but partner is unknown.
<: the nucleotide i is paired to a nucleotide j with $j > i$.
>: the nucleotide i is paired to a nucleotide j with $j < i$.
`.`: no constraint for this nucleotide.

For an example on usage of such a "hard" constraint string see Sect. 3.2.2. "Hard" constraints are implemented today not by addition of large bonus or malus energies, but by direct restrictions in the folding algorithm [56]. "Soft" constraints bias the ensemble of secondary structures to either favor or discourage certain structures or structural features by bonus and/or malus energies [56]; this allows a user, for example, to force sequence ranges to be paired or unpaired, prohibit or force a helix at a given position, or to add bonus energies if a sequence motif appears in a special structural context and thus is available for binding to a ligand [56].

Further options include the following:

`--noLP`: Avoid structure(s) with lonely, non-stacked basepairs. The stability of basepairs is mostly due to stacking on neighboring basepairs, but not due to their hydrogen bonds; each base can have almost as many hydrogen bonds to water molecules as to the pairing partner. Consequently, a basepair not stacking on a neighboring basepair should not appear in a structure, but will be predicted without this option if for example a loop of size $(a + b + 2)$ nts is energetically less favorable than two loops of size a and b separated by a lonely pair.

`-d INT` (or `--dangles=INT`): This option determines how energies due to (non-paired) nucleotides adjacent to helices in free, dangling ends and junctions are handled.

- `-d0`: Ignore any energies that might arise from dangling ends or in junction loops.

`-d1`: Account only for energies of an unpaired nucleotide in at most one dangling end.

`-d2`: The check from `-d1` is ignored; that is, an unpaired nucleotide can participate in stacking on both sides. This is the default for mfe and partition folding.

`-d3`: Allow for coaxial stacking of adjacent helices in junctions. This works only for mfe but not for partition folding.

Nearly no experiment-based parameters for junction loops are known. The standard approximation is $\Delta G = a + n_1 \cdot b + n_2 \cdot c$ depending on the number of unpaired nucleotides n_1 and helices n_2 in the loop with empirical constants a, b, and c. This approach is mainly chosen to keep low the algorithmic complexity. Thus options `-d2` or `-d3` are good choices, but the latter works only for mfe folding.

`--shape=<filename>;`

`--shapeMethod=[D/Z/W];`

`--shapeConversion=[M/C/S/L/O]`: Use "selective 2'-hydroxyl acylation analyzed by primer extension" (SHAPE) reactivity data—given in `filename`—to guide structure predictions by various methods, as detailed in the manual page of `RNAfold`. Guiding structure prediction by data from SHAPE or other structure-probing data is also implemented in `RNAstructure/Fold` [113] and `Rsample` [93]. But be careful with the interpretation of SHAPE data [50].

3.2 The Consensus Structure

Homologous ncRNAs, as the members of a viroid species, should have a common structure for functional reasons, but their sequences usually are different due to the evolutionary distance of the RNAs and further biological constraints. The Watson-Crick (WC) basepairs—and to a lower extent also wobble pairs—are isosteric to each other and thus can form regular helices with any WC basepair combination; in addition, members of certain groups of non-WC pairs are also isosteric to each other [51, 98]. Thus, a mutation of a nucleotide in a helix can be compensated for either by a reversion to the original nucleotide or by an additional mutation of the partner nucleotide to restore WC pairing; similarly a non-WC-pair can be exchanged by a different non-WC-pair from the same isostericity family [52]. This argument based in biophysics and in phylogeny helps to establish a statistically significant secondary and higher structure. Even the tertiary structure of ribosomal RNA was (nearly perfectly) predicted by covariation analysis prior to the x-ray structure of ribosomes [33, 72]. With PSTVd, isostericity of non-WC pairs helped to establish structural models for several loops [95, 100, 117, 119, 120]. Note, however, that such a covariation analysis works best if the sequences can still be aligned by a sequence alignment program but have only a low overall sequence identity (≤80%). If these conditions are met, a few (≤10) sequences are usually sufficient to improve sensitivity and specificity of structure prediction. With sequences from a single viroid species, which have a sequence identity ≥85% by definition [15], many more sequences are necessary for improvement of a consensus structure in comparison to a single-sequence prediction. `R-scape` [82] is a good choice to test for statistical significance of covarying basepairs.

Already in 1985 Sankoff published a comparative approach, in which sequence alignment and mfe folding was performed simultaneously [84]. The main drawback of this algorithm—besides the scoring problem of nucleotide matching and basepairing probability—is its time complexity of $O(n^{3k})$ with space requirement $O(n^{2k})$ for sequence length n and k sequences, which makes the algorithm unfeasible. In the mean time, however, several programs are available that are able to handle the simultaneous aligning and folding using some restrictions; these are for example `Dynalign` [36, 64], `foldalign` [27], `LocARNA` [109, 110], and `StrAl` [12]. `RNAcast` avoids the multiple alignment completely but predicts a shape class common to a sequence set [77]. Other tools rely on a multiple alignment produced by a more-or-less standard alignment program and then try to infer a consensus secondary structure from the evolutionary and energetic information contained in the alignment [22]. In the following we will show the use of two tools from the latter class. `RNAalifold` uses sequence covariation, which is extracted from a given sequence alignment, as bonus energies in a "soft" constrained computation [5, 40]. `ConStruct` is a semi-automatic, graphical tool suitable for improving RNA alignments correct in terms of both consensus sequence and consensus structure by a combination of sequence alignment, thermodynamic data, and various measures of covariation; a user may correct the sequence alignment to optimize structural aspects and is guided in this step by a consensus dotplot, which displays all thermodynamically optimal basepairs and the corresponding covariation [60, 111].

3.2.1 Sequence Alignment

A prerequisite for many consensus structure prediction programs, like the two tools `RNAalifold` and `ConStruct` used in Sects. 3.2.2 and 3.2.3, is an alignment of the sequences. This alignment should be as good as possible not only in terms of a sequence alignment but also in terms of a structure-aware alignment. We assume that the user has downloaded the viroid sequences from, for example, GenBank from the taxonomy browser or via a `BLAST` search [116]. This set of sequences might contain partial sequences and sequences of wrong polarity, and the sequences might not have a common start position. This latter problem is due to the circular nature of viroids, but only linear sequences are stored in GenBank and only these can be handled by most alignment programs. At first a rough alignment is necessary to make a decision on removal of unwanted, partial sequences and on change of sequence polarity and start position. For this first alignment, the `mafft` web service is excellent [45]: with default options it is fast, output is available in FASTA and `Clustal` format, respectively, and sequences with presumed wrong polarity are highlighted. The programs `MARS` [2] and `cyclope` [69], respectively, might be used to "rotate" the sequences to have a common start position, which however might not be the conventional one. At last a time-consuming but close-to-perfect alignment should be produced with the cleaned sequence set. If only a few sequences are in the set, `LocARNA` [109, 110] might be the optimal choice. With a larger sequence set, `MAFFT-X-INS-i` [44] is preferable because it works for such larger sets and still takes structural information into account, however, on a simpler level than `LocARNA`. `MAFFT-X-INS-i` has to be locally installed and is not available on the web server. Corresponding command lines are seen in Fig. 3a line 1 and Fig. 4a line 1. Useful options of `MAFFT-X-INS-i` and the other `MAFFT` variants are:

`--maxiterate` <#>:	Maximum number of iterative refinement.
`--retree` <#>:	The guide tree is built the given number of times during the progressive stage; default is 2.
`--reorder`:	Sort output according to similarity of aligned sequences; the default without this option is order of the input file.
`--clustalout`:	Output in `Clustal` format; default is output in FASTA format.
`--thread` <#>:	Number of threads to use.

In the two examples, the output to STDOUT is redirected into a file in FASTA format.

3.2.2 `RNAalifold`

`RNAalifold` calculates for each sequence in an alignment the mfe structure and partition function, as `RNAfold` for single sequences, but then adds a covariance term

$$\gamma_{i,j} = \frac{1}{2} \sum_{\substack{\alpha,\beta \in A \\ \alpha \neq \beta}} x \cdot R(\alpha_i \alpha_j, \beta_i \beta_j)$$

for each consensus pair in sequences α and β in columns i and j of the alignment A; x is a scaling factor and R is either a simple distance measure

$$\begin{aligned} R(\alpha_i\alpha_j,\beta_i\beta_j) &= h(\alpha_i,\beta_i)+h(\alpha_j,\beta_j) \\ \text{with} \quad h(\alpha_k,\beta_k) &= 0 \quad \text{if } \alpha_k=\beta_k,\ k\in[i,j] \\ h(\alpha_k,\beta_k) &= 1 \quad \text{if } \alpha_k \neq \beta_k \\ \text{and} \quad (\alpha_i\alpha_j) &\in \mathcal{B} = [\text{AU, UA, CG, GC, GU, UG}] \\ (\beta_i\beta_j) &\in \mathcal{B} \end{aligned}$$

or incorporates a weighting scheme for all possible substitutions based on a RIBOSUM matrix [5, 46]; both measures additionally include penalties if a "basepair" contains gaps. Use of the RIBOSUM matrix leads to slight improvements in comparison to the simple distance measure [5, 46].

The ratio of the energy of the consensus structure (including the covariance "energies") E_{cons} to the average free energy of the single sequence structures $\bar{E}_{single}$ is the structure conservation index [31, 104]:

$$SCI = \frac{E_{cons}}{\bar{E}_{single}}$$

If `RNAalifold` does not find a consensus structure, the *SCI* is close to zero; if all single structures are close to the consensus structure, the *SCI* is close to 1; with bonus energies due to covarying basepairs the *SCI* is above 1.

Figure 3 shows the prediction of alternative foldings in the terminal left region of PSTVd sequences. The file `PSTVd_TL.vie` contains the sequence regions nucleotides 325–358/1–35 from 274 different PSTVd sequences, which were aligned by `MAFFT-X-INS-i`. The standard input to `RNAalifold` is a sequence file in `Clustal` format; thus the conversion of `MAFFT`'s output in FASTA format to `Clustal` format via the `ClustalX` GUI (Fig. 3a line 2); an alternative to the use of `ClustalX` would be, for example, `sreformat` from the `SQUID` package [17], `seqret` from the EMBOSS package [10], or to use directly the `--clustalout` option of `MAFFT`. A part of the resulting multiple sequence file is shown in lines 4–11. Then `RNAalifold` was run with the following options (line 12 + 13):

`-p`	Calculate the partition function and produces a corresponding dotplot (Fig. 3b)
`--sci`	Compute the *SCI*
`--ribosum_scoring`	Use RIBOSUM scoring
`-d2`	Add dangling end energies for nucleotides adjacent to a helix on both sides
`--noLP`	Avoid lonely pairs
`--color`	Produce a colored version of the consensus structure plot (Fig. 3d)
`--aln`	Produce a structure-annotated alignment in file `aln.ps` (not shown)
`--aln-EPS-cols=85`	Number of columns in `aln.ps`

`RNAalifold` writes a file `alifold.out` that contains a list of likely pairs sorted by credibility; this output file is overwritten by each run of `RNAalifold`. A few lines from `alifold.out` are shown in Fig. 3a, lines 17–20, each line shows a position *i*, *j* of a basepair, the number of non-pairs at this position, the probability of pairing, the entropy of this position, and a list of types and numbers

```
1 $ mafft-xinsi --maxiterate 1000 PSTVd_TL.vie > PSTVd_TL.fa
2 $ clustalx PSTVd_TL.fa [-> Save as PSTVd_TL.aln in ClustAl format]
3 $ more PSTVd_TL.aln
4 CLUSTAL 2.1 multiple sequence alignment
5 PTVSEQC          GGGUGUUUAGCCCUUGGAACC-GC-AGUUGGUUCC-UCGGAACUAAACUC
6 AY360446         GGGUGUUUAGCCCUUGGAACC-GC-AGUUGGUUCC-UCGGAACUAAACUC
7 ...
8 HM107846_rev     GGGUGUUUAGCCCUUGGAACC-GC-AGUUGGUUCC-UCGGAACUAAACUC
9 HM107848_rev     GGGUGUUUAGCCCUUGGAACC-GC-AGUUGGUUCC-UCGGAACUAAACUC
10                 *****    ***  *****  ** **   ***      *  * *  **
11 ...
12 $ RNAalifold -p --sci --ribosum_scoring -d2 --noLP --color --aln \
13     --aln-EPS-cols=85 < PSTVd_TL_bif.aln > PSTVd_TL_bif_aln_woC.log
14 $ more rnaalifold.out
15 274 sequence; length of alignment 72
16 alifold output
17    21    53  0 100.0%   0.000 CG:273  UG:1
18    33    41  0 100.0%   0.001 UG:1    UA:273
19 ...
20    29    46  0  35.7%   1.757 UA:272  --:2
21 ...
22 $ more PSTVd_TL_bif_aln_woC.log
23 GGGUGUUUAGCCCUUGGAACC_GC_AGUUGGUUCC_UCGGAACUAAACUCGUGGUUCCUGUGGUUCACACCU
24 ((((((..((((...((((((.((.((((((((((...)))))).)))).))))))))...)))).)))))) (-51.61=-34.09-17.52) [sci=1.5059]
25 ((((((..((((...((((((.((.(((((((((((...))))))))))).))))))))...)))).)))))) [-52.75]
26 ((((((..((((...((((((.((.((((((((((...))))))).))).))))))))...)))).)))))) {-51.61=-34.10-17.50 d=2.18}
27  frequency of mfe structure in ensemble 0.993251; ensemble diversity 3.20
28 $ more constraint.txt
29 ....................(..........)...........(..........)..................
30 $ cat constraint.txt | \
31  RNAalifold -p --sci --ribosum_scoring -d2 --noLP --color --aln \
32    --aln-EPS-cols=85 \
33    --enforceConstraint -C PSTVd_TL_bif.aln > PSTVd_TL_bif_aln_wC.log
34 $ more PSTVd_TL_bif_aln_wC.log
35 GGGUGUUUAGCCCUUGGAACC_GC_AGUUGGUUCC_UCGGAACUAAACUCGUGGUUCCUGUGGUUCACACCU
36 ((((((..((((...((((((........))))))...(((((((......)))))))...)))).)))))) (-40.23=-25.66-14.57) [sci=1.1739]
37 ((((((..((((...((((((........))))))...(((((((......)))))))...)))).)))))) [-40.77]
38 ((((((..((((...((((((........))))))...(((((((......)))))))...)))).)))))) {-40.23=-25.66-14.57 d=2.33}
39  frequency of mfe structure in ensemble 0.996824; ensemble diversity 4.06
```

(a)

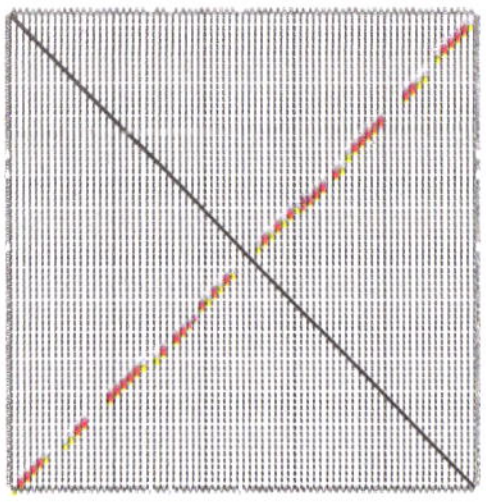

(b)

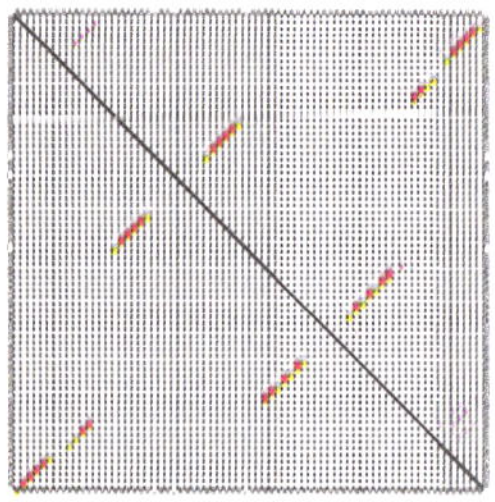

(d)

(c)

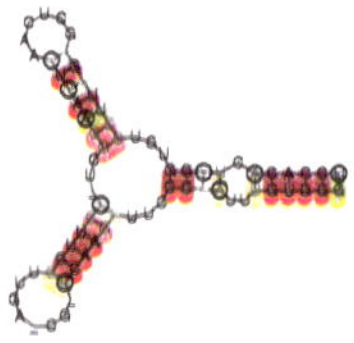

(e)

Fig. 3 Consensus structure prediction using `RNAalifold`. (**a**) Commands to produce the structure representations shown in (**b**)–(**e**). The alignment consists of terminal left domains of 274 PSTVd sequences. (**b**), (**d**) Dotplots from `alidot.ps` produced by commands shown in part (**a**), lines 12, 13 and 30, 32, respectively. The bottom left triangle shows the consensus basepairs in the mfe structure; the top right triangle shows the consensus basepairs from the partition folding. (**c**), (**e**) Structure drawings from `alirna.ps` produced by commands shown in part (**a**), lines 12, 13 and 30, 32, respectively. Colors encode base-pair probabilities; compensatory mutations are marked by a circle around the variable base(s); pairs that cannot be formed by some of the sequences are shown gray instead of black of pairs or gaps (line 20) occurring at this position. This information is also shown in the structure plot (Fig. 3c) by a color code and circles: compensatory mutations are marked by a circle around the variable base(s); pairs that cannot be formed by some of the sequences are shown gray instead of black; colors encode base-pair probabilities, conservation, or positional entropy.

`RNAalifold` writes information on predicted structures and their energies to STDOUT, which was re-directed to the file `PSTVd_TL_bif_aln_woC.log`. Its content consists of the consensus sequence (Fig. 3a line 23), the consensus mfe structure and its energy (line 24), the consensus structure obtained from partition folding and its energy (line 25), a highly probable structure from the partition folding (line 26), and a note on frequency of the mfe structure in the ensemble and the diversity of the ensemble (line 27). At the end of each structure line is given the respective energy as a sum of the (mean) $\Delta G^0_{37°C}$ and the covariance term. In the example, the structure of line 26 is identical to the structure of line 24, which has not to be the case.

Furthermore, `RNAalifold` produces a consensus dotplot in file `alidot.ps` (Fig. 3b) and a secondary structure drawing in `alirna.ps` (Fig. 3c); both output files are overwritten by each run of `RNAalifold`.

Next a bifurcated structure of the terminal-left region was forced by a constraint on two basepairs. Thus the file `constraint.txt` contains a single line with the length of the alignment; the bracket pairs denote the forced basepairs, while the dots denote no constraints for the corresponding nucleotide. `RNAalifold` reads `constraint.txt` from STDIN; additional options show to `RNAalifold` what to do:

`--enforceConstraint`	Enforce given basepairs
`-C <filename>`	Calculate structures subject to constraints and read the alignment from `<filename>`

For both types of calculation—without and with constraints—`RNAalifold` predicts consensus structures with quite negative energies, and both structures are supported by covariations as judged by the respective covariance terms and *SCI* values. The rod-like structure, however, has a lower ΔG^0, a lower covariance term, and a higher *SCI* than the bifurcated structure. Taking into account only the biophysically relevant ΔG^0 values for the rod-shaped and the bifurcated structure, it follows for the equilibrium between these structures at $T = (273.15 + 37)$ K:

$$\Delta\Delta G_T^0 = \Delta G_{T,\mathrm{rod}}^0 - \Delta G_{T,\mathrm{bif}}^0 = (-34.09 + 25.66)\ \mathrm{kcal/mol}$$

$$K_T = \frac{c_{\mathrm{rod}}}{c_{\mathrm{bif}}} = \exp\left(-\frac{\Delta\Delta G_T^0}{R \cdot T}\right) = 8.79 \times 10^5$$

That is, the concentration of the bifurcated structure will be about 1.15×10^{-6} times less than that of the rod-like structure in conditions of thermodynamic equilibrium.

3.2.3 `ConStruct`

`ConStruct` [111] consists of two graphical user interfaces (GUI). The first, named `cs_fold`, uses `RNAfold` to predict basepairing dotplots for each individual sequence of an alignment. The second, named `cs_dp`, inserts the gap positions from the alignment into the dotplots, which then are all of identical size, and superimposes them. Basepairs from the individual sequences are shown as green dots with size proportional to pairing probability; the probability of basepairs at identical positions are averaged and shown as red dots in the upper right triangle of a dotplot (Figs. 4b and c). The lower left triangle of the dotplot shows a covariation score (expected mutual information content I and variants thereof; for an introduction see Sect. 3.3.1 and Eq. (3)). The consensus structure prediction is done by dynamic programming maximizing the probability of consensus basepairs, or the covariation score, or a weighted average of both; results are depicted graphically as structure plot (Fig. 4d) or as structure-annotated alignment (Fig. 4e).

Additionally, the user is able to modify the alignment via the GUI; single nucleotides, sequence regions of a single sequence or of a set of sequences can be moved to a previous gap position. The dotplot is immediately updated to reflect the new alignment. Thus the user is able to correct the alignment in terms of sequence identity and structure conservation. The effort for the dotplot update is proportional to number of sequences and basepairs, which limits the interactive use of `cs_dp` to a few hundred sequences of viroid length. For further details and options we refer to the manual [111].

Figure 4 outlines the use of `ConStruct` and shows a few outputs based on a `MAFFT-X-INS-i` alignment of 842 PLMVd sequences. Sequences are selected to be full-length, without duplicates, and are not from the deep-sequencing studies.

Figure 4b is the dotplot: the upper right triangle shows the overlay of the basepairing probabilities of all individual sequences (green dots), the consensus pairs (reddish dots), and gaps of the alignment (bars with color from white to blue corresponding to sequences with gaps at that position). Most sequences have gaps close to the right end of the plot; for example, sequences with the "peach calico insertion" have extended their sequences in this region [70, 96]. In the rod-like part (P11 stem-loop) most

```
$ mafft-xinsi --maxiterate 10000 PLMVd.vie > PLMVd.fa
$ cs_fold -f PLMVd.fa [-> Save Project File; Fold; Exit]
$ cs_dp -f PLMVd.proj
```

(a)

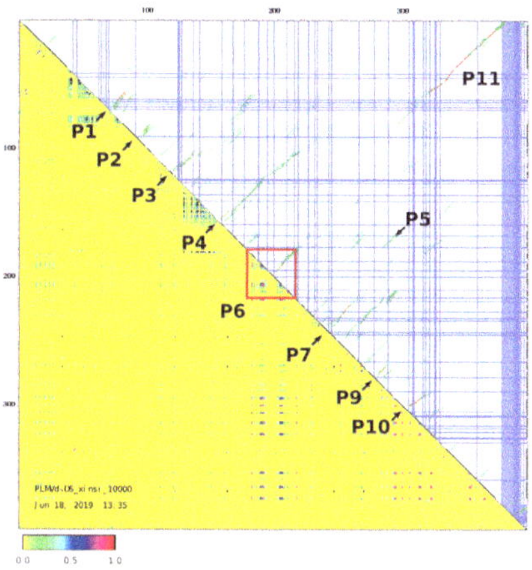

(b)

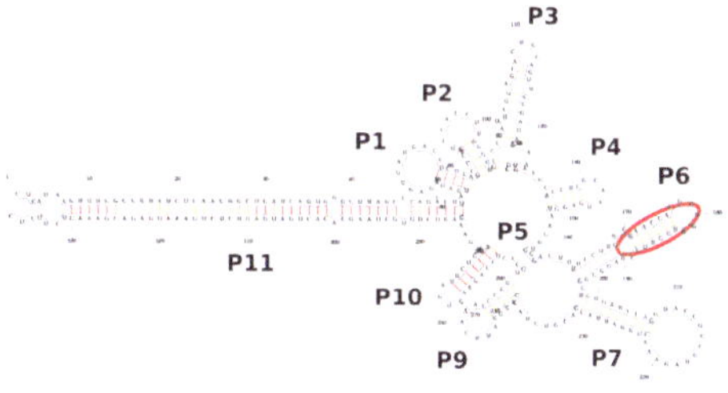

(d)

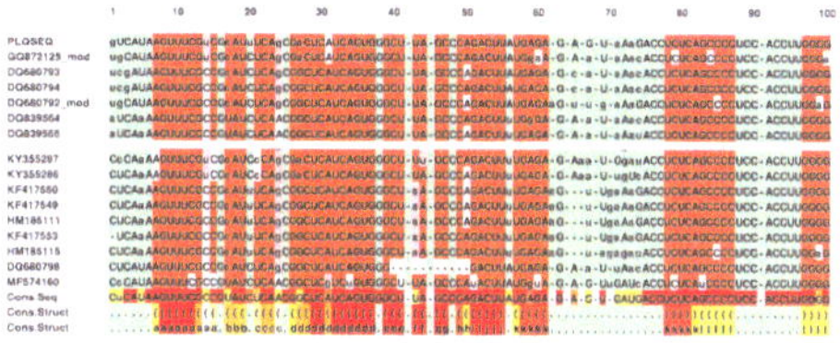

(e)

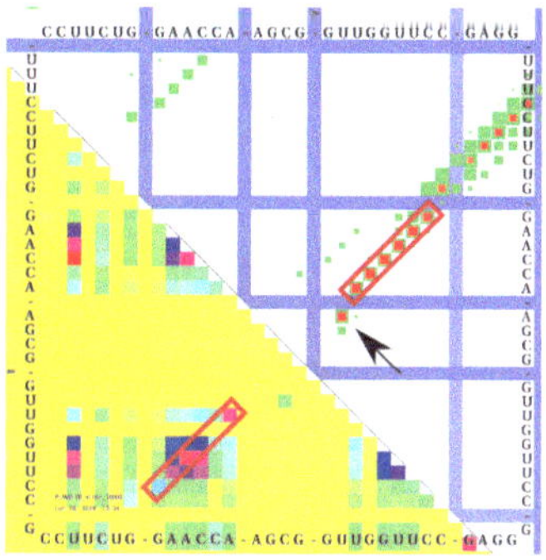

(c)

BP	NoBP	BP	CsBP	P(TD)	I(x,y)	chi^2(df,p)
G:C	41	699	102	0.380	0.262	>= 0.023=x2(16,.999)
A:U	3	642	197	0.442	0.493	>= 0.023=x2(16,.999)
A:U	4	637	201	0.458	0.584	>= 0.017=x2(9,.999)
C:G	0	674	168	0.474	0.414	>= 0.017=x2(9,.999)
C:G	1	840	1	0.474	0.002	>= 0.002=x2(3,.10)
A:U	10	653	179	0.452	0.573	>= 0.017=x2(9,.999)
A:U	4	838	0	0.408	0.000	=> not significant

(f)

Fig. 4 Consensus structure prediction using `ConStruct`. (**a**) `MAFFT-X-INS-i` was used to align 842 PLMVd sequences, `cs_fold` predicted the individual structures, and `cs_dp` was used to view the superimposed dotplots and to extract consensus sequence and structure. (**b**) Superimposed dotplots; the top right triangle shows individual basepair probabilities (green squares), consensus basepair probabilities (red squares), and gap positions and frequencies (bluish bars); the bottom left triangle shows mutual information content (color code below the dotplot); the red box surrounds the part that is shown enlarged in (**c**). Position of stem-loops P1–P7, and P9–P11 are marked; P8, a pseudoknot between hairpin loops of P6 and P7, is not part of the secondary structure. The arrow in (**c**) points to a lonely pair that is not stacked because a single sequence has an insertion. (**d**) Consensus sequence and structure; colors of lines connecting basepairs are from yellow to red proportional to basepairing probability; the helix circled in red corresponds to the region boxed in (**c**). (**e**) Part of the alignment with consensus sequence and structure in the three bottom lines; the consensus sequence is colored from yellow to red proportional to sequence conservation; the consensus structure is colored green for non-paired regions and from white to red proportional to basepairing probability; in the sequences, uppercase letters coincide with the consensus, non-paired regions are on green background; red, pink, and white background are for conserved and paired, non-conserved and paired (showing covariation), and non-paired nucleotides (contradicting the predicted basepair). (**f**) Part of a table listing features of the helix boxed in (**c**) and (**d**): type of basepair (BP), number of non-pairs (NoBP), WC pairs (BP), covarying pairs (CsBP), sum of thermodynamic basepairing probability (P(TD)), mutual information content (I), and χ^2 value with respective degrees of freedom (df) and significance level (p) sequences have coinciding basepairs (many red dots); this is also true for the remaining hairpin stems P1–P7, P9, and P10; the pseudoknot P8 between hairpin loops of P6 and P7 cannot be found in a secondary structure prediction. However, many gaps are found especially in regions of the hairpin loops, which are due to single insertions or deletions.

The lower left triangle of the dotplot (Fig. 4b) shows the covariation score. Despite the many sequences in the alignment, only a part of P6 is supported by both thermodynamics and covariation score (see red boxes in Fig. 4c, red ellipse in Fig. 4d). Six of its seven basepairs are supported by statistically significant mutual information content (Fig. 4f). The distal basepair (see arrow in Fig. 4c) is not stacked on the next basepair of P6 because a single sequence has an insertion.

3.3 Drawing

Graphical representations of a secondary structure based on a single sequence or of a consensus secondary structure based on a multiple alignment are required at least for publications but are already helpful during experimental verification steps. All such, usually planar representations have their individual merits but also common drawbacks because a secondary structure is not planar, basepairs are mostly suggested as perfectly closed, and single-stranded regions are always shown as completely open. However, each helix is already a three-dimensional object, basepairs might have a probability less than 1 for their formation (see Figs. 2a and c versus Figs. 2e–g), and nucleotides in single-stranded regions might form non-Watson-Crick pairs or might be strongly stacked [51, 95, 119]. Taken together, be careful with the simplifications of most graphical representations. For a more general overview of structure drawing programs we refer to Ponty and Leclerc [74]; here we will concentrate to a few programs that are useful for viroid structures and sequences.

The raw data formats used by most programs are quite different, but interconversion of formats is possible, for example, by the help of `sreformat` [17]. Quite common formats are listed below:

Vienna format with additional dot-bracket line(s): In addition to header and sequence line(s) (see lines 3 and 4 in Fig. 1a) the format has an additional line consisting of dots for unpaired nucleotides and matching pairs of brackets for basepairs. This line has the identical length as the sequence.

Stockholm format is able to represent a single sequence as well as a multiple alignment. The usual file extension is `.sto` or `.stk`. Sequence lines start with an ID followed by the respective sequence, which might include gap symbols. Lines starting with `#=` contain additional information; `#=GC` lines contain annotations per column of an alignment; and `#=GF` lines contain free text. For example, `#=GC SS_cons` is used to indicate a consensus structure by brackets of any type and even may contain pseudoknots represented by character combinations. These lines are used by `R2R` to obtain drawing commands (see Sect. 3.3.6); for an example of a Stockholm file used by `R2R` see Fig. 9a. For more details of the Stockholm format, the reader is referred to the internet [92, 108].

CONNECT format: This format is used by `Mfold` to represent a secondary structure of a single sequence; the usual file extension is `.ct`. The file starts with a line consisting of the sequence length n followed by a comment (e.g., the free energy of the structure obtained by `Mfold`); each of the following n lines consist of six fields: position of the nucleotide in the sequence, a nucleotide character, position of the previous nucleotide (in 5'–3' order; 0 is used if the description refers to the first position), position of the next nucleotide (in 5'–3' order; 0 is used if the description refers to the last position), position of the basepairing partner (or 0 if unpaired), and again the position.

PDB format is used to describe three-dimensional models from x-ray or NMR analysis or from modeling studies. It is the format used in the Protein data bank [4]. None of the other formats can be converted into PDB format, but `rnaview` [115] or `x3dna-dssr` [58] are able to extract, for example, information on Watson-Crick and non-canonical basepairs.

As output formats, we strongly recommend vector representations (like pdf, eps, or svg (scalable vector graphics)) over bitmap formats (like png (portable network graphics) or jp(e)g (Joint Photographic Experts Group)). The latter make it difficult to change color, to modify character size and line width, or to rescale certain parts or full figures. With vector formats any modifications are performed much easier, and conversion to bitmap formats, if required by the publisher, can be done by many programs (like `gimp` or Adobe Photoshop®).

3.3.1 Sequence and Structure Logos

Consensus sequences and structures are mainly used in the viroid field to depict certain recurrent motifs like terminal conserved region (TCR) [47], loop E [7, 21], RY motif for binding of VirP1 [28, 95], or extrastable hairpins [37, 96]. The major problem with the usual consensus presentations (like CUNGNGGUUCCU-GUGGU for the TCR of *Pospiviroidae* members) is the lack of information on the degree of conservation of individual nucleotides or basepairs. These disadvantages are circumvented by using sequence or structure logos that are based on information theory [9, 34, 86–88, 91]. Assume one has assembled an alignment of (many) sequences, what will be the base b in a certain column C_i of the alignment if one would determine the next sequence? The answer is given by Shannon's entropy

$$H(C_i) \quad = - \sum_{b \in [A,C,G,U]} f(C_i = b) \log_2 f(C_i = b)$$

with the base frequency f (or, more correctly, its probability). The unit of H is bit, nit, and digit if the logarithm is to base 2, e, and 10, respectively. If only a certain nucleotide appears in a column, the uncertainty about the nucleotide at the same position in the next, new sequence is $H_i = -1 \log_2 1 + 3 \cdot 0 \log_2 0 = 0$ bit; that is, one expects again this nucleotide. If all four nucleotides appear in a column with equal frequency, the uncertainty is maximal with $H_i = -4 \cdot \frac{1}{4} \log_2 \frac{1}{4} = 2$ bit.

If one would compare an alignment of randomly selected genomic regions, then the uncertainty about what base would be the next would be:

$$H_g \quad = - \sum_{b \in [A,C,G,U]} p_b \log_2 p_b$$

which will be close to 2 bit for a genome with random base composition ($p_b = \frac{1}{4}$). For viroids, this value is also close to 2 bit: $H_{\text{PSTVd}} = 1.98$ bit with $p_A = 0.20$, $p_C = 0.30$, $p_G = 0.28$, $p_U = 0.22$; $H_{\text{ASBVd}} = 1.95$ bit with $p_A = 0.28$, $p_C = 0.17$, $p_G = 0.21$, $p_U = 0.34$. Consequently, a pattern, as those mentioned above, decreases the uncertainty below that of randomly aligned fragments; that is, this decrease

$$R(C_i) \quad = H_g - H(C_i) \tag{2}$$

is simply the difference between entropy of genomic background and that of the pattern. For an example, see Sect. 3.3.2 and Figs. 5a and b.

So far, no dependence of different alignment columns is assumed: on the contrary, columns forming basepairs have to show dependence of their nucleotides. The joint entropy of columns C_i and C_j is described by

$$H(C_i, C_j) = - \sum_{b \in [A,C,G,U]} \sum_{b \in [A,C,G,U]} f(C_i = b, C_j = b) \log_2 f(C_i = b, C_j = b)$$

with the frequency $f(C_i = b, C_j = b)$ of the joint appearance of base b in column C_i and base b in column C_j. The expected mutual information I or rate of information transmission

$$I(C_i, C_j) = H(C_i) + H(C_j) - H(C_i, C_j) = \sum_{b \in [A,C,G,U]} \sum_{b \in [A,C,G,U]} f(C_i = b, C_j = b) \log_2 \frac{f(C_i = b, C_j = b)}{f(C_i = b) \cdot f(C_j = b)} \quad (3)$$

is defined as the entropy of the different columns decreased by the joint entropy of both columns. If columns C_i and C_j contain each a single type of nucleotide, then $I(C_i, C_j) = 0$; for example, if column C_i consists only of `G` and column C_j consists only of `C`, one gains no information on the dependence of both columns. The information is maximal with $I = 2$ bit if each column has a random base composition ($H(C_i) = H(C_j) = 2$ bit) but contains only pairs of `G:C`, `C:G`, `A:U`, and `U:A` ($H(C_i, C_j) = 2$ bit). Note, however, that $I = 2$ bit results also from two columns of random base composition but only pairs of `A:G`, `G:A`, `C:U`, and `U:C`; information theory has no concept of WC pairs and thus highlights also column pairs with non-WC pairs. For an example, see Sect. 3.3.3 and Fig. 5c.

The handling of gaps is a problem in the formulas shown above. One possibility is to handle them as a fifth "nucleotide." This might lead to negative "information" if sufficiently many gaps are present at a given position; such negative values, however, are not shown in the corresponding pictures.

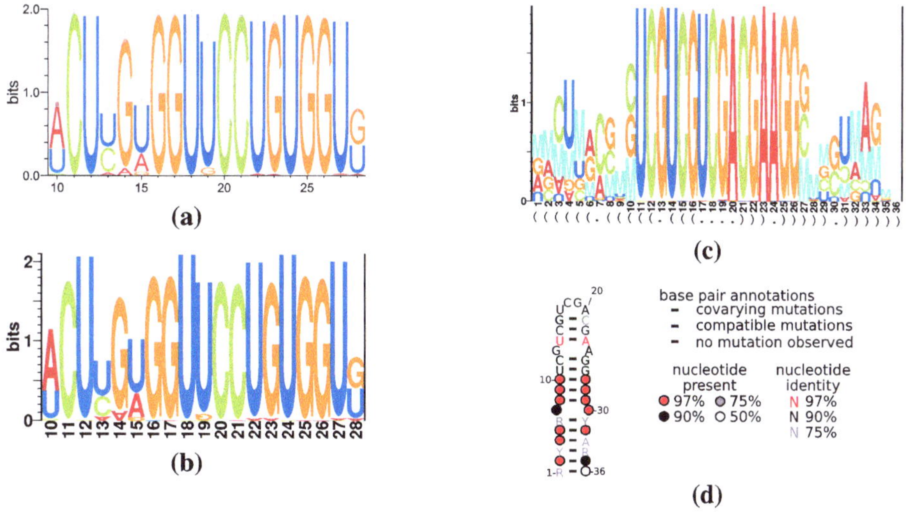

Fig. 5 Sequence and structure logos. (**a**), (**b**): Sequence logos of the terminal conserved region (pos 11–27) of 912 *pospiviroid* members created by WebLogo server (**a**) and RNA structure logo server (**b**), respectively. (**c**) Structure logo of hairpin I of 507 *Apscaviroid* members created by the RNA structure logo server [26]; the letters M denote the mutual information content of potentially basepaired positions, which are given in dot-bracket notation below the x-axis. (**d**) Secondary structure of hairpin I drawn by `R2R` from the same data set as the structure logo shown in (**c**); right: general legend to `R2R` drawings. For further details see text

3.3.2 Sequence Logo

With sequence logos as shown in Fig. 5a and b, each position of the alignment is represented by a column of stacked nucleotide characters; the total height of the column reflects the information content of this position (see (2)); the height of characters reflects their individual contribution to the position's information. Such logos are produced easily using one of the available web services (Table 1).

Figure 5a was produced with the WebLogo server [11]. It needs as input a multiple alignment file in FASTA, `ClustalW`, flat format, and others; of course all sequences have to be of identical length. For background composition $f_{GC} = 0.57$ was given; only the stand-alone program allows a user to give the full background distribution. The error bars indicate an approximate Bayesian 95% confidence interval.

Figure 5b was produced with the RNA structure logo server [26]. The input is an alignment in an unusual format:

```
1 > ...................
2 > ACUCGUGGUUCCUGUGGUU
3 > UCUUGAGGUUCCUGUGGUG
4 ...
```

All lines start with a '> '. Line 1 uses a string to denote a consensus structure (see Sect. 3.3.3); to produce a pure sequence logo, this line has to consist of dots of sequence length. Each further line contains a sequence. A full background distribution may be given. Here the nucleotide distribution of 913 full-length sequences of *Pospiviroidae* members was given ($p_A = 0.1983$, $p_C = 0.2947$, $p_G = 0.2795$, $p_U = 0.2275$). When a nucleotide appears less than expected, the symbol will be displayed upside-down (see positions 13–15, 19 in Fig. 5b).

The difference between the two logos (Fig. 5a and b) is quite obvious: the background correction in (b) leads to lower information by C and G (smaller height of C and G) because C and G are overrepresented in the genomic background. Otherwise both logos are quite similar. In comparison to the standard consensus representation by CUN$_{13}$GN$_{15}$GGUUCCUGUGGU, the logos show clearly that positions 13 and 15 are H (not G) and D (not C), respectively, and that both positions are as non-informative as positions 10 and 28 surrounding the TCR. Furthermore, the Uracils—as less common nucleotides in the genomic background—are more critical to the motif in comparison to the Guanines and Cytosines.

3.3.3 Structure Logo

Structure logos [26] are an extension of sequence logos (see previous section): they indicate mutual information (see Eq. (3)) on potentially basepaired positions in RNA in addition to the sequence logo.

The sequence plus structure logo (Fig. 5c) was produced with the RNA structure logo server [26] for hairpin I (HPI) of 507 *Apscaviroid* members with a random background distribution of nucleotides. The input is in an unusual format as mentioned in the previous section:

```
1 > ((((((.(((((.(((....))).))))).))))))
2 > UCCUGGCGACUCGUCGUCGACGAAGGGUCCUCAGGA
3 > AGCUGACUGGUCGUCGUCGACGAAGGCUCGUCAGCU
4 ...
```

Line 1 indicates the basepaired positions in dot-bracket notation, for which the mutual information content is shown by characters 'M' in Fig. 5c. Basepaired positions can be given with any bracket ("()", "[]", "{}", "<>") or letter pairs; this allows a user to give, for example, pseudoknotted structures.

The central part of the HPI sequence is highly conserved: entropies at positions 11–26 are close to 2 bit. The same is seen in the secondary structure depicted in Fig. 5d: identities of nucleotides at positions 11–26 are ≥90%. Consequently the mutual information content of basepairs in this range (like U_{11} : G_{26}) is close to 0 bit. In contrast, basepairs between positions 2–10 and 27–35 have nucleotide identities below 90%, and, despite a lot of gaps, their mutual information content is clearly obvious.

3.3.1 `RNAplot`

`RNAplot` belongs to the ViennaRNA package [55] and is internally used for example by `RNAfold` to produce the standard graphics output (Fig. 1b), but it is also available as a standalone program. Figure 6 shows structure calculation for the terminal left region of a PSTVd by `RNAfold` and two versions of graphics output by `RNAplot`.

```
$ more PSTVd_TL.vie
> PSTVd_TL
GGGUGUUUAGCCCUUGGAACCGCAGUUGGUUCCUCGGAACUAAACUCGUGGUUCCUGUGGUUCACACCU
$ RNAfold < PSTVd_TL.vie > PSTVd_TL.log
$ more PSTVd_TL.log
> PSTVd_TL
GGGUGUUUAGCCCUUGGAACCGCAGUUGGUUCCUCGGAACUAAACUCGUGGUUCCUGUGGUUCACACCU
((((((..((((...(((((((((((((((((....))))).)))).))))))))...)))).)))))) (-34.50)
$ RNAplot < PSTVd_TL.log
$ mv PSTVd_TL_ss.ps PSTVd_TL_ss_noOpt.eps
$ rotate_ss.pl -a -90 -m PSTVd_TL_ss_noOpt.eps > PSTVd_TL_ss_noOpt_m.eps

$ RNAplot --layout-type=0 < PSTVd_TL.log
$ mv PSTVd_TL_ss.ps PSTVd_TL_ss_t0.eps
$ rotate_ss.pl -m PSTVd_TL_ss_t0.eps > PSTVd_TL_ss_t0_m.eps
```

(a)

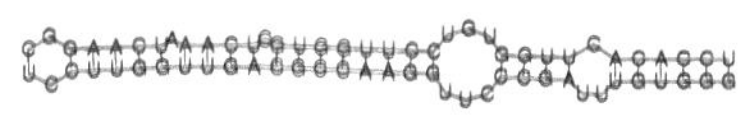

(b)

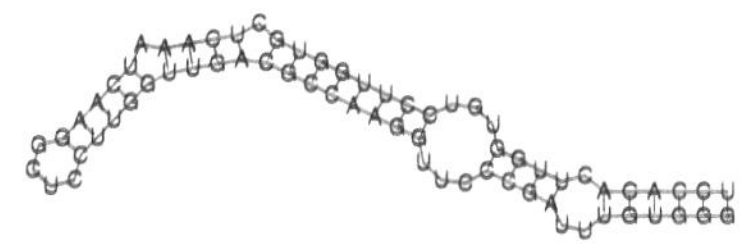

(c)

Fig. 6 `RNAplot`. (**a**): Commands to produce the structure representations shown in (**b**) and (**c**). (**b**) Standard output of `RNAplot`. (**c**) Output of `RNAplot` with option `--layout-type=0` or short `-t 0`

4 Predict the structure by RNAfold. The contents of the input file PSTVd_TL.vie is shown in lines 2 and 3; the output is redirected to PSTVd_TL.log.

6–8 The output file PSTVd_TL.log contains in addition to the input file the mfe structure in bracket dot notation and the $\Delta G^0_{37\,^\circ\mathrm{C}}$ in kcal/mol (line 8).

9 RNAplot produces a secondary structure drawing using PSTVd_TL.log as input; the additional energy data at the end of the structure line do not bother the program. The output drawing is in PSTVd_TL_ss.ps; this filename is created from the heading line (see line 6) of the input.

10 Rename the output file to PSTVd_TL_ss_noOpt.eps.

11 Rotate the drawing by an angle of -90° and mirror it;output is stored in PSTVd_TL_ss_noOpt_m.eps. This final drawing is shown in Fig. 6b.

12–14 RNAplot with the option --layout-type=0 (or -t 0) produces a secondary structure layout with constant spacing of backbone-consecutive nucleotides and basepairs; thus loop nucleotides are positioned along an imaginary circle (Fig. 6c).

3.3.5 VARNA

VARNA (Visualization Applet for RNA) [13] draws RNA secondary and some tertiary structures in several different layouts; these layouts include "radial" (similar to Fig. 6c), "naview" (named according to the program naview [8]; similar to Fig. 6b), "circular" (not shown; the sequence is drawn on a circle and basepairs are connected by lines inside of the circle), and "linear" (Fig. 7b). VARNA accepts several file format as input (e.g., Vienna dot-bracket notation (.dbn) and connect (.ct)). Output formats are, for example, EPS and SVG.

The example in Fig. 7 shows the use of VARNA to produce a "linear" representation of two alternative structures of the terminal left region of a PSTVd. The used command and options (Fig. 7a) are as follows:

1 Let java load the java archive file (the path has to be adjusted to your local installation).

2,3 Input of structure in dot-bracket notation; pseudoknots can be specified using the bracket types "{}" and "[]".

4,5 Input of sequence.

6–9 Highlight given sequence positions by color in hexadecimal notation (#006ab3, blue; #0e4067, dark blue).

10 Set the default color for basepairs.

11 Draw nucleotide letters.

12 Set the distance between letters.

```
1 $ java -cp ~/VARNAv3-93-src.jar fr.orsay.lri.varna.applications.VARNAcmd \
2   -structureDBN \
3    "((((((..((((...((((((((((((((((((..)))))).)))).))))))))...)))).))))))" \
4   -sequenceDBN \
5    "GGGUGUUUAGCCCUUGGAACCGCAGUUGGUUCCUCGGAACUAAACUCGUGGUUCCUGUGGUUCACACCU" \
6   -highlightRegion "1-6:fill=#006ab3,outline=#0e4067;\
7    64-69:fill=#006ab3,outline=#0e4067;\
8     9-12:fill=#006ab3,outline=#0e4067;\
9    59-62:fill=#006ab3,outline=#0e4067;" \
10  -bp "#006ab3" \
11  -drawBases True \
12  -spaceBetweenBases 1.0 \
13  -bpStyle line \
14  -algorithm line \
15  -o PSTVd_TL_woC_cons.eps
16 $ java -cp ~/VARNAv3-93-src.jar fr.orsay.lri.varna.applications.VARNAcmd \
17  -structureDBN \
18   "((((((..((((...((((((......))))))..(((((((......)))))))...)))).))))))" \
19   ... # as lines 4--11
20  -o PSTVd_TL_wC_cons.eps
```

(a)

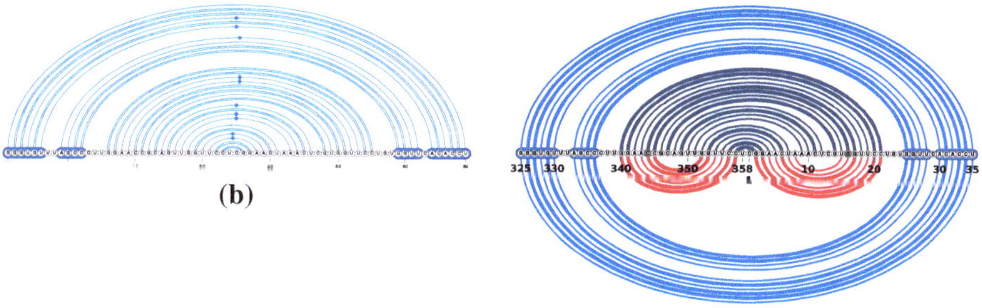

(b)

(c)

Fig. 7 VARNA. Sequence and mfe structure are for the terminal left region of a PSTVd as in Fig. 6a, lines 6–8. (**a**) Lines 1–15 show the command to produce the structure representation shown in (**b**). (**b**) Graphics output from the command shown in (**a**), lines 1–15. (**c**) The command in (**a**), lines 16–20 was used to draw a bifurcated structure of the same sequence; this second drawing was combined with the drawing in (**b**) in a graphics program (inkscape); the numbering was adjusted, and colors of lines connecting basepairs were modified to highlight the basepairs differing in both structures

13 Connect basepairs by lines.

14 Draw the sequence in a linear layout.

15 Define output file; the format is guessed from the file extension.

For further options and details we refer to the internet [73]. In addition VARNA has a graphical user interface called via java--jar ˜/VARNAv3-93-src.jar.

3.3.6 R2R

R2R is a program that allows a user to draw consensus secondary structures from an alignment in Stockholm format, but is also able to produce structures for a single sequence. The program lacks a graphical user interface. Instead it relies on a Stockholm-formatted file with added R2R commands. A simple example of such a file (with name TL_PTVA.sto) is shown in Fig. 8a:

```
1  # STOCKHOLM 1.0

2  PTVA                        GGGUGUUUAGCCCUUGGAACCGCAGUUGGUUCCUCGGAACUAAACUCGUGGUUCCUGUGGUUCACACCU
3  #=GC SS_cons                ((((((..((((...((((((((((((((((((....))))))).))).))))))))...)))).))))))
4  #=GC R2R_XLABEL_outline     .............................................TTTTTTTTTTTTTTTTT.......
5  #=GC R2R_XLABEL_labels      ....................................................T................

6  #=GF R2R set_dir pos0 180

7  #=GF R2R outline_nuc outline:T
8  #=GF R2R tick_label labels:T TCR
9  #=GF R2R nuc_color all rgb:0,0,0
10 #=GF R2R nobpannot

11 #=GF R2R_consensus tick_label_regular_numbering 0 10 firstNucNum 325
12 #=GF Makefile skeleton
13 #=GF Makefile skeleton-with-pairbonds
14 //
```

(a)

```
1  $ mkdir test
2  $ cd test
3  $ cp ~/somewhere/R2R-1.0.5/demo/MetamakeDemos.pl ./r2r_make

4  # my $r2r="../src/r2r";
5  # my $subfam="../src/SelectSubFamilyFromStockholm.pl";
6  my $r2r="~/somewhere/R2R-1.0.5/src/r2r";
7  my $subfam="~/somewhere/R2R-1.0.5/src/SelectSubFamilyFromStockholm.pl";

8  $ perl r2r_make --disableR2RUsageWarning
9  Running find . -name "*.sto" -print
10         done (found 1 files)
11 Apparently this is your first run of this script in this directory;
12 making 'intermediate' subdirectory
13 found 1 motifs

14 $ make all-pdf
15 $ make all-svg
```

(b)

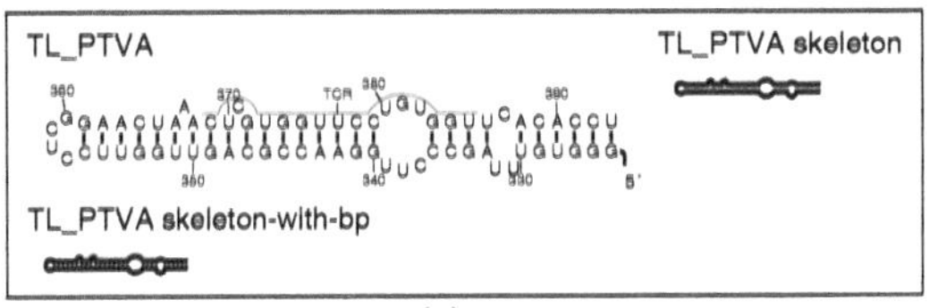

(c)

Fig. 8 A simple drawing by R2R. (**a**) A Stockholm file `TL_PTVA.sto` of a single sequence supplemented with R2R commands. (**b**) Commands to modify (lines 3–7) a script from the R2R installation, which is then used to write a `Makefile`. Execution of this file by `make` (lines 14, 15) produces output in pdf or svg format shown in (**c**)

1,14 Begin and end lines of a Stockholm file

2 Sequence name plus sequence

3 Structure line

4,5,7,8 Definition and use of labels: a line starting with `#=GC R2R_XLABEL_`*name* defines *name*, which is `outline` and `labels`, respectively, in the example. The following character string, which has the length of the sequence or the alignment, consists of characters naming columns and dots otherwise. These label lines can be used with different commands:

`#=GF R2R_outline_nuc` *`name:x`* outlines nucleotides of columns that are marked in the line `#=GC R2R_XLABEL_`*`name`* by *x*; `#=GF R2R_tick_label` *`name:x text`* annotates the nucleotide in the column(s) marked by *x* with a tick mark and a label with *text*.

6 Sets the drawing orientation of the first, 5' nucleotide: 0° is default, resulting in the second nucleotide drawn to the right of the first nucleotide; with 180° the second nucleotide is drawn to the left of the first nucleotide.

9 Set font color of all nucleotides to black; `R2R` usually colors nucleotide letters according to percentage of nucleotide identity in a multiple alignment, which would result here with only a single sequence in red letters (see legend in Fig. 5d).

10 No basepair annotation by background shading; `R2R` usually colors basepairs according to their covariation in a multiple sequence file, which would result here with only a single sequence in a red background shading (see legend in Fig. 5d).

11 Command used to label nucleotides with their positions. The syntax is

`tick_label_regular_numbering` *`start skip`* `[firstNucNum]` *num*

meaning that nucleotide position *`start`* as well as *`start`*+ *`skip`*, *`start`*+ 2 ·*`skip`*, etc. are labeled; the 5' nucleotide is not labeled but automatically labeled with 5'; if `firstNucNum` is given, then the first nucleotide gets the number *num*.

12, 13 Produce a structure drawing of only the backbone without and with lines denoting basepairs, respectively.

Execution of `R2R` is simple using a provided `perl` script that has to be slightly modified as outlined in Fig. 8b:

1 Create a new directory, for example, `test`

2 Change to the new directory

3 Copy and rename the file `MetamakeDemos.pl` to `r2r_make`

Modify lines 8 and 9 of the file `r2r_make` to fit to your local installation of `R2R`; that is, comment the original lines by a # (see 4 and 5) and add two appropriate new lines (see 6 and 7). This modification has to be done only once; the script is then usable in any directory.

Create a STOCKHOLM file with content as shown in Fig. 8a; the extension has to be `.sto`.

8 Execute the script; the option `--disableR2RUsageWarning` (or shorter `disabler`) disables inclusion of the following remark into the graphics output that is worth to be read:

`R2R is not intended to evaluate evidence for covariation or RNA structure where this is in question. It is not appropriate to use R2Rs covariation markings to declare that there is evidence of structural conservation within an alignment. R2R is a drawing program. As the original paper, Weinberg and Breaker, 2011 wrote:`

`"This automated R2R annotation [of covariation] does not reflect the extent or confidence of covariation. While such information can be useful, we believe that thorough evaluation of covariation evidence ultimately requires analysis of the full sequence alignment. For example, misleading covariation can result from an incorrect alignment of sequences, or from alignments of sequences that do not function as structured RNAs. Unfortunately, there is no accepted method to assign confidence that entirely eliminates the need to analyze the full alignment."`

The script scans the current directory (and all subdirectories) for presence of files with extension `.sto`, writes a `Makefile`, creates the subdirectories `intermediate`, `output-pdf` and `output-svg`, and writes some remarks (lines 9–13) to the terminal.

14, 15 The program `make` with option `all-pdf` and `all-svg`, respectively, repeatedly executes `R2R` with all necessary options; the final output files are stored in `output-pdf/` and `output-svg/`, respectively. Files in `intermediate/` are superfluous.

In Fig. 9 is demonstrated that `R2R` is able to produce different structural views from a single alignment:

1,59 Begin and end lines of a Stockholm file.

2–4 The alignment part, of which only two lines are shown, consists of the terminal left domain of 274 PSTVd sequences.

5 The "standard" consensus structure; this line is referred to by the label "primary." This consensus structure was obtained with the help of `ConStruct` without using constraints for the `RNAfold` prediction.

6 An alternative consensus structure; its name is `_1`. This consensus structure was obtained with the help of `ConStruct` and `RNAfold` with forced basepairs marked by 'F' in the structures at the bottom of Fig. 9b.

7,8 These lines are used to outline and to label certain nucleotides; see lines 10–19.

9 Define drawing direction.

```
# STOCKHOLM 1.0
PTVA                              GGGUGUUUAGCCCUUGGAACC.GC.AGUUGGUUCC.UCGGAACUAAACUCGUGGUUCCUGUGGUUCACACCU
AB623143                          GGGUGUUUAGCCCUUGGAACC.GC.AGUUGGUUCC.UCGGAACUAAACUCGUGGUUCCUGUGGUUCACACCU
...
#=GC SS_cons                      ((((((..((((...((((((.((.(((((((((.....)))))).))).))))))))...)))).))))))
#=GC SS_cons_1                    ((((((..((((...((((((........))))))...(((((((......)))))))...)))).))))))
#=GC R2R_XLABEL_Region            ...............................................TTTTTTTTTTTTTTTTT........
#=GC R2R_XLABEL_RegionLabel       .....1.........2...f.......3..f.....45....f...6......f..7...T.....8.....
#=GF R2R set_dir pos0 180
#=GF R2R outline_nuc Region:T
#=GF R2R tick_label RegionLabel:T TCR
#=GF R2R tick_label RegionLabel:1 330
#=GF R2R tick_label RegionLabel:2 340
#=GF R2R tick_label RegionLabel:3 350
#=GF R2R tick_label RegionLabel:4 359
#=GF R2R tick_label RegionLabel:5 1
#=GF R2R tick_label RegionLabel:6 10
#=GF R2R tick_label RegionLabel:7 20
#=GF R2R tick_label RegionLabel:8 30
#=GF R2R SetDrawingParam nucTickLabel_fontSize 12
#=GF R2R SetDrawingParam nucTickLabel_tickLen 0.2
#=GF R2R SetDrawingParam outlineNucPenWidth 0.03
#=GF R2R SetDrawingParam outlineNucExtraRadius 0.1
#=GF R2R SetDrawingParam indicateOneseqWobblesAndNonCanonicals false

#=GF SUBFAM_PERL_PRED BIF return 1
#=GF SUBFAM_BIF_R2R subst_ss _1 primary
#=GF SUBFAM_BIF_R2R set_dir pos0 180
#=GF SUBFAM_BIF_R2R SetDrawingParam nucTickLabel_fontSize 12
#=GF SUBFAM_BIF_R2R SetDrawingParam nucTickLabel_tickLen 0.2
#=GF SUBFAM_BIF_R2R SetDrawingParam outlineNucPenWidth 0.03
#=GF SUBFAM_BIF_R2R SetDrawingParam outlineNucExtraRadius 0.1
#=GF SUBFAM_BIF_R2R outline_nuc Region:T
#=GF SUBFAM_BIF_R2R tick_label RegionLabel:T TCR
#=GF SUBFAM_BIF_R2R tick_label RegionLabel:f F
#=GF SUBFAM_BIF_R2R tick_label RegionLabel:1 330
...
#=GF SUBFAM_BIF_R2R tick_label RegionLabel:8 30

#=GF SUBFAM_PERL_PRED BIFone return ($hitId eq "AB623143")
#=GF SUBFAM_BIFone_R2R subst_ss _1 primary
#=GF SUBFAM_BIFone_R2R set_dir pos0 180
#=GF SUBFAM_BIFone_R2R outline_nuc Region:T
#=GF SUBFAM_BIFone_R2R tick_label RegionLabel:f F
#=GF SUBFAM_BIFone_R2R SetDrawingParam nucTickLabel_fontSize 12
#=GF SUBFAM_BIFone_R2R SetDrawingParam nucTickLabel_tickLen 0.2
#=GF SUBFAM_BIFone_R2R SetDrawingParam outlineNucPenWidth 0.03
#=GF SUBFAM_BIFone_R2R SetDrawingParam outlineNucExtraRadius 0.1
#=GF SUBFAM_BIFone_R2R SetDrawingParam nucShrinkWithCircleNuc 0.7
#=GF SUBFAM_BIFone_R2R nuc_color all rgb:0,0,0
#=GF SUBFAM_BIFone_R2R nobpannot
#=GF CC AB623143           GGGUGUUUAGCCCUUGGAACC.GC.AGUUGGUUCC.UCGGAACUAAACUCGUGGUUCCUGUGGUUCACACCU
#=GC R2R_XLABEL_mapBzCN    =====---=....--..................=.---....----.=-......=..=====-=.....=
#=GF SUBFAM_BIFone_R2R circle_nuc mapBzCN:= rgb:251,233,60
#=GF SUBFAM_BIFone_R2R circle_nuc mapBzCN:- rgb:240,99,93

#=GF CC AB623143           GGGUGUUUAGCCCUUGGAACC.GC.AGUUGGUUCC.UCGGAACUAAACUCGUGGUUCCUGUGGUUCACACCU
#=GR AB623143 CLEAVAGE:BzCN =====---=....--..................=.---....----.=-......=..=====-=.....=
#=GR AB623143 DEL_COLS:BzCN ....................-..-..........-...................................
#=GF R2R_oneseq AB623143:BzCN tick_label_disable_default_numbering
#=GF R2R_oneseq AB623143:BzCN SetDrawingParam nucShrinkWithCircleNuc 0.7
//
```

(a)

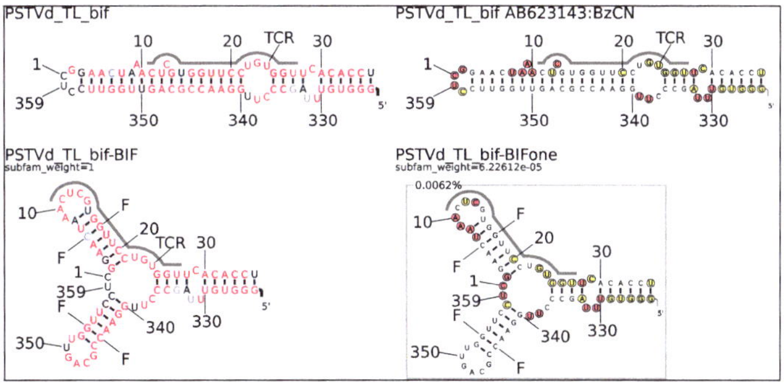

(b)

Fig. 9 Multiple structures created by `R2R` from one Stockholm file. (**a**) The STOCKHOLM file `PSTVd_TL_bif.sto` used to create the output shown in (**b**)

10 The nucleotides marked by 'T' in line 7 are marked by a line on the outside of the structure; this marks the TCR.

11 The nucleotide marked by 'T' in line 8 is marked by 'TCR'.

12–19 The sequence positions are labeled by the consensus positions of PSTVd. The standard numbering by `R2R` (e.g., with

`tick_label_regular_numbering` *start skip*

see Fig. 8a line 11) would only allow for a continuous numbering.

20–23 These lines demonstrate some options used to change font size, length of ticks, width of outlines, and distance of outlines from the nucleotide characters, respectively.

24 By default, WC and wobble pairs are indicated by a line and a dot, respectively, as bond symbols. If set to false, `R2R` uses the same WC symbol for all pairings.

The lines 1–24 (without line 6) are sufficient to produce the top left structure of Fig. 9b.

25–37 These lines define a "subfamily," the structure of which is drawn separately (Fig. 9b bottom left).

25 This line with syntax

`SUBFAM_PERL_PRED` *predicate-name perl-code*

is a call to `SelectSubFamilyFromStockholm.pl` (see Fig. 8b line 7) and defines a name, which here is `BIF`, and the `perl` statement `return 1`, which is evaluated for each sequence. The '`1`' means simply `true`; that is all sequences will be included in the 'family' named `BIF`. The following lines 26–37 are only valid for this family due to the mentioning of `BIF` in the statements.

26 The primary consensus structure (line 5) is substituted by the alternative structure `_1` (line 6) for the family `BIF`.

27–37 These lines repeat the statements of lines 9–23 for family `BIF`. Line 34 is used to mark the basepairs forced for the `RNAfold` prediction.

The next part (lines 38–53) is used to draw a single sequence (GenBank AC AB623143) in the suboptimal structure annotated with SHAPE data [25] (Fig. 9b right bottom). Note that the SHAPE data was obtained with the full-length sequence.

38 Define the name `BIFone` and the `perl` statement`return ($hitId eq "AB623143")`. The variable `$hitId` is defined in `SelectSubFamilyFromStockholm.pl`. Only if the content of `$hitId` is equal to the string AB623143, the statement returns true; that is, the family consists only of the

single sequence with AC AB623143. The following lines 39–53 are only valid for this family due to the mentioning of `BIFone` in the statements.

47 Shrink the font size of the nucleotide characters by a factor of 0.7 to allow for their circling.

48 Draw the nucleotides in black instead of the usual color annotation.

49 Avoid the usual background shading of basepairs.

50 A comment line showing the sequence of AB623143 for comparison to the SHAPE data.

51 This label line represents the SHAPE data using the symbols '-', '=', and '.' for nucleotides with high, intermediate, and low reactivity, respectively [25].

52,53 Mark the respective nucleotides (see line 51) by background colors red and yellow. The colors are given as RGB values.

The last block of instructions (lines 54–58) annotates the single sequence drawing of AB623143 (`oneseq` mode) in the optimal consensus structure with SHAPE data [25] (Fig. 9a top right).

55 Define cleavage data for the sequence AB623143. The syntax is `#=GR` *`ID`* `CLEAVAGE:` *`label SHAPE symbols`*
The pre-defined SHAPE symbols and the associated background colors are = (yellow), - (red), + (green), and ? (gray).

56 `R2R` has to know the positions that are not present in the selected sequence but in the alignment; these positions are marked by '-'.

57 Disable the standard numbering while drawing the structure of AB623143.

Here the SHAPE data fit quite well. All nucleotides of the hairpin loop (nts 358, 359, 1, 2) are flexible, while one of the hairpin loops (nts 346–351) of the bifurcated structure would be inflexible. The A nucleotides at position 8–10 are flexible and not only A_9, which is drawn as bulged, because any of these A nucleotides can be paired without any energy difference between the different structures; note that the same nucleotides have intermediate entropy and probability of pairing in Fig. 2e and f.

References

1. Andronescu M, Condon A, Hoos H, Mathews D, Murphy K (2007) Efficient parameter estimation for RNA secondary structure prediction. Bioinformatics 23: i19–28. https://doi.org/10.1093/bioinformatics/btm223
2. Ayad L, Pissis S (2017) MARS: improving multiple circular sequence alignment using refined sequences. BMC Genom. 18:86. https://doi.org/10.1186/s12864-016-3477-5

3. Benson D, Cavanaugh M, Clark K, Karsch-Mizrachi I, Lipman D, Ostell J, Sayers E (2012) GenBank. Nucleic Acids Res 41: D36–42. https://doi.org/10.1093/nar/gks1195
4. Berman H, Westbrook J, Feng Z, Gilliland G, Bhat T, Weissig H, Shindyalov I, Bourne P (2000) The protein data bank. Nucleic Acids Res 28:235–242. https://doi.org/10.1093/nar/28.1.235
5. Bernhart S, Hofacker I, Will S, Gruber A, Stadler P (2008) RNAalifold: improved consensus structure prediction for RNA alignments. BMC Bioinf. 9:474. https://doi.org/10.1186/1471-2105-9-474
6. Berry K (2019) epstopdf – Convert EPS to PDF using Ghostscript. https://ctan.org/pkg/epstopdf
7. Branch A, Benenfeld B, Robertson H (1985) Ultraviolet light-induced crosslinking reveals a unique region of local tertiary structure in potato spindle tuber viroid and HeLa 5 S RNA. Proc Natl Acad Sci U S A 82:6590–6594. http://www.pnas.org/content/82/19/6590.full.pdf+html?sid=bd413635-93ac-4ab7-b4a6-a55a78ae6f34
8. Bruccoleri R, Heinrich G (1988) An improved algorithm for nucleic acid secondary structure display. Comp Appl Biosci 4:167–173. https://doi.org/10.1093/bioinformatics/4.1.167
9. Chiu D, Kolodziejczak T (1991) Inferring consensus structure from nucleic acid sequences. Comp Appl Biosci 7:347–352. https://doi.org/doi:10.1093/bioinformatics/7.3.347
10. Chojnacki S, Cowley A, Lee J, Foix A, Lopez R (2017) Programmatic access to bioinformatics tools from EMBL-EBI update: 2017. Nucleic Acids Res 45:W550–W553. https://doi.org/10.1093/nar/gkx273
11. Crooks G, Hon G, Chandonia J, Brenner S (2004) WebLogo: a sequence logo generator. Genome Res 14:1188–1190. https://doi.org/10.1101/gr.849004
12. Dalli D, Wilm A, Mainz I, Steger G (2006) STRAL: progressive alignment of non-coding RNA using base pairing probability vectors in quadratic time. Bioinformatics 22:1593–1599. https://doi.org/10.1093/bioinformatics/btl142
13. Darty K, Denise A, Ponty Y (2009) VARNA: interactive drawing and editing of the RNA secondary structure. Bioinformatics 25:1974–1975. https://doi.org/10.1093/bioinformatics/btp250
14. Ding Y, Chan C, Lawrence C (2005) RNA secondary structure prediction by centroids in a Boltzmann weighted ensemble. RNA 11:1157–1166. https://doi.org/10.1261/rna.2500605
15. Di Serio F, Flores R, Verhoeven J, Li S, Pallás V, Randles J, Sano T, Vidalakis G, Owens R (2014) Current status of viroid taxonomy. Arch Virol 159:3467–3478. https://doi.org/10.1007/s00705-014-2200-6
16. Do C, Woods D, Batzoglou S (2006) CONTRAfold: RNA secondary structure prediction without physics-based models. Bioinformatics 22:e90–e98. https://doi.org/10.1093/bioinformatics/btl246
17. Eddy S (2002) SQUID – C function library for sequence analysis. https://eddylab.org/software.html
18. Flamm C, Fontana W, Hofacker I, Schuster P (2000) RNA folding at elementary step resolution. RNA 6:325–338. https://doi.org/10.1017/S1355838200992161
19. Flamm C, Hofacker I, Stadler P, Wolfinger M (2008) Barrier trees of degenerate landscapes. Z Phys Chem 216:155. https://doi.org/10.1524/zpch.2002.216.2.155
20. Forsdyke D (2007) Calculation of folding energies of single-stranded nucleic acid sequences: conceptual issues. J Theor Biol 248:745–753. https://doi.org/10.1016/j.jtbi.2007.07.008
21. Freidhoff P, Bruist M (2019) In Silico Survey of the Central Conserved Regions in Viroids of the Pospiviroidae Family for Conserved Asymmetric Loop Structures. RNA. https://doi.org/10.1261/rna.070409.119
22. Gardner P, Giegerich R (2004) A comprehensive comparison of comparative RNA structure prediction approaches. BMC Bioinf 5:140. https://doi.org/10.1186/1471-2105-5-140
23. Giegerich R, Voss B, Rehmsmeier M (2004) Abstract shapes of RNA. Nucleic Acids Res 32:4843–4851. https://doi.org/10.1093/nar/gkh779
24. Giguère T, Perreault J (2017) Classification of the *Pospiviroidae* based on their structural hallmarks. PLoS ONE 12:e0182536. https://doi.org/10.1371/journal.pone.0182536
25. Giguère T, Adkar-Purushothama C, Perreault J (2014) Comprehensive secondary structure elucidation of four genera of the family *Pospiviroidae*. PLoS ONE 9:e98655. https://doi.org/10.1371/journal.pone.0098655
26. Gorodkin J, Heyer L, Brunak S, Stormo G (1997) Displaying the information contents

of structural RNA alignments: the structure logos. Comput Appl Biosci 13:583–586. https://doi.org/10.1093/bioinformatics/13.6.583
27. Gorodkin J, Heyer L, Stormo G (1997) Finding the most significant common sequence and structure motifs in a set of RNA sequences. Nucleic Acids Res 25:3724–3732. https://doi.org/10.1093/nar/25.18.3724
28. Gozmanova M, Denti M, Minkov I, Tsagris M, Tabler M (2003) Characterization of the RNA motif responsible for the specific interaction of potato spindle tuber viroid RNA (PSTVd) and the tomato protein Virp1. Nucleic Acids Res 31:5534–5543. https://doi.org/10.1093/nar/gkg777
29. Griffiths-Jones S (2005) RALEE–RNA ALignment editor in Emacs. Bioinformatics 21:257–259. https://doi.org/10.1093/bioinformatics/bth489
30. Gross H, Domdey H, Lossow C, Jank P, Raba M, Alberty H, Sänger H (1978) Nucleotide sequence and secondary structure of potato spindle tuber viroid. Nature 273:203–208. https://doi.org/10.1038/273203a0
31. Gruber A, Bernhart S, Hofacker I, Washietl S (2008) Strategies for measuring evolutionary conservation of RNA secondary structures. BMC Bioinf 9:122. https://doi.org/10.1186/1471-2105-9-122
32. Gruber A, Lorenz R, Bernhart S, Neuböck R, Hofacker I (2008) The Vienna RNA websuite. Nucleic Acids Res 36:W70–W74. https://doi.org/10.1093/nar/gkn188
33. Gutell R (2014) Ten lessons with Carl Woese about RNA and comparative analysis. RNA Biol 11:254–272. https://doi.org/10.4161/rna.28718
34. Gutell R, Power A, Hertz G, Putz E, Stormo G (1992) Identifying constraints on the higher-order structure of RNA: continued development and application of comparative sequence analysis methods. Nucleic Acids Res 20:5785–5795. https://doi.org/10.1093/nar/20.21.5785
35. Hamada M, Kiryu H, Sato K, Mituyama T, Asai K (2009) Prediction of RNA secondary structure using generalized centroid estimators. Bioinformatics 25:465–473. https://doi.org/10.1093/bioinformatics/btn601
36. Harmanci A, Sharma G, Mathews D (2007) Efficient pairwise RNA structure prediction using probabilistic alignment constraints in Dynalign. BMC Bioinf 8:130. http://www.biomedcentral.com/1471-2105/8/130
37. Henco K, Sänger H, Riesner D (1979) Fine structure melting of viroids as studied by kinetic methods. Nucleic Acids Res 6:3041–3059. https://doi.org/10.1093/nar/6.9.3041
38. Hofacker I, Fontana W, Stadler P, Bonhoeffer S, Tacker M, Schuster P (1994) Fast folding and comparison of RNA structures. Monatsh Chem 125:167–188. https://doi.org/10.1007/BF00818163
39. Hofacker I, Schuster P, Stadler P (1998) Combinatorics of RNA secondary structures. Discrete Appl Math 88:207237. https://doi.org/10.1016/S0166-218X(98)00073-0
40. Hofacker I, Fekete M, Stadler P (2002) Secondary structure prediction for aligned RNA sequences. J Mol Biol 319(5):1059–1066. https://doi.org/10.1016/S0022-2836(02)00308-X
41. Hogeweg P, Hesper B (1984) Energy directed folding of RNA sequences. Nucleic Acids Res 12:67–74. https://doi.org/10.1093/nar/12.1Part1.67
42. Janssen S, Giegerich R (2014) Abstract shape analysis of RNA. Methods Mol Biol 1097:215–245. https://doi.org/10.1007/978-1-62703-709-9_11
43. Janssen S, Giegerich R (2014) The RNA shapes studio. Bioinformatics 31:423–425. https://doi.org/10.1093/bioinformatics/btu649
44. Katoh K, Toh H (2008) Improved accuracy of multiple ncRNA alignment by incorporating structural information into a MAFFT-based framework. BMC Bioinf. 9:212. https://doi.org/10.1186/1471-2105-9-212
45. Katoh K, Rozewicki J, Yamada K (2017) MAFFT online service: multiple sequence alignment, interactive sequence choice and visualization. Brief Bioinf. https://doi.org/10.1093/bib/bbx108
46. Klein R, Eddy S (2003) RSEARCH: finding homologs of single structured RNA sequences. BMC Bioinf. 4:44. https://doi.org/10.1186/1471-2105-4-44
47. Koltunow A, Rezaian M (1988) Grapevine yellow speckle viroid: structural features of a new viroid group. Nucleic Acids Res 16:849–864. https://doi.org/10.1093/nar/16.3.849
48. Langowski J, Henco K, Riesner D, Sänger H (1978) Common structural features of different viroids: serial arrangement of double helical sections and internal loops. Nucleic Acids Res 5:1589–1610. https://doi.org/10.1093/nar/5.5.1589

49. Larkin M, Blackshields G, Brown N, Chenna R, McGettigan P, McWilliam H, Valentin F, Wallace I, Wilm A, Lopez R, Thompson J, Gibson T, Higgins D (2007) Clustal W and Clustal X version 2.0. Bioinformatics 23:2947–2948. https://doi.org/10.1093/bioinformatics/btm404
50. Leonard C, Hajdin C, Karabiber F, Mathews D, Favorov O, Dokholyan N, Weeks K (2013) Principles for understanding the accuracy of SHAPE-directed RNA structure modeling. Biochemistry 52:588–595. https://doi.org/10.1021/bi300755u
51. Leontis N, Stombaugh J, Westhof E (2002) The non-Watson-Crick base pairs and their associated isostericity matrices. Nucleic Acids Res 30:3497–3531. https://doi.org/10.1093/nar/gkf481
52. Lescoute A, Leontis N, Massire C, Westhof E (2005) Recurrent structural RNA motifs, isostericity matrices and sequence alignments. Nucleic Acids Res 33:2395–2409. http://nar.oxfordjournals.org/cgi/content/full/33/8/2395
53. López-Carrasco A, Flores R (2017) Dissecting the secondary structure of the circular RNA of a nuclear viroid *in vivo*: a "naked" rod-like conformation similar but not identical to that observed *in vitro*. RNA Biol 14:1046–1054. https://doi.org/10.1080/15476286.2016.1223005
54. López-Carrasco A, Flores R (2017) The predominant circular form of avocado sunblotch viroid accumulates in planta as a free RNA adopting a rod-shaped secondary structure unprotected by tightly bound host proteins. J Gen Virol 98:1913–1922. https://doi.org/10.1099/jgv.0.000846
55. Lorenz R, Bernhart S, Höner Zu Siederdissen C, Tafer H, Flamm C, Stadler P, Hofacker I (2011) ViennaRNA Package 2.0. Algorithms Mol Biol 6:26. https://doi.org/10.1186/1748-7188-6-26
56. Lorenz R, Hofacker I, Stadler P (2016) RNA folding with hard and soft constraints. Algorithms Mol Biol 11:8. https://doi.org/10.1186/s13015-016-0070-z
57. Loss P, Schmitz M, Steger G, Riesner D (1991) Formation of a thermodynamically metastable structure containing hairpin II is critical for infectivity of potato spindle tuber viroid RNA. EMBO J 10:719–727. https://doi.org/10.1002/j.1460-2075.1991.tb08002.x
58. Lu X, Olson W (2008) 3DNA: a versatile, integrated software system for the analysis, rebuilding and visualization of three-dimensional nucleic-acid structures. Nat Protoc 3:1213–1227. https://doi.org/10.1038/nprot.2008.104
59. Lu Z, Gloor J, Mathews D (2009) Improved RNA secondary structure prediction by maximizing expected pair accuracy. RNA 15:1805–1813. https://doi.org/10.1261/rna.1643609
60. Lück R, Gräf S, Steger G (1999) *ConStruct*: a tool for thermodynamic controlled prediction of conserved secondary structure. Nucleic Acids Res 27:4208–4217. https://doi.org/10.1093/nar/27.21.4208
61. Markham N, Zuker M (2005) DINAMelt web server for nucleic acid melting prediction. Nucleic Acids Res 33:W577–W581. https://doi.org/10.1093/nar/gki591
62. Markham N, Zuker M (2008) UNAFold: software for nucleic acid folding and hybridization. Methods Mol Biol 453:3–31. https://doi.org/10.1007/978-1-60327-429-6_1
63. Mathews D (2014) RNA secondary structure analysis using RNAstructure. Curr Protoc Bioinf. 46:12.6.1–25. https://doi.org/10.1002/0471250953.bi1206s46
64. Mathews D, Turner D (2002) Dynalign: an algorithm for finding the secondary structure common to two RNA sequences. J Mol Biol 317:191–203. https://doi.org/10.1006/jmbi.2001.5351
65. Mathews D, Sabina J, Zuker M, Turner D (1999) Expanded sequence dependence of thermodynamic parameters improves prediction of RNA secondary structure. J Mol Biol 288:911–940. https://doi.org/10.1006/jmbi.1999.2700
66. Mathews D, Disney M, Childs J, Schroeder S, Zuker M, Turner D (2004) Incorporating chemical modification constraints into a dynamic programming algorithm for prediction of RNA secondary structure. Proc Natl Acad Sci U S A 101:7287–7292. https://doi.org/10.1073/pnas.0401799101
67. Mathews D, Turner D, Watson R (2016) RNA secondary structure prediction. Curr Protoc Nucleic Acid Chem 67:11.2.1–11.2.19. https://doi.org/10.1002/cpnc.19
68. McCaskill J (1990) The equilibrium partition function and base pair binding probabilities for RNA secondary structure. Biopolymers 29:1105–1119. https://doi.org/10.1002/bip.360290621
69. Mosig A, Hofacker I, Stadler P (2006) Comparative analysis of cyclic sequences: viroids and other small circular RNAs. In: German conference on bioinformatics (GCB), pp

93–102. http://subs.emis.de/LNI/Proceedings/Proceedings83/article5487.html

70. Navarro B, Gisel A, Rodio M, Delgado S, Flores R, Di Serio F (2012) Small RNAs containing the pathogenic determinant of a chloroplast-replicating viroid guide the degradation of a host mRNA as predicted by RNA silencing. Plant J 70:991–1003. https://doi.org/10.1111/j.1365-313X.2012.04940.x
71. OnlineConvertercom (2019) Online Converter; EPS to PDF. https://www.onlineconverter.com/eps-to-pdf
72. Pace N, Thomas B, Woese C (1999) Probing RNA structure, function, and history by comparative analysis. In: Gesteland R, Cech T, Atkins J (eds) The RNA World. Cold Spring Harbor Laboratory Press, New York, pp 113–141
73. Ponty Y (2019) VARNA: visualization applet for RNA. http://varna.lri.fr/index.php?lang=en&page=command&css=varna
74. Ponty Y, Leclerc F (2015) Drawing and editing the secondary structure(s) of RNA. Methods Mol Biol 1269:63–100. https://doi.org/10.1007/978-1-4939-2291-8_5, https://hal.inria.fr/hal-01079893/document
75. Project TI (2019) Inkscape. https://inkscape.org/
76. Qi Y, Pélissier T, Itaya A, Hunt E, Wassenegger M, Ding B (2004) Direct role of a viroid RNA motif in mediating directional RNA trafficking across a specific cellular boundary. Plant Cell 16:1741–1752. https://doi.org/10.1105/tpc.021980
77. Reeder J, Giegerich R (2005) Consensus shapes: an alternative to the Sankoff algorithm for RNA consensus structure prediction. Bioinformatics 21:3516–3523. https://doi.org/10.1093/bioinformatics/bti577
78. Repsilber D, Wiese U, Rachen M, Schröder A, Riesner D, Steger G (1999) Formation of metastable RNA structures by sequential folding during transcription: time-resolved structural analysis of potato spindle tuber viroid (−)-stranded RNA by temperature-gradient gel electrophoresis. RNA 5:574–584. https://doi.org/10.1017/S1355838299982018
79. Reuter J, Mathews D (2010) RNAstructure: software for RNA secondary structure prediction and analysis. BMC Bioinf. 11:129. https://doi.org/10.1186/1471-2105-11-129
80. Riesner D, Henco K, Steger G (1991) Temperature-gradient gel electrophoresis: a method for the analysis of conformational transitions and mutations in nucleic acids and proteins. In: Chrambach A, Dunn M, Radola B (eds) Advances in electrophoresis, VCH Verlagsgesellschaft, Weinheim, vol 4, pp 169–250
81. Rivas E, Eddy S (2000) Secondary structure alone is generally not statistically significant for the detection of noncoding RNAs. Bioinformatics 16:583–605. https://doi.org/10.1093/bioinformatics/16.7.583
82. Rivas E, Clements J, Eddy S (2016) A statistical test for conserved RNA structure shows lack of evidence for structure in lncRNAs. Nat Methods 14:45–48. https://doi.org/10.1038/nmeth.4066
83. Sänger H, Klotz G, Riesner D, Gross H, Kleinschmidt A (1976) Viroids are single-stranded covalently closed circular RNA molecules existing as highly base-paired rod--like structures. Proc Natl Acad Sci U S A 73:3852–3856. https://doi.org/10.1073/pnas.73.11.3852
84. Sankoff D (1985) Simultaneous solution of the RNA folding, alignment and protosequence problems. SIAM J Appl Math 45:810–825. https://doi.org/10.1137/0145048
85. Schmitz M, Steger G (1996) Description of RNA folding by "simulated annealing". J Mol Biol 255:254–266. https://doi.org/10.1006/jmbi.1996.0021
86. Schneider T (2013) Information theory primer with an appendix on logarithms. https://doi.org/10.13140/2.1.2607.2000
87. Schneider T, Stephens R (1990) Sequence logos: a new way to display consensus sequences. Nucleic Acids Res 18:6097–6100. https://doi.org/10.1093/nar/18.20.6097
88. Schneider T, Stormo G, Gold L, Ehrenfeucht A (1986) Information content of binding sites on nucleotide sequences. J Mol Biol 188:415–431. https://doi.org/10.1016/0022-2836(86)90165-8
89. Schröder A, Riesner D (2002) Detection and analysis of hairpin II, an essential metastable structural element in viroid replication intermediates. Nucleic Acids Res 30:3349–3359. https://doi.org/10.1093/nar/gkf454
90. Seibel P, Müller T, Dandekar T, Wolf M (2008) Synchronous visual analysis and editing of RNA sequence and secondary structure alignments using 4SALE. BMC Res Notes 1:91. https://doi.org/10.1186/1756-0500-1-91
91. Shannon C (1948) A mathematical theory of communication. Bell Syst Tech J 27:379–423. http://math.harvard.edu/

~ctm/home/text/others/shannon/entropy/entropy.pdf, https://doi.org/10.1002/j.1538-7305.1948.tb01338.x

92. Sonnhammer E (2019) Stockholm format. http://sonnhammer.sbc.su.se/Stockholm.html
93. Spasic A, Assmann S, Bevilacqua P, Mathews D (2018) Modeling RNA secondary structure folding ensembles using SHAPE mapping data. Nucleic Acids Res 46:314–323. https://doi.org/10.1093/nar/gkx1057
94. Spasic A, Berger K, Chen J, Seetin M, Turner D, Mathews D (2018) Improving RNA nearest neighbor parameters for helices by going beyond the two-state model. Nucleic Acids Res 46:4883–4892. https://doi.org/10.1093/nar/gky270
95. Steger G (2017) Modelling the three-dimensional structure of the right-terminal domain of pospiviroids. Sci Rep 7:711. https://doi.org/10.1038/s41598-017-00764-x
96. Steger G, Perreault JP (2016) Structure and associated biological functions of viroids. Adv Virus Res 94:141–172. https://doi.org/10.1016/bs.aivir.2015.11.002
97. Steger G, Hofmann H, Förtsch J, Gross H, Randles J, Sänger H, Riesner D (1984) Conformational transitions in viroids and virusoids: comparison of results from energy minimization algorithm and from experimental data. J Biomol Struct Dyn 2:543–571. https://doi.org/10.1080/07391102.1984.10507591
98. Stombaugh J, Zirbel C, Westhof E, Leontis N (2009) Frequency and isostericity of RNA base pairs. Nucleic Acids Res 37:2294–2312. https://doi.org/10.1093/nar/gkp011
99. Sun T, Zhao C, Chen S (2018) Predicting cotranscriptional folding kinetics for riboswitch. J Phys Chem B 122:7484–7496. https://doi.org/10.1021/acs.jpcb.8b04249
100. Takeda R, Petrov A, Leontis N, Ding B (2011) A three-dimensional RNA motif in *Potato spindle tuber viroid* mediates trafficking from palisade mesophyll to spongy mesophyll in *Nicotiana benthamiana*. Plant Cell 23:258–272. https://doi.org/10.1105/tpc.110.081414
101. Tinoco I Jr, Uhlenbeck O, Levine M (1971) Estimation of secondary structure in ribonucleic acids. Nature 230:362–367. https://doi.org/10.1038/230362a0
102. Turner D, Mathews D (2009) NNDB: the nearest neighbor parameter database for predicting stability of nucleic acid secondary structure. Nucleic Acids Res 38:D280–2. https://doi.org/10.1093/nar/gkp892
103. Wang Y, Zirbel C, Leontis N, Ding B (2018) RNA 3-dimensional structural motifs as a critical constraint of viroid RNA evolution. PLoS Pathog 14:e1006801. https://doi.org/10.1371/journal.ppat.1006801
104. Washietl S, Hofacker I, Stadler P (2005) Fast and reliable prediction of noncoding RNAs. Proc Natl Acad Sci U S A 102:2269–2270. https://doi.org/10.1073/pnas.0409169102
105. Waterman M (1978) Secondary structure of single-stranded nucleic acids. In: Studies on Foundations and Combinatorics, Advances in Mathematics Suppl. Studies, Academic Press, New York, pp 167–212
106. Waterman M, Smith T (1978) RNA secondary structure: a complete mathematical analysis. Math Biosci 42:257–266. https://doi.org/10.1016/0025-5564(78)90099-8
107. Weinberg Z, Breaker R (2011) R2R – software to speed the depiction of aesthetic consensus RNA secondary structures. BMC Bioinf. 12:3. https://doi.org/10.1186/1471-2105-12-3
108. Wikipedia (2019) Stockholm format. https://en.wikipedia.org/wiki/Stockholm_format
109. Will S, Reiche K, Hofacker I, Stadler P, Backofen R (2007) Inferring noncoding RNA families and classes by means of genome-scale structure-based clustering. PLoS Comput Biol 3:e65. https://doi.org/10.1371/journal.pcbi.0030065
110. Will S, Joshi T, Hofacker I, Stadler P, Backofen R (2012) LocARNA-P: accurate boundary prediction and improved detection of structural RNAs. RNA 18:900–914. https://doi.org/10.1261/rna.029041.111
111. Wilm A, Linnenbrink K, Steger G (2008) ConStruct: improved construction of RNA consensus structures. BMC Bioinf. 9:219. https://doi.org/10.1186/1471-2105-9-219
112. Wolfinger M, Svrcek-Seiler W, Flamm C, Hofacker I, Stadler P (2004) Efficient computation of RNA folding dynamics. J Phys A 37:4731–4741. https://doi.org/10.1088/0305-4470/37/17/005
113. Wu Y, Shi B, Ding X, Liu T, Hu X, Yip K, Yang Z, Mathews D, Lu Z (2015) Improved prediction of RNA secondary structure by integrating the free energy model with restraints derived from experimental probing data. Nucleic Acids Res 43:7247–7259. https://doi.org/10.1093/nar/gkv706

114. Wuchty S, Fontana W, Hofacker I, Schuster P (1999) Complete suboptimal folding of RNA and the stability of secondary structures. Biopolymers 49:145–165. https://doi.org/10.1002/(SICI)1097-0282(199902)49:2<145::AID-BIP4>3.0.CO;2-G
115. Yang H, Jossinet F, N L, Chen L, Westbrook J, Berman H, Westhof E (2003) Tools for the automatic identification and classification of RNA base pairs. Nucleic Acids Res 31:3450–3460. https://doi.org/10.1093/nar/gkg529
116. Zhang Z, Schwartz S, Wagner L, Miller W (2000) A greedy algorithm for aligning DNA sequences. J Comput Biol 7:203–214. https://doi.org/10.1089/10665270050081478
117. Zhong X, Ding B (2008) Distinct RNA motifs mediate systemic RNA trafficking. Plant Signal Behav 3:58–59. https://doi.org/10.4161/psb.3.1.4890
118. Zhong X, Leontis N, Qian S, Itaya A, Qi Y, Boris-Lawrie K, Ding B (2006) Tertiary structural and functional analyses of a viroid RNA motif by isostericity matrix and mutagenesis reveal its essential role in replication. J Virol 80:8566–8581. https://doi.org/10.1128/JVI.00837-06
119. Zhong X, Leontis N, Qian S, Itaya A, Qi Y, Boris-Lawrie K, Ding B (2006) Tertiary structural and functional analyses of a viroid RNA motif by isostericity matrix and mutagenesis reveal its essential role in replication. J Virol 80:8566–8581. http://jvi.asm.org/cgi/reprint/80/17/8566
120. Zhong X, Tao X, Stombaugh J, Leontis N, Ding B (2007) Tertiary structure and function of an RNA motif required for plant vascular entry to initiate systemic trafficking. EMBO J 26:3836–3846. https://doi.org/10.1038/sj.emboj.7601812
121. Zuker M (1989) On finding all suboptimal foldings of an RNA molecule. Science 244:48–52. https://doi.org/10.1126/science.2468181
122. Zuker M (2003) Mfold web server for nucleic acid folding and hybridization prediction. Nucleic Acids Res 31:3406–3415. https://doi.org/10.1093/nar/gkg595
123. Zuker M, Stiegler P (1981) Optimal computer folding of large RNA sequences using thermodynamics and auxiliary information. Nucleic Acids Res 9:133–148. https://doi.org/10.1093/nar/9.1.133

INDEX

Ayala L. N. Rao et al. (eds.), *Viroids: Methods and Protocols*, Methods in Molecular Biology, vol. 2316,
https://doi.org/10.1007/978-1-0716-1464-8, © Springer Science+Business Media, LLC, part of Springer Nature 2022

MIX
Papier aus verantwortungsvollen Quellen
Paper from responsible sources
FSC® C105338

If you have any concerns about our products,
you can contact us on
ProductSafety@springernature.com

In case Publisher is established outside the EU,
the EU authorized representative is:
Springer Nature Customer Service Center GmbH
Europaplatz 3, 69115 Heidelberg, Germany

Printed by Libri Plureos GmbH
in Hamburg, Germany